Food, Culture and Identity in the Neolithic and Early Bronze Age

Edited by

Mike Parker Pearson

BAR International Series 1117
2003

Published in 2016 by
BAR Publishing, Oxford

BAR International Series 1117

Food, Culture and Identity in the Neolithic and Early Bronze Age

ISBN 978 1 8417 1495 0

BAR Publishing is the trading name of British Archaeological Reports (Oxford) Ltd. British Archaeological Reports was first incorporated in 1974 to publish the BAR Series, International and British. In 1992 Hadrian Books Ltd became part of the BAR group. This volume was originally published by Archaeopress in conjunction with British Archaeological Reports (Oxford) Ltd / Hadrian Books Ltd, the Series principal publisher, in 2003. This present volume is published by BAR Publishing, 2016.

Printed in England

BAR
PUBLISHING

BAR titles are available from:

BAR Publishing
122 Banbury Rd, Oxford, OX2 7BP, UK
EMAIL info@barpublishing.com
PHONE +44 (0)1865 310431
FAX +44 (0)1865 316916
www.barpublishing.com

CONTENTS

LIST OF CONTRIBUTORS

Paul Budd	Dept. of Archaeology, University of Durham, South Road, Durham, DH1 3LE, UK
Andrew Chamberlain	Dept. of Archaeology, University of Sheffield, Northgate House, Sheffield, S1 4ET, UK
Carolyn Chenery	NERC Isotope Geosciences Laboratory, British Geological Survey, Keyworth, Nottingham, NG12 5GG, UK
Susan Cox	Dept. of Archaeology, University of Sheffield, Northgate House, Sheffield, S1 4ET, UK
Oliver Craig	Dept. of Fossil Fuels, University of Newcastle, Newcastle upon Tyne, UK
Jane Evans	NERC Isotope Geosciences Laboratory, British Geological Survey, Keyworth, Nottingham, NG12 5GG, UK
Gill Hey	Oxford Archaeology, Janus House, Osney Mead, Oxford, OX2 0ES, UK
Eva Koch	National Museum, Copenhagen, Denmark
Liliana Janik	George Pitt-Rivers Laboratory for Bioarchaeology, Dept. of Archaeology, University of Cambridge, Cambridge, UK
Andrew Jones	Dept. of Archaeology, University of Southampton, Southampton, UK
Janet Montgomery	Dept. of Archaeological Sciences, University of Bradford, Bradford, BD7 1DP, UK
Jacqui Mulville	School of History & Archaeology, Cardiff University, PO Box 909, Cardiff, UK
Pia Nystrom	Dept. of Archaeology, University of Sheffield, Northgate House, Sheffield, S1 4ET, UK
Mike Parker Pearson	Dept. of Archaeology, University of Sheffield, Northgate House, Sheffield, S1 4ET, UK
Keith Ray	Herefordshire Archaeology, Conservation and Environmental Planning, Herefordshire County Council, PO Box 230, Hereford, HR1 2ZB, UK
Colin Richards	School of Art History & Archaeology, University of Manchester, Oxford Road, Manchester, M13 9PL, UK
Mike Richards	Dept. of Archaeological Sciences, University of Bradford, Bradford, BD7 1DP, UK

Mark Robinson	Natural History Museum, Parks Road, Oxford, UK
Julian Thomas	School of Art History & Archaeology, University of Manchester, Oxford Road, Manchester, M13 9PL, UK
Soultana-Maria Valamoti	Dept. of Archaeology, University of Thessaloniki, Thessaloniki, Greece
Annsofie Witkin	Oxford Archaeology, Janus House, Osney Mead, Oxford, OX2 0ES, UK

LIST OF FIGURES AND TABLES

Figures

Tables

FOOD, IDENTITY AND CULTURE: AN INTRODUCTION AND OVERVIEW

Mike Parker Pearson

Food is as indispensable as the air that we breathe. Rich in metaphor, symbolism and meaning, it is much more than the nutritional substance which provides the calorific input necessary to sustain life. As both artefact and ecofact, food bridges the awkwardly constructed gap between culture and nature. The 'natural' world out there must be brought into our 'cultured' bodies through the ambiguous zone of the mouth. Animals feed but humans eat, in a myriad different cultural forms - feasting, dining, working lunches, power breakfasts, snacks, holy communion, tea breaks and takeaways. Meals are one of life's most regular and sustained rituals, often but not always involving commensality and supernatural blessing. Food is also deeply implicated in politics and the construction of identity and culture. Its temporality, spatiality and materiality all have profound effects on how people have lived their lives and how our worlds have been transformed. As Klaus Eder expresses it, 'culinary operations dominate the reproduction of modern social self-construction in every field of social power' (1996: 138) at every level from the state banquet to Sunday lunch. Food is also at the heart of how we construct our bodies and ourselves, physically and emotionally (Lupton 1996). Moralities and symbolic meanings - and the experiences and sensations through which these are lived - are all articulated through food in ways that are contradictory, conflicting and ever-changing (Lupton 1996: 154).

In the last twenty years historians and social scientists have seen a veritable explosion of research into food and its consumption and social context.[1] And yet archaeology has been slow to catch on. This is all the more surprising since the 'bread and butter' of archaeology are the residues of food preparation and consumption - animal bones, pottery and other containers, cooking places and other technologies of preparation, plant remains (micro and macro), landscapes and settlements, grave goods, chemical residues, human bones and human coprolites. The many specialists in these fields have, of course, produced quantities of information but their work has been largely directed towards answering questions about subsistence, production and economy. Such questions were staples of the culture-historical, palaeoeconomic and processual archaeologies whose materialist concerns tended to emphasise production over consumption. Food studies are often the side dish of an environmental archaeology that is primarily concerned with human-environment relations on a wider scale.

The very constitution of an 'environmental' archaeology has led to divisions of the discipline, each producing research of an empirical nature on distinct categories of material evidence. This necessary professional fragmentation into archaeological specialisms has created centres and boundaries which have worked against the re-combination of archaeological evidence into an integrated and coherent category of 'food residues'. In a strange, almost double leap of inference, 'environmental' archaeologists have studied animal bones, for example, in order to understand not so much the consumption of the meat that they bore but the patterns of husbandry and stock-raising which produced that meat. The same could be said for studies of botanical residues in which the main aim has been to investigate agricultural production, stepping beyond the apparently more prosaic concerns of how the plant materials were prepared, eaten and disposed of. The situation has been worse in traditional pottery studies, in which typologies, culture-historical affiliations, chronologies, trade networks - in fact, almost anything other than pottery's significance for investigating food practices - have been foregrounded.

Whilst this 'ghettoisation' has created the academic structure necessary for the development of expertise and methodologies, we must not lose sight of the crucial fact that it is only through the integrated application of environmental, ceramic, chemical, osteological and other evidence that a social archaeology of food is possible.

[1] Adams 1990; Appadurai 1981; 1993; Arnott 1975; Bell and Valentine 1997; Charles and Kerr 1988; Cook *et al.* 1999; Counihan and Kaplan 1998; Counihan and Van Esterik 1997; Elias 1978; Falk 1994; Farb and Armelagos 1980; Fenton and Kisban 1986; Fenton and Myrdal 1988; Fenton and Owen 1981; Fiddes 1991; Fieldhouse 1986; Fischler 1988; Forster and Ranum 1979; Goodman *et al.* 1995; Goody 1982; Hage 1979; James 1996; Kahn 1986; Kiple and Ornelas 2000; Korsmeyer 1999; Kuper 1997; Lentz 1999; Lupton 1994; 1996; MacClancy 1992; McCracken 1982; Meigs 1984; 1997; Mennell 1985; Mennell *et al.* 1992; Messer 1984; Ohnuki-Tierney 1993; Okely 1983; Paston-Williams 1993; Pullar 1970; Revel 1982; Richards 1939; Scholliers 2001; Shanklin 1985; Simoons 1979; 1994; Tannahill 1973; Visser 1986; Walcher *et al.* 1976; Warde 1997; Wiessner and Schiefenhovel 1996.

PART ONE: CULTURAL APPROACHES TO FOOD

This volume of papers arises out of a conference held in Sheffield in 1999, organised jointly by The Prehistoric Society and the Sheffield University Archaeology Society, on 'Food, Identity and Culture in the Neolithic and Early Bronze Age'. The aim was to bring together the different archaeological interests - from archaeological science and humanities perspectives - in food as cultural artefact/ecofact, to examine the potential of the new and developing scientific techniques for reconstructing prehistoric food habits, and to foster an integrated approach to the archaeology of food regardless of different researchers' specialisms. The deep divide between the humanities and the scientific, exemplified in the archaeology of food, has been described as 'both productive and uncomfortable' (Gosden 1999: 8) and there is no doubt that many speakers and members of the audience went away from the Sheffield conference with their feet still placed securely on one side or other of this divide. One of the conference's outcomes was the growing awareness within the Archaeology department at Sheffield that food, culture and identity were very much topics of intellectual interest and that this research theme may provide the synergy to further integrate the 'two cultures' - theoretical/humanistic and technical/scientific - which still divide archaeology both practically and institutionally.

Several recent archaeological conferences have included sessions on food and culture. In 1999 Fokke Gerritsen organised an enjoyable symposium 'From Calorie to Culture' in Leiden (Gerritsen 2000) and in 2001 the Society for American Archaeology's annual conference in New Orleans commenced with a plenary session on food and feasting. From the stew of the World Archaeology Congress in New Delhi in 1996 came a large (and almost prohibitively expensive) volume on the prehistory of food (Gosden and Hather 1999) and food has also been part of the subject of other conferences in recent years (see Cumberpatch and Blinkhorn 1997; Miracle and Milner 2002). In the same period, a number of books have come out on prehistoric and ancient food (Crowe 2000; Davidson 1997; Garnsey 1999; Vaughan and Coulson 1999; Wiessner and Schiefenhövel 1996; Wilson 1988; Wood 2001). Compare this with archaeologists' lack of interest and low output on such matters in the 1970s and 1980s. Jane Renfrew's book on prehistoric food (1985) was the only work of its kind and even this was considered by some to be a light-hearted and 'popular' work. The study of food was simply not thought to be an important aspect of knowing the past unless it was related to adaptation, carrying capacity, nutritional stress, optimal foraging, resource availability, risk buffering or some such conception of 'food as fuel'.[2] So what changed?

[2] See Jerome *et al.* 1980 for anthropological and archaeological approaches along adaptationist lines, although there were cultural historical and *annaliste* writings by this time (*e.g.* Tannahill 1973 and papers in Forster and Ranum 1979). The excellent book by Farb and Armelagos (1980) had a foot in both camps.

The development of a social archaeology of food

One of the earliest archaeological investigations of 'foodways' ('the whole inter-related system of food conceptualisation, procurement, distribution, preservation, preparation, and consumption') was contained in James Deetz's remarkable book *In Small Things Forgotten* (1977: 49-61), a historical archaeology which stood alone for many years in its detailed consideration of food. The development of post-processual approaches in archaeology has been considered to be the catalyst for consumption-based approaches to food, amongst other aspects of material culture (Sherratt 1997).

Yet there is a strong case to be made that, as the early post-processualists sifted through material culture, food was largely ignored (Gerritsen 2000: 169). Whereas topics such as rubbish, burial practices, craft production, and pottery design were the empirical categories of evidence tackled by Hodder's 'first wave' of post-processualist research students, food was not on that early menu. Ethnoarchaeological studies within this vein in the late 1970s and 1980s, however, certainly did look at food but within broader social themes of gender, caste and ethnicity (Braithwaite 1982; Hodder 1982; Miller 1985; Moore 1982; 1986) but it did not feature as a major ingredient in later post-processual projects until the late 1990s.

This lack of interest is all the more remarkable given the impact in social anthropology of structuralist and post-structuralist writings about food. Lévi-Strauss' culinary triangle had been widely discussed, re-interpreted and criticised by structuralists and cultural materialists alike (Leach 1964; Douglas 1972; Harris 1979; 1985) and there was a new generation of post-structuralist anthropologists who were developing more contextualised and subtle applications which framed food within the various spatialities and temporalities of gender and other social processes of life in traditional societies around the world, in such places as Algeria, Papua New Guinea, Greece, Amazonia and Thailand, for example (Bourdieu 1977; 1990; Bulmer 1967; Danforth 1982; Hugh-Jones 1979; Tambiah 1969). Even at archaeological conferences during the 1980s, such as WAC at Southampton in 1986, the resulting publications on themes tackling human meanings and practices concerning animals and food were almost entirely written by anthropologists rather than archaeologists (Ingold 1989; Willis 1990).

In the simplest of terms, an archaeology of food is made difficult by the fact that the object of study is consumed and only its residues remain. But this is almost certainly not the reason why it has been so late in coming. No doubt part of archaeologists' reluctance to engage directly with the topic was the problem of 'the two cultures'. The post-processualists, mostly research students, were not trained to learn the 'science' (whether it be archaeozoology, palaeoethnobotany, chemistry or human osteology) in order to apply empirically the

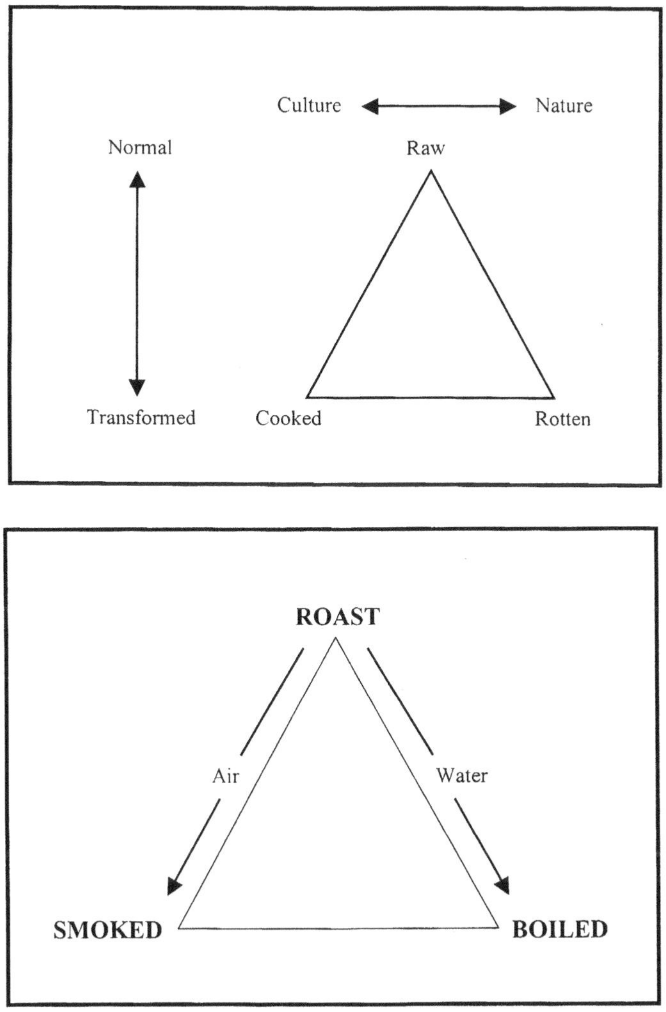

Figure 1.1. *Lévi-Strauss's culinary triangle.*

theoretical approaches, whilst the archaeological scientists remained unconvinced that these post-processual approaches had any validity. Only now is the gap being closed as a new generation pursues the theoretical agendas of post-processual archaeology with their own expertise in the scientific techniques which were previously harnessed to answering different kinds of empiricist and ecological questions (A. Jones 2002). At the same time, a few of the first-generation post-processualists were able to carry out field projects in which they could encourage the direct involvement of archaeological scientists to address these issues (Hodder 1999; Craig *et al.* 2000; Richards 2002).

In Britain the social archaeology of food really developed in studies of the first millennium BC. Barrett's papers on ceramics and on food, metal and gender in the Late Bronze Age (1980; 1989) were followed by studies of published, well-excavated Iron Age settlements in Wessex and the Thames valley (Hill 1989; 1995; Parker Pearson 1996), more recently carried through into the Roman period (Meadows 1997). The Iron Age, but in France and Germany, was also the scene for Dietler's research into the social and political dimensions of drinking in Late Hallstatt society (1990). Hodder's *Domestication of Europe* (1990), as Gosden has pointed out (1999: 9), was probably the first book-length archaeological attempt to explain prehistoric social change through, *inter alia*, cultural attitudes to food.[3]

Contemporary approaches to the social archaeology of food

The social and cultural anthropologically-driven approach to food studies in archaeology is just one of several. Experimental archaeology has been a long-standing field of enquiry which has probably never been more popular. Experiments with boiling meat in skins or heating water with hot stones (such as O'Kelly 1954) have been regular accompaniments to reconstructing roundhouses and other attempts to lead a prehistoric life in the shadow of suburbia. The sophistication of experimental archaeology in this field has, however, reached new heights with the work of field centres such as the Cornwall Celtic Village (Wood 2000; 2001). Using the food resources which would have been available in prehistory, Jacqui Wood and her team - like Napoleon's chef at Marengo - have created a dazzling and even mouth-watering menu of over a hundred possible recipes from disparate and apparently unpalatable ingredients. They have also explored the

[3] To appreciate something of the variety of archaeological approaches to food, see Campbell 2000, Curtis 2001, Dietler and Hayden 2001, Hastorf 1998, Jones 1998 and 1999, M. Jones 2002; Legge *et al.* 1998, Milner and Miracle 2002, Richmond 1999, Thomas 2000 and Verpoorte 2000.

many ways of cooking in water, with hot stones, in clay, in ovens and over open fires.

This research is informed by anthropological perspectives on food preparation in many different societies (see Kuper 1997) and it combines a degree of ethnoarchaeological knowledge with practical experiment. It is an engaging and enjoyable way of finding out what might have been possible but, like all experimental archaeology, it suffers from the inescapable limitations imposed by our own ethnocentric perspective. What may seem most logical, efficient, effective, pragmatic - or appetising in this case - to us is not necessarily how it was in prehistory. The past is still that other country where people do things differently, in ways that may not have a comparison in any society - traditional or otherwise - today. Our own cultural valorisation of cuisine, as a self-contained field of endeavour in which innovation is privileged, is a major potential stumbling block of which we have to be aware.

Ethnoarchaeology offers another range of possibilities of how people might have prepared and eaten their food. One of its important lessons is that the very category of 'edible' may be delimited quite differently to our own (see Kuper 1997 for some examples). Fried insects or half-digested contents of sheep's intestines are considered as delicacies in some cultures but do not make the grade in others. Learning what is considered 'edible' or even 'tasty' in other cultures is a useful way of removing the cultural blinkers from our own unacknowledged ethnocentric interpretations of what food is, and how it should be prepared, served and eaten (Farb and Armelagos 1980; Parker Pearson 2000b). Recreated prehistoric foods can produce outraged reactions from contemporary tasters; on the panel game show *Animal, Vegetable, Mineral*, broadcast in the 1950s, Mortimer Wheeler famously expressed the sentiment that Tollund man was probably glad to die having eaten the gruel whose residues were found in the bog body's gut. As noted above, ethnoarchaeology only opens up possibilities, widening our horizons rather than providing precise parallels to what happened in prehistory.

Empirical research into the residues of past foodways is, of course, the means by which we learn directly about the past. We may identify the species of plants and animals that were consumed and the manners in which they were prepared and eaten, and how and where their waste was discarded. Yet we must be mindful of the problems of a 'blind empiricism' which is undirected, atheoretical and prone to ethnocentric assumptions about human foodways. Some of us may wish to hide behind the label of 'technician', merely producing results for others to interpret - an attitude criticised long ago by Wheeler as unacceptable (1956). Others may consider theory to be no more than speculation and best left out. Yet theory is the rationalisation of our experiences, directing our questions, defining our research topics, and linking these to wider issues of human life.

Food and theory: 'good to think'

The starting point for any discussion of theory has to be Claude Lévi-Strauss and his culinary triangle, a concept developed from his readings on South American indigenous mythologies in which he situated food between two axes of transformation, one of nature-culture and the other of normal-transformed (1966a; 1969; 1974; 1978). The raw could be set in opposition to the cooked and both placed in opposition to the rotten (Figure 1.1). He considered each to resonate with particular forms of cooking - respectively roasting/grilling, smoking and boiling. Within Lévi-Strauss' universalising model, he considered roasting as wasteful and high-status whereas boiling was conserving and low-status. Edmund Leach pointed out that boiled may equate with rotten in South American mythology but nowhere else, and was also critical of Lévi-Strauss' identification of 'rotten' as a major node, noting that only a few foods (such as, for Europeans, blue cheese, sour cream and pickled fish) came under this category (Leach 1964). Marvin Harris also demonstrated that Lévi-Strauss' generalisation about South American endogenous and exogenous cannibalism - that kin ('cultural') were eaten boiled whilst strangers ('non-cultural') were eaten roasted (Harris 1968) - was an empirical fallacy.

Lévi-Strauss and other structuralists (see also Barthes 1972) hoped to discover 'how, in any particular society, cooking is a language through which that society unconsciously reveals its structure, unless - just as unconsciously - it resigns itself to using the medium to express its contradictions' (Lévi-Strauss 1978: 495). Food and drink could be analysed as systems of classification which expressed relationships between people. Cooking and cuisine were not categories separate to other forms of experience but were divisible into series of oppositions which could be detected in the social, economic, aesthetic and religious aspects of life such as men/women, family/society, village/bush, thrift/ extravagance, nobles/commoners and sacred/profane.

The British structuralists Edmund Leach and Mary Douglas also made major contributions to the anthropology of food. Leach refined Lévi-Strauss' binary opposition of edible/inedible into a tripartite scheme of edible/edible-but-taboo/edible-but-not-recognised-as-food (unconsciously tabooed) (1964: 32; see also Leach 1976). He also conceived of edibility according to the social distance between the eater and the eaten (see also Sahlins 1976 and Tambiah 1969). In Britain one does not eat one's pet cat (too close) or remote wild animals such as birds or badgers (too far) but those animals in the middle (domesticates and game, although the latter are surrounded with ritual rules). Lévi-Strauss categorised taboo food as an anomaly in between nature and culture; Leach placed it within the nature-culture gradient but marked by a confusion of the distinction between eater and food.

Mary Douglas' conception of taboo was not tied to the nature-culture gradient but to the anomalous within cultural categorisations. It derived in part from her

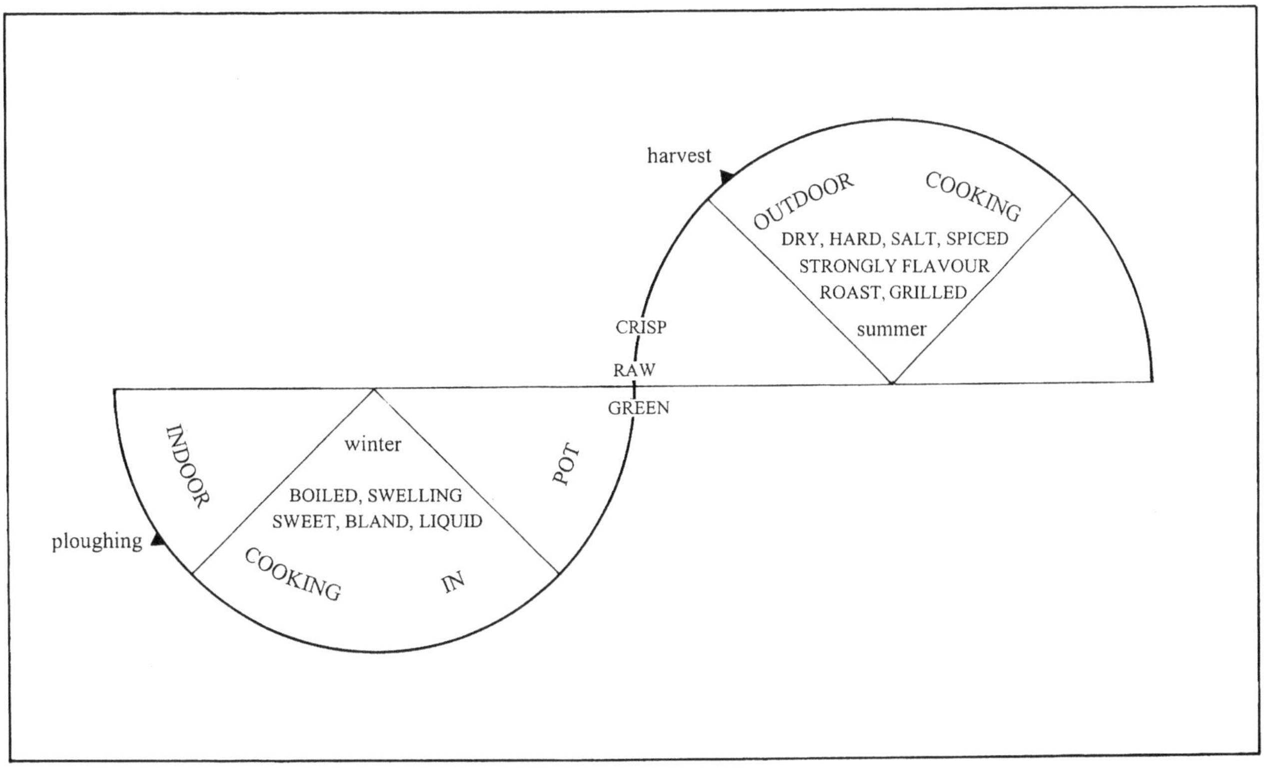

Figure 1.2. *Bourdieu's conception of the structuring principles of Kabyle Berber life and cuisine.*

studies of Lele attitudes towards the pangolin and Jewish prohibitions on the pig, which brought out the ways in which these creatures do not fit the wider classificatory system and thus were anomalous and therefore dangerous - the pangolin lives in the forest like a forest creature, yet has scales like a fish, walks upright on its hind legs and also gives birth singly like people (1975). Her reading of the prohibition on the pig in Jewish culture was based not on its literal 'dirtiness' but on its anomalous status within the classifications of animals (1966; see also Soler 1979). Along with camels and hyraxes, the pig has cloven hooves but does not chew the cud. By not conforming to categories, it is impure, dangerous and symbolically uncontrollable.

In our own Western society, food can of course be dangerous to sufferers of allergies - nuts, gluten and oysters, for example - but it may also be dangerous in Mary Douglas' terms. To ingest food is to risk the body's integrity by polluting it (Lupton 1996: 114): '"Raw" food is fraught with danger, a "wilderness" that is tamed by culinary treatment.' (Fischler 1988: 287). People do not need recourse to formal taboos to describe certain foods as equivalent to excrement (as in 'I'm not eating that foreign muck').

Mary Douglas also devised an ingenious analysis of British foodways of the post-war years in which the structure of the main dish, 'meat and two veg' (formalised as A+2B), is replicated in the meal itself (starter, main course, pudding), and the day's three meals (main one and two smaller meals), a scheme that was embellished at Christmas, Sunday lunch and other celebrations (Douglas 1972; Douglas and Nicod 1974). Each meal was thus a structured event that structured others in its own image and formed a homology of the daily fare. She also pointed out that the 1960s meal incorporated a series of oppositional categories: hot/cold, bland/spiced, liquid/ solid and savoury/sweet.

The structuralist method was flawed, however, with its dubiously universal structures, its static systems of thought which could not account for change, and its creation of an unengaged cultural 'mind' instead of human practice as the subject of study. Yet Lévi-Strauss demonstrated that both animals and food were 'good to think' (1966b), as opposed to being merely 'good to eat', thereby opening the way for approaches which recognised that food habits were formed by cultural as well as environmental/ecological factors. Virulent debates raged in the 1970s between the ecological functionalists and materialists on the one hand and their structuralist/post-structuralist counterparts on the other about the reasons why cows were sacred in India, why pigs were dirty in Judaism and Islam, and why sacrificed warriors were eaten in Aztec Mesoamerica (Diener and Robkin 1978; Harris 1975; 1978; 1985; Harris and Ross 1987; Soler 1979; Simoons 1979; 1994).

'Good to think about', 'good to think with' as well as 'good to taste'

Post-structuralist anthropologists rejected the mentalist and universalist aspirations of Lévi-Strauss' formulation - food was not just 'good to think' since 'cultures are not only thought but also lived' (Tambiah 1969). Bourdieu's analysis of Kabyle Berber food practices revealed a complex web of inter-relationships between seasonality, gender and cooking (Figure 1.2): the moist, boiled and swelling foods of winter and spring (linked to the indoor world of women and the pregnancy of people and animals) were replaced in summer by the spiced, roasted and dry foods prepared by men out-of-doors (1977; 1990). Hugh-Jones' study of life in a Barasana longhouse (1979) demonstrated a similar spatiality of food associations in which women (other than the elderly) were deprived of 'anti-foods' (the nourishment for the souls of the dead) of coca, beer, tobacco and hallucinogenic plants derived from the forest areas which were the preserve of men. Morris' study of Malawian foods demonstrates that explanations of dietary taboos simply in terms of cosmological or classificatory schemes is inadequate - there are wide individual variations in species eaten which relate, in part, to religious affiliation, avoidance of clan totem animals, gender and ethnicity (1998: 209). He further criticises Lévi-Strauss' nature-culture transformation by explaining that, for the society at the centre of his research, meat-eating is not a domination over nature but a way of harnessing the powers of wild nature, particularly of the animals and ancestral spirits that dwell in the woodlands whence fertility derives in the form of rain and semen.

Descola also deconstructs the ethnocentric basis of this 'conceptual stand-off' between nature and culture which he traces back to the ancient Greeks (1994: 93; see Valamoti this volume for its possible prehistoric ancestry in Greece). In his work on the Amazonian Achuar, he points out that this (for us) deep-rooted dualism has no place in their world in which there is a sliding continuum between human beings and nature's beings, wherein humans, most plants, animals and meteors are persons with souls whereas most insects, fish, poultry and some plants are not (1994: 325). Closest to people are the other highly socialised creatures (other humans) or those that follow the same marriage rules (peanuts, manioc, toucans, woolly monkeys). Further down the scale are the sociable but incestuous creatures such as howler monkeys and dogs.

From a different direction, Eder illuminates the contradictory ways in which Western society has culturally constructed the concept of nature as a separate domain to be conquered or embraced and contrasts case studies of Jivaro and ancient Jewish food moralities (as embodied through taboos) with those of contemporary Europe (1996). In the Jivaro case he identifies structuring principles of social/physical 'openness' and 'closedness' (Jivaro proscribe the eating of animals which are considered to have either no anus or an anus that is too open) whilst those in the Jewish case concern violations of the boundaries of categorisation and social identity (*cf* Douglas 1966; Soler 1979). Both are different from the contemporary British and European structuring principle of the nature-culture divide, identified by Leach (1964), and help to objectify its relative status as a cultural and social construction of the modern world which might have had little validity in prehistory.

Tastescapes and the senses

Food is 'good to think with' and 'good to think of' in terms of 'good to taste' (Falk 1994: 68). Yet, like all phenomenological approaches, taste experience does not provide a simple key to understanding how people of the past experienced their worlds. As Pasi Falk explains, cultural categorisations are the basic principles of food behaviour and thus affect food and taste preferences. We may well all taste (approximately) the same things but whether we like or loathe them depends largely on our cultural categories. 'At the sensory level taste preferences are necessarily also related to and even determined by the symbolic principles which translate the material universe into *representations* of the edible vs. inedible, which are then further specified into different sub-categories according to taboos and ritual rules.' (Falk 1994: 68).

The experiencing of food and drink is not restricted to the domains of smell and taste. How it looks is extremely important (blue-dyed food is deeply unattractive to many), as is how it feels (few of us can tolerate sand in our food) and even the way that it sounds (the cork coming out of a bottle, the bubbling of soup) can really matter. Westerners' concerns with obsessive hygiene, household cleanliness, and distaste for oozing and slithering food (identified by Lupton [1996: 114] as redolent of our bodily fluids of excrement, semen, saliva, vomit *etc.*) are not shared universally. Food - and its end products of excrement and garbage - is, however, among the more prominent markers within the 'smellscape'.

Food, drink and narcotics are also the dominant stimulators of taste. As Marcel Proust made clear over a century ago, smell and taste can also stimulate intense memory recall and intimate personal experiences and reactions. David Sutton's recent social anthropological study of food in contemporary Greece has developed this theme to explore the significance of taste and smell in constructing memory (2001). In the modern Western world our sensory experiences of both smell and taste are severely skewed; 'bad' smells are masked or hidden where possible and we are rarely assaulted by the smells of rotten meat, decaying corpses or excrement and sewers. Our tastebuds and sense of smell are also titillated to a level hitherto undreamed of, by a food and drinks industry which has developed beyond the stimulants of early capitalism - sugar, chocolate, coffee, tea and spices - to a bewildering array of chemicals and E-numbers.

The sociology and anthropology of the senses are now established fields of inquiry (Classen 1993; Classen *et al.* 1994; Kus 1992) with a specialised sub-field for smell (Howes 1987; 1991; Synnott 1991). It has been claimed that archaeologists privilege the sense of sight above all others and fail to consider the significance of experience derived from all the senses combined that make up human life experience. When we turn to prehistory, the strongest smells were most likely the 'bad' ones! Yet the role of smell in social life might have been very significant. David Howes points to what he considers a universal association between olfaction and transition (1987) and, although we may question such universality, there are many historical and anthropological examples of burnt offerings, ritual purifications, incense and decomposition as 'smelly symbols' and embodiments of liminality, transition and supernatural communication.

At a more mundane level, the smell of cooking food - for ordinary meals or feasts – would also have served as a marker of transition even if Neolithic cooking was, to our pampered noses and tastebuds, bland and perhaps even disgusting. In a world which might have been largely devoid of strong tastes, substances such as honey would have been highly prized. The smell of roasting meat at feasts may have evoked strong feelings and memories, not least because it so closely resembles the smell of a cremating corpse. The senses of smell and taste fade with age and we can perhaps presume further impairment after a lifetime spent inhaling smoke from indoor hearths and fires.

Food and time

Food and drink are the structuring elements of everyday life (Braudel 1981); they are used to divide up the day, the week and the year, punctuating the passing of the seasons and the life cycle with consumption events and festivities that create time itself. The food of infants is different from that of children and adults. The very elderly may also have to be fed with substances that can pass easily through toothless mouths.

'In any socially regulated set of consumption practices, those that centre around the body, and especially around the feeding of the body, take on the function of structuring temporal rhythm, of setting the minimum temporal measure (by analogy to musical activity) on which much more complex, and "chaotic" patterns can be built. Pushing the analogy a step further, the small habits of consumption, typically daily food habits, can perform a percussive role in organising large-scale consumption patterns, which may be contrived of much more complex orders of repetition and improvisation.' (Appadurai 1993: 13).

Practices centred around food and drink can also be used to create 'timeless' traditional representations of the past fixed in the present. Proust's gastronomical sense of nostalgia has in recent years been hitched to the capitalist reinvention of tradition, as 'heritage' cheeses and other traditional foodstuffs have been marketed as 'original' products (though with different tastes and health standards to days gone by). This sense of rewriting is also found in non-literate, non-Western societies such as that of the Sabarl islanders of Melanesia. Here the mortuary feast cycle *segaiya* commemorates dead individuals but also reworks people's relationships to the deceased, to create a new future for the dead, a new sense of order in the face of mortality and a new and reinforced perspective on cultural values (Battaglia 1990: 155). The mortuary feasts are as much about learning to forget the deceased as they are about remembering them in new ways: *segaiya* makes ancestors as it makes memories (Battaglia 1990: 96).

Taboos and how they change

The conundrum of change in traditional societies, a question of great importance to archaeologists, has been tackled by relatively few anthropologists. Among the most important work is that of Sahlins on the cosmological collapse of the Hawaiian kingdom - linked to the misinterpreting of Captain Cook as a returning deity – in which he highlights the constant reinterpretation of myths in changing and undermining cosmologies (1985). In a similar vein, Barth stresses that cosmologies are always in the making, being modified and transformed in the wake of historical events and interactions with other communities (1987). Falk concludes that the adoption of one culture's food by another must rest not on its innate taste being judged as superior but on its positive representation, as something that stands for good and valued aspects of life (1994: 79-85). As a result, and only as a result, is it considered to taste good. Falk suggests that such foods may be incorporated into controlled situations of taboo or feasting - available at certain times to certain people and not to others. Deemed as powerful, dangerous or uncontrollable, such foods also have a dual character since they are at the same time enticing and exotic. Examples of this include the introductions of bananas and tomatoes into sixteenth-century England when they were sought after initially for their supposed aphrodisiac qualities (Farb and Armelagos 1980). In southern Madagascar, the introduction of rice for funerary feasts in recent years has been possible because of its 'good' and sacred connotations (Parker Pearson 2000b: 222).

Consuming bodies

Recent social theorists of food (Lupton 1996; Falk 1994) have emphasised the corporeality of the human body in consumption. Falk poses a heuristic distinction between 'primitive' and modern society in which the 'primitive' body is characterised by openness within a society that is a sharing, 'eating-community' (Falk 1994: 20). He stresses the sense in which, in traditional societies, the eating of food incorporates the eater into the community: 'eating into one's body/self and being eaten into the community' (Figure 1.3). Shared ritual meals serve to further insulate members within such a community. They locate themselves at the centre of their perceived universe, communing with the supernatural, often through sacrifice (normally of an animal whose meat is then eaten), which sacralises the partakers in the ritual. The use of narcotics can further enhance such experiences to emphasise their distinctiveness and imprint their power on the participants.

Falk also perceives a historical transformation in the abandonment of the food taboos of 'primitive' societies. In their place are the weaker system of distinction - Bourdieu's social judgement of taste (1984) - in which class and background inform people's preferences and dislikes. The search for innovation, which has characterised culinary art in the last three centuries, has eroded the rigid classifications of food taboo amongst all but the most orthodox.

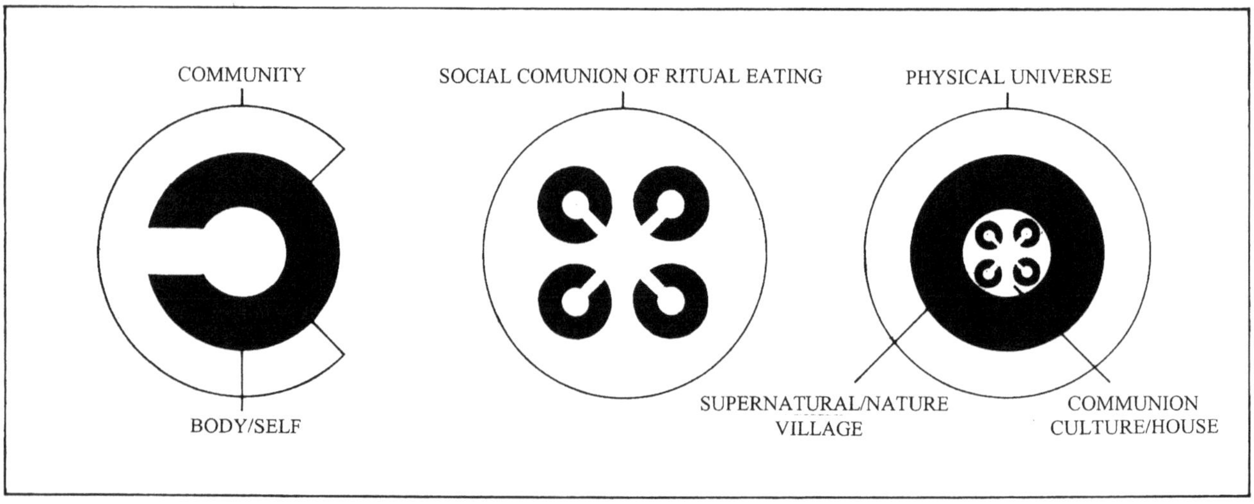

Figure 1.3. *Pasi Falk's model of pre-modern eating communities.*

Food, gender and bodies

In a recent study of British households, the production, serving and consumption of food were found to be gendered at all levels - shopping, choice of dishes, portion size, clearing up. Gender inequality was a conspicuous dimension of the domestic politics of food irrespective of social class (Charles and Kerr 1988). In many non-Western societies women are also in charge of the food-work, whether it be cooking the daily meals of a household or providing the labour for growing and raising the crops and animals which are to be consumed at ceremonial feasts. There are, of course, exceptions and feasting may also be the one time when men actually do the cooking, as among the Akha of northern Thailand (Clarke 2001) and the Gimi of Papua New Guinea (Gillison 1980), in a rite of reversal that serves to highlight and naturalise the order of everyday activity.

Women's bodies are profoundly implicated with the consumption of food, not only because they physically provide food for their infants but also because, historically, food is 'inscribed' into the shape of the body itself. Fat is, historically, a sign of well-being, prosperity, beauty, insulation from hunger, sociability and fertility. In the twentieth/twenty-first century white Western world, obesity has been twisted into a psychological problem, especially for women, and is associated with mental illness and other negative imagery (Counihan 1999). Foods take on strong moral values - 'naughty but nice' or 'tasteless but good for you' - and create anxieties and guilt about our bodies and selves. Featherstone describes how taste, in Bourdieu's sense of cultural capital (1984), is embodied by being not only inscribed on the body but also by being made apparent in body size, demeanour, ways of eating and drinking, walking, sitting, speaking and gesturing (1987: 123).

Meigs' remarkable study of the concept of *nu* among the Hua of Papua New Guinea illustrates how food and eating are principal means by which vital essences are transferred between organisms and objects, to the extent of altering gender (1984). Foods (except for wild species) contain the *nu* of their producer, the substance that provides nourishment and growth, and each person has only a finite amount of *nu* which is dispersed in their bodily fluids as well as in food transactions. Old women who have expended their *nu* become as men whereas old men, who have absorbed women's *nu* through food and sex, become as women. Meigs points out that food contains the self and feelings of the producer; it is not an alienable item that may be separated from its provider (1997). Thus eating is an emotional and mystical event in which the self blends with the surrounding world. 'Bad' food is that which comes from 'bad' people rather than from any intrinsic properties of the food itself.

Food and identity

'Tell me what you eat and I will tell you what you are' wrote Anselme Brillat-Savarin in 1825. Yet Ludwig Feuerbach's pithy aphorism from 1850 'Der Mensch ist was er isst' (We are what we eat) is the one that is remembered today. The meanings of this phrase have a kaleidoscopic quality. It can be taken literally, as it is in some societies where ingestion is considered to cause people to take on physically the attributes of certain foods (as Meigs and others have illustrated among various New Guinea communities in which men must avoid food that resembles female attributes and women must avoid that which looks like male genitalia; 1984; 1997). Yet there is also a wide field in which our food and how we eat it serve to build our experiences and identities, both individual and collective. Food is both a highly condensed social fact and a marvellously plastic kind of collective representation (Appadurai 1981: 494). There is now a large literature on food and identity, in terms of gender, nationalism, class and other group identities (see references in footnote 1).

In addition to the set-piece ethnographies, a series of studies of particular foodstuffs within or between cultures have made important contributions. Nick Fiddes' perceptive book about meat in Western society draws out its multivalent connotations and associations with sex, masculinity, power, barbarity, superiority and domination over animals (1991). Others have shown how even the most basic of staples - rice, bread, yams, maize - have powerful mythologies embedded in their preparation and consumption (Mack 1986; Farb and Armelagos 1980; Visser 1986; Bowden 1983; Fortune 1932; Taube 1996; 2000). Even when our staples do not carry heavy semantic loadings (such as the 'staff of life' or the sacrificial body of Christ) or are not imbued with complex meanings (such as confetti or hot-cross buns[4]), they are important and comforting reference points which remind us of who we are and what we value. Such concerns are thrown into high relief when travelling abroad, to the extent that the lack of mundane items like Marmite and soft bread emphasise the distressing dislocation from one's everyday national, ethnic and cultural milieu.

Foodscapes: we are where we eat

Of particular interest to prehistoric archaeology is the notion that 'we are *where* we eat' (Bell and Valentine 1997). Food is involved in practices that affect our sense of place and identity. In the world before the supermarket and before mercantile capitalism, the landscape of any community was essentially its 'foodscape', its source of continuing nourishment and reproduction. To leave it or to destroy it was to lose the very means of survival. Of course, the self-contained, self-sufficient subsistence-level community with its domestic level of production is something of a mythologised fiction. Exchange in animals and foodstuffs is probably as deeply embedded in human societies as the sharing of food, animals in particular being important exchange items in traditional societies (Parker Pearson 2000b). The pre-global foodscape was a complex lattice of human relationships rather than simply the local resource of fields, plants and animals. The 'foodscape' is thus not a restricted territory to which the subsistence farmer is tied but a sometimes extensive network of exchanges, hospitality and obligations which may involve a variety of locales and varying rights over what may be taken, when and by whom.

Food and politics

'The French will only be united under the threat of danger. Nobody can simply bring together a country that has 265 kinds of cheese' Charles de Gaulle 1951.

'I want good vibes in my food, but I'm eating Vietnam for breakfast' Elaine Sundancer, American commune farmer (cited in Belasco 1993: 80).

The six billionth inhabitant of the planet was born in 1999/2000. As population numbers continue to rise and wealth remains unequally distributed so we will have increasing difficulty feeding the world in the next few centuries. Food will continue to be a political weapon, withheld and deployed as world powers see fit. Ever since the potato famine, when the surplus of English grain was denied to the starving Irish, the global politics of food have been linked closely with famine and war. Even in the Bronze Age, the biblical tale of Joseph's warning to Pharaoh that he should construct stores for the produce of seven fat years, to be followed by seven lean ones, provides an indicator of the political power behind the control of foodstuffs.

Food is used to dominate and to resist. The caste system in India, complex and variable as it is, affects not only who eats what (Brahmins are traditionally supposed to eat raw and superior cooked foods whereas meat is eaten only by lower castes) but on who gives food to whom, in that givers must be of equal or higher status to receivers (Appadurai 1981; Miller 1985). The personal politics of vegetarianism and other forms of ethical eating in the Western world articulate with broader political concerns about factory farming, animal rights, rain forest despoliation, and world over-population, to mention the main issues (Bell and Valentine 1997: 109-12; Fiddes 1991: 195-223). Macdonalds and Coca-Cola have become totemic representations of the insidious effects of economic globalisation (or American hegemony), subject to boycotting of their products and destruction of their premises during rallies and riots in America, Europe and the Middle East.

Food is also an important element of identity politics. Chaudhuri's broad-brush account of south Asian history highlights three material discourses of identity - architecture, dress and food - and shows how different communities in the region, together with European colonial incomers, were prepared to maintain, integrate or surrender their identities along these dimensions. He concludes that food was the dimension along which most were prepared to adopt new ways whereas architectural style was the most resistant (Chaudhuri 1990). Food also provides something of a microcosm of globalisation. Allison James has characterised British food preferences as four different discourses on identities (1996), in which re-authentication of local identities through global food, expatriate appreciation of local foreign foods, heritage-isation of local British foods, and creolisation of British and foreign foods are all strategies by which the British use food to reshape both local and global identities in a complex interplay of meanings and intentions relating to who we are.

Cuisine and class

It is a truism in our male-dominated society that the social elevation of a craft into an art is accompanied by the emergence of male practitioners in what was hitherto predominantly women's work. Just as knitting and pottery have spawned 'master' artists, male chefs are household names whose books and television programmes are consumed by millions. Cooking is today a complex meshing of aspirational values, aestheticism and constant innovation, carried out by charismatic cooks with culinary finesse and skill who demonstrate that their food is especially good to think. If cooking is a language then cuisine is the high-culture Latin or Mandarin spoken to empower, impress and control.

[4] Confetti was traditionally grain which was thrown at the bride and groom; hot-cross buns are thought to have been associated with fertility linked to the pagan deity Ostara; Farb and Armelagos 1980.

Jack Goody's classic study of the social significance of cuisine and its development and spread provides archaeologists with a useful model to interrogate (1982). Cuisine can be defined not simply as the elevation of cooking into an art form requiring specialists and esoteric knowledge but also as the preparation of multiple dishes which may be served in complex arrangements especially during ceremonial occasions. As such, it has been a feature of class-based societies for thousands of years, especially in China, south Asia and Europe. Goody considered that there was no cuisine within sub-Saharan Africa on the basis that, even at ceremonial feasts, it was volume that counted rather than diversity of dishes. He also made the case for sub-Saharan Africa having no deeply developed pre-colonial class systems and argued that globally, cuisine and class have gone hand in hand for millennia. In retrospect, Goody's classificaton of African societies, ancient and modern, as having no evolved classes is far too sweeping as a generalisation. Among the many sub-Saharan ancient civilisations, those such as Kerma, Great Zimbabwe, Jenne-Jenno and Axum seem to have supported profound and long-term class differences (Connah 2001). Despite this lack of perfect fit, the class/cuisine idea is still of interest for prehistorians comparing, for example, the Neolithic and Bronze Age Aegean with the British Isles in the same period.

Feasting, politics and power

If the foods that people choose are bound up with political and moral issues, then taking part in public consumption events also has a major political dimension. Dietler and Hayden's edited volume on feasting (2001) is an explicit recognition of the political contexts and opportunities for social change which may be embodied in the kinds of large-scale ritualised consumption events defined as feasts. Feasting has for some time been identified as a means by which status and power are negotiated and the social order transformed (Friedman 1975; Friedman and Rowlands 1978; Rappaport 1968; Bloch 1999). Dietler's recent theorisation of feasting has, however, rescued the concept from these earlier structural Marxist and functionalist concerns: 'it is essential for archaeologists to come to grips with the arenas of social action in which, and the sets of practices by which, the micropolitics of daily life are played out... we need to think seriously and realistically about political life as it is lived and experienced if we are to fill our analytical categories with meaningful content and advance beyond mechanistic structural correlations, vague pronouncements about overdetermined social processes, and sweeping evolutionary teleologies' (2001: 66).

Feasts, then, are occasions when large-scale hospitality creates debts and obligations, when reputations are made and lost, when the social order is exhibited, challenged and reformulated, when the work of many may be claimed by the few, and when new factions, mobilisations, alliances and other relationships are formed or dissolved. Bender has suggested that the transition from gathering to a food-producing way of life in the Near East might have been effected through the competition engendered by feasting (Bender 1978). The contexts for feasting may be multiple and varied: to extract tribute, to recruit military aid, to honour the ancestors, to bury the dead, to extend hospitality to outsiders, to make alliances between lineages, to reward work parties, and many more (Hayden 2001: fig. 2.1).

The ways in which archaeologists can differentiate feasting from the remains of day-to-day consumption are problematic. Hayden has produced a set of criteria for feasts (2001: tab. 2.1) which includes not only the expected food residues (mainly large items in large quantities) but also the enhanced architectural arrangements (specially constructed facilities) and special locations (such as at a tomb). In some cases, the size of the feasting midden itself becomes a mark of the status of the feast-givers (Huffman 1996: 20-2). Yet Hayden's criteria perhaps need to be turned on their head. The real difficulty that archaeologists face is whether it is possible to identify the remains of *day-to-day* food consumption. The archaeology of everyday food practices is perhaps rather more difficult to recognise than we may think. It is not impossible that *most if not all* animal bones on a prehistoric site may be the result of feasting.

Without markets and refrigerators, the killing of large domestic animals in agricultural societies is often reserved for special occasions of feasting - sacrifice of animals is not often a daily activity. Where the bones are deposited in the aftermath of feasting (and sometimes the meat, blood and marrow may be conserved for days or even weeks after) may not show up as a feasting midden, since the bones may be chewed by dogs or scattered around a settlement (Parker Pearson 2000b: 227-30). Equally, the separation of pottery broken in day-to-day contexts from pottery broken in feasts or other rituals (see Brück 2001) requires careful study of formation processes and temporalities of midden accumulation. The frugalities of day-to-day life (in which even cracked and broken pots may be carefully reused) often form a marked and even deliberate contrast to the wasteful consumption involved in feasting.

PART TWO: NEOLITHIC AND EARLY BRONZE AGE BRITAIN - THE CULINARY BASIS

There has been surprisingly little discussion or analysis of cooking practices in Neolithic and Early Bronze Age Britain. One of the notable exceptions, and among the earliest examples, is John Hedges' description for Neolithic Orkney: '[Meat, grain and wild seeds] may have ended up as gruel or stews - some species of boiled sustenance - no doubt supplemented by wild turnips and other edible roots, fruits and leaves of plants. Pots are frequently burnt on the outside from having been placed in embers and some... have knobs or holes by which they could have been suspended. There were, however, other methods of cooking available. Meat could obviously have been roasted - this being suitable for large joints - and a variety of victuals from gulls' eggs to unleavened bread can be cooked on a stone slab laid on a glowing fire. There were also ovens... It is not to be supposed that food would necessarily have been very basic. Given eggs, cheese, butter, milk, yoghurt, honey, flour, fat, nuts, oil, meat, fish, game, seafood and a whole range of vegetable products a great variety of dishes *could* have been prepared.' (1984: 217).

In terms of culinary (pre-)history, the European and west Asian Neolithic and Early Bronze Age are of profound significance for establishing some of the fundamental transformations in human lifestyles, a few of which - in the form of lactose and alcohol tolerance - have had biological impacts on the human body. Amongst the world-transforming innovations of this period were:

- monumental architecture in which to stage feasting events,
- ceramic and sheet-metal technology for the manufacture of feasting and drinking paraphernalia,
- the consumption of milk as a secondary animal product,
- the incorporation into foodstuffs of yeasts to leaven bread and make alcohol,
- the use of bacteria for making cheese, curds and whey, and yoghurt,
- the controlling of fermentation processes (other than those involving yeasts),
- the beginning of long-distance transport of luxuries such as salt.

Furthermore, a series of innovations which had been made among earlier hunter-gatherer communities were now adopted widely in western Europe:

- the development and widespread use of ceramic containers for boiling and storage,[5]
- ovens for baking, broiling and roasting,[6]
- storage facilities to break out of the tyrannical grip of seasonal availability,[7]
- the use of grinding stones to make porridge and bread.[8]

The inception of the Neolithic way of life has, since Gordon Childe's time, been classically understood as a move from food gathering to food production (1928), although we increasingly recognise that this transition to agro-pastoral farming was a long-term and geographically diverse process (Dennell 1985: 184-7; Edmonds 1999; Zvelebil 1994). The literature has been dominated until recently by questions about production, subsistence, husbandry and ecology, with only scant attention paid to the fact that culture is constructed through consumption and not just through production. Foodways, along with other forms of material culture consumption, serve to materialise, temporalise, spatialise and substantiate human culture. Andrew Sherratt has done more than most to raise awareness of consumption as a useful if neglected field of study in this period: 'The perspective offered here would deny that population growth (or even agrarian improvement) is an independent variable; instead it would assert that societies grow and thrive as they successfully interact with their neighbours, and not least in the mutual provision of consumable commodities.' (1999: 33).

Innovations and transformations in food technology and consumption also provide people with new metaphors for thinking about the world and acting within it. There is growing evidence that the introduction of cereals was associated with new metaphors for conceptions of life and the hereafter (A. Jones 2000; Parker Pearson 1999: 91-2; Tilley 1996: 188-9). Cooking in its new forms might have been linked to human gestation and growth (Jones 2000). Entirely new categories of food experience ('rotted' products, intoxicating substances) might have opened up new social categories and dimensions of practice. Pottery, for example, may facilitate not only the use of new foodstuffs like cereals and allow more intensive use of existing foods but also enable more complex sequences of food storage and production, such as slow heating of stews, porridge, broth and weaning foods (Gibson and Woods 1990: 58; Jones 1996: 296; Thomas 1999: 96). Julian Thomas (1999: 225) has argued that these new ceramic foodways involving storage, cooking, combination, presentation and consumption would have promoted an episodic and punctuated sense of time, transforming the rhythms of social life.

Cooking

The advent of ceramics is a momentous event in human history, associated with the earliest farming communities and also with hunter-gatherer societies in Japan and northern Europe. In the context of the latter, Chris Tilley has discussed how Scandinavian Ertebølle pots made boiling possible as a mode of food preparation,

[5] In fact, pottery seems to have developed first in the Upper Palaeolithic of Japan, China and the far east of Russia around 13,000-11,000 BC. Conversely it was not used in the Near East until after the aceramic Neolithic.

[6] Ovens are known from the Upper Palaeolithic around 25,000 years ago at Dolní Vestonice in the Czech Republic and also in Mesolithic contexts.

[7] Pits and other storage facilities are known from the Mesolithic (Rowley-Conwy and Zvelebil 1989; Zvelebil 1994).

[8] Grinders are known from the Upper Palaeolithic and Mesolithic, presumably for processing seeds and nuts into an edible state (Wright 2000: 92-8; Zvelebil 1994).

transforming foods most completely in a Lévi-Straussian way from their natural condition into a decomposed and artificial state (1996: 66). He surmised that these highly nutritious boiled foods might have been prized in ceremonial rituals and that this new style of cooking food might have coincided with an increased 'cooking of the land' by widespread burning in order to transform the land itself from nature to culture. That pots are themselves 'cooked' is a point of potential interest since food can be boiled in nothing more than an animal skin suspended over the fire.

Ranges and shapes of British Neolithic pottery provide food for thought on the manner of cooking, storage and distribution. The vast majority of vessels were under a litre in volume (Thomas 1999: figs 5.6, 5.13) and can be classified as personal cups and small bowls. Very few were bigger than eight litres and, of these, a small number were capable of holding volumes over 20 litres. The rounded bases of Early Neolithic pottery make these vessels suitable for holding or for setting in soft beds of ashes. Their shapes are predominantly open or 'neutral' (Thomas 1999: 103), sufficiently wide-mouthed to allow the sharing-out of the food contained within. The same is also true of the later Neolithic impressed Peterborough Wares and Late Neolithic-Early Bronze Age Beakers but not of the Late Neolithic Grooved Ware whose decoration and shape are reminiscent of basketry containers. These flat-bottomed Grooved Ware skeuomorphs hint at the existence of organic containers in the Early Neolithic and also suggest the possibility that closed vessels might have been synonymous with collecting and storing food, as opposed to the open-mouthed vessels from which boiled food could be shared and consumed (L. Hurcombe pers. comm.).

Ann Woodward has recently suggested that British prehistoric pottery (at least, prior to the Middle Iron Age) does not constitute 'domestic' ware since its contexts of deposition, even when associated with houses, are so often special deposits of smashed or whole vessels (1998). She argues that Neolithic and Early Bronze Age pottery was made and used specifically for consumption at communal gatherings and feasting events. The proposal is deliberately provocative and may be overstating the case: everyday household settings in which ceramic refuse may have been produced rarely survive from these periods and are thus seriously under-represented among the archaeological remains. Yet it does highlight the significance of ceremonial circumstances in which most of the ceramics seem to have been deposited.

The hearth, oven and cooking pit are also aspects of cooking technology that have social implications. Katherine Wright (2000) has shown how the place of cooking shifted during the Pre-Pottery Neolithic of the Near East, a fluid and unstructured arrangement (both indoors and outside) in the Natufian and PPNA giving way to milling, cooking and storage areas near house entrances (visible and semi-public) and then, towards the end of the PPNB, to more secluded and cloistered settings within the village. Investigating cooking in Late Neolithic Greece, Halstead similarly points to changes along a public/privacy: hospitality/private consumption gradient in which cooking areas (formerly in courtyards between houses) became enclosed behind kitchen and yard walls, fine decorated tableware disappeared, and surplus food might have been stored rather than shared in communal feasting (2001: 86; but see Valamoti this volume). Halstead (1989) also contrasts the fine painted pottery and shared open-area cooking and neighbourly hospitality of Early Neolithic Greece with their absence in the LBK of central Europe in that same period, implying a lack of food-sharing and hospitality in that region,though Thorpe has argued that the Greek pattern may be interpreted as inter-household competition within communities whereas the LBK societies were more co-operative at the community level (1996: 32).

Cooking facilities are by no means limited to household settings. From beneath an Early Bronze Age cairn at Stannon Downs, Cornwall, a stone-lined pit with carbonised wood and a cover of a flat stone and rings of stones has been interpreted as a cooking pit (Harris *et al.* 1984; Wood 2000: 93). Strangely, there was no burial under this cairn and it is assumed to have been erected as part of the funerary ceremonies for burials under other cairns. The baking of meat in this earth oven might well have accompanied the 'cooking' of the corpse as it was cremated prior to burial in a pot under an adjacent mound.

Recent work on animal bones points to potential methods of establishing just how meat was cooked in the Neolithic. Albarella and Serjeantson (2002) have identified barbecuing or roasting of pigs at Durrington Walls from the burnt distal ends of lower limb bones. Pearce and Luff's experimental work on fresh, boiled and roasted bones (1994) highlighted differences in fragmentation but the degree to which differentiation is possible is problematic (see Montón Subías 2002 for a recent summary).

Salt and boiling

Two of the great transformations in culinary practices that do not, in the main, occur until after the Early Bronze Age are the archaeologically identifiable use of salt and the boiling of large meat joints as a mode of feasting. It is not until the later Bronze Age that we have evidence for large-scale production of salt - from coastal boiling sites or mined from underground seams - and for the use of cauldrons or the construction of outdoor boiling pits surrounded by mounds of heated rocks. In central Europe there is evidence for salt exploitation as early as the Middle Neolithic and briquetage from the Early Bronze Age although it is predominantly from the Late Bronze Age onwards that briquetage and extraction facilities have survived (Harding 2000: 249-54).

Salt is a useful preservative which extends the period of edibility of meat, fish and milk products in particular and could have been used as such in prehistory. It is also an important flavourer and an essential part of our diet. Additionally, it might have been associated with special spiritual powers like those described for the plant-derived potassium salt of the Baruya of Papua New Guinea, which is used only in ritual contexts and not in daily consumption. Reserved mainly for rites of passage, salt is regarded by the Baruya as a source of strength which accumulates in the liver (the centre of a person's force) and it is associated with male strength in the form of semen (Godelier 1999: 139).

Meat provides a certain amount of salt in its own right but we can expect that mineral and sea salt would have been prized and exchanged in small quantities over great distances before the Bronze Age, valued as a necessary luxury well before the long-distance trade in exotic spices in the historical period. Of course, its archaeological identification as a trade item or a food is supremely difficult in the absence of distinctive non-perishable containers.

The boiling mounds or *fulachta fiadh* found in Britain, Ireland, Scandinavia and central Europe (Harding 2000: 23) are mainly Late Bronze Age in date or later, though some are known from the end of the Early Bronze Age (Burgess 1980: 148). Their large stone or wooden troughs were capable of holding large volumes of water which could have boiled substantial portions of meat. These sites are therefore considered to have been outdoor feasting locations yet those that have been excavated have yielded little other material culture and, on account of the acidity of the soil, no animal bones (O'Kelly 1954; Burgess 1980: 223-5).

Seasonality and metaphor

The capacity for storage that is evident from Neolithic pottery undoubtedly enabled communities to insulate themselves to a minor degree from the fluctuations in food availability during the year, particularly important during the spring and early summer, those least productive seasons in terms of human dietary needs. The cereals brought to Britain were wheat (einkorn, emmer and bread wheat) and barley (six-row hulled and naked); although they were just a limited part of the Neolithic diet these cereals, together with hazel nuts, are the most visible 'tip of the iceberg' in terms of available plant foods. Bread could be produced from stored grain long after the harvest was over. Its remains occasionally survive in Neolithic and Early Bronze Age contexts from the north Italian and Swiss lake sites and it is recorded at a Hungarian cave at Aggtelek (Harding 2000: 132). In Britain an Early Neolithic pit at Yarnton dating to 3620-3350 BC contained a small piece of barley or mixed cereal 'bread' or 'cake', of coarsely ground grains, together with an apple core, hazelnut shell fragments and wheat grains (Robinson 2000: 89; Hey *et al.* this volume).

From our twenty-first century Western perspective, however, it is hard to appreciate just how enthralled these prehistoric farmers were to the seasonal rhythms of food availability. Other than stored products only a few foodstuffs, such as meat and the hundred or so types of edible fungi, were available all year round (Figure 1.4). Seasonality models have been proposed for the Early Mesolithic (Clark 1954; Rowley-Conwy and Zvelebil 1989; Zvelebil 1994) and for the Iron Age (Giles and Parker Pearson 1999; Parker Pearson and Sharples 1999: fig. 1.11). The pre-farming seasonal round into which the domesticated cereals had to be imported was one which followed, to some extent, an unfolding process of growth: the main plant foods available in the winter and spring would have been edible roots and tubers (though they are best eaten after summer flowering of the plant; *cf* Mabey 1972: 60-74; Wood 2001: 141-52), followed by stems and shoots and young leaves in the early summer. In late summer and autumn the foods available included seeds, nuts and fruits. Thus the annual round of plant foods consumed perhaps mirrored the cycle of growth, from root to stem to leaves to fruit/seed (Table 1.1). Within such a cycle the use of stored nuts, cereals and dried fruits/vegetables would have provided contrast, punctuation and tempo, particularly through the provision of feasts in which such stored foods could be consumed with meat.

As in earlier periods – and, indeed, until modern times - the leanest season of the year was spring and early summer (Figure 1.4). The bounty of late summer and autumn included cereals, nuts, fruits, seeds and fungi. Honey was available in the summer and autumn months whilst milk would have been plentiful from late spring onwards. Roots were obtainable throughout the year but key species such as pignut, salsify and herb bennet are best in midsummer whilst dandelion and wild parsnip are tastiest in the autumn (Mabey 1972). Stems of plants such as burdock, asparagus and sea kale are also best eaten between May and September. Greens (including nettle, bistort and mallow) were at their most abundant between the spring and autumn but certain species such as goosegrass, chickweed and sorrel could be gathered and eaten in winter and early spring. If seafood and river fish (see below) were not consumed in the lean season, then there would have been a strong reliance on stored foodstuffs and fresh meat in that half of the year between winter and early summer. Thus the annual cycle might have been conceptualised as a revolving dualism of fresh and stored foods, no doubt with attendant symbolism relating to the life-death cycle and the staging of ceremonies and rites of passage.

Roots and tubers were probably an important part of the diet, capable of being eaten over most of the year either raw, boiled or roasted, yet they hardly ever survive archaeologically (but see Moffett 1991). In more recent centuries roots were treated as animal food, a poor substitute for bread, and the cause of wind, melancholy and diarrhoea (Mabey 1972: 60). The tooth wear noted from human remains within the Early Neolithic long cairn at Whitwell quarry (Chamberlain and Witkin this volume; Nystrom and Cox this volume) is consonant with the eating of tough roots and tubers rather than softer processed cereals.[9] This 'Mesolithic dietary

[9] Further evidence for the importance of root crops in the Neolithic diet may also be present in Denmark where low levels of dental alveolar reduction in the Neolithic period have been attributed to the consumption of coarse, hard and tough food such as root crops and meat (Bennike 1985: 156).

Table 1.1. *Some of the vegetable foods available in prehistoric Britain. Those shown in bold have been identified on British Neolithic sites.*

Edible roots & tubers	Stems & shoots	Young leaves	Seeds	Nuts	Fruits
Burdock *Arctium minus*	Burdock *Arctium minus*	Bedstraw *Galium* sp.	**Barley** ***Triticum* sp.**	Beech *Fagus sylvatica*	**Crab apple** ***Malus sylvestris***
Dandelion *Taraxacum officinale*	Marsh-thistle *Cirsium palustre*	Beech *Fagus sylvatica*	**Fat hen** ***Chenopodium album***	**Hazel** ***Corylus avellana***	Elderberry *Sambucus nigra*
Early purple orchis *Orchis mascula*	Nettle *Urtica dioica*	Bistort *Polygonum bistorta*	**Wheat** ***Hordeum* sp.**	Oak *Quercus robur*	**Hawthorn** ***Crataegus* sp.**
Pignut ***Conopodium majus***	Sea kale *Crambe maritime*	Burdock *Arctium minus*	Wild celery *Apium graveolens*		**Raspberry/ blackberry** ***Rubus* sp.**
Sea beet *Beta vulgaris*	Wood-garlic *Allium ursinum*	Chickweed *Stellaria media*			**Sloe** ***Prunus spinosa***
Silverweed *Potentilla anserina*		Dandelion *Taraxacum officinale*			
Water parsnip *Sium latifolium*		**Fat hen** ***Chenopodium album***			
White water-lily *Nymphaea alba*		**Hawthorn** ***Crataegus* sp.**			
Wood-garlic *Allium ursinum*		Hedge-garlic *Alliaria petiolata*			
		Nettle *Urtica dioica*			
		Samphire *Crithmum maritimum*			
		Sheep's sorrel *Rumex acetosella*			
		Shepherd's purse *Capsella bursa-pastoris*			
		Sow-thistle *Sonchus asper*			
		Wild celery *Apium graveolens*			
		Yarrow *Achillea millefolium*			

signature'[10] within a Neolithic setting demonstrates something of just how important the 'wild' or native plant foods continued to be even in the presence of cereals and probably pulses. The latter are not certainly documented from Neolithic Britain although peas and beans were consumed by LBK farmers in the Netherlands (Bakels 2000).

The structuring principles of Neolithic foodways in Britain might thus have been a real and metaphorical cycle of annual growth, an understanding of native *versus* outsider foods, the seasonal contrast between fresh and stored produce, and also the physical experience of softness/hardness. The latter would have been embedded into the growing process as infants graduated from breast

[10] The 'Mesolithic dietary signature' amongst pre-farming communities across Europe was further complicated by a dual pattern of dietary intake (Meikeljohn and Zvelebil 1991; Schulting and Richards 2000; Bonsall *et al.* 1997; Lillie 1997). Recent isotope results from the bone of a Late Mesolithic woman, found in the River Trent in inland Britain, indicate that she ate almost nothing but meat (Davies 2002).

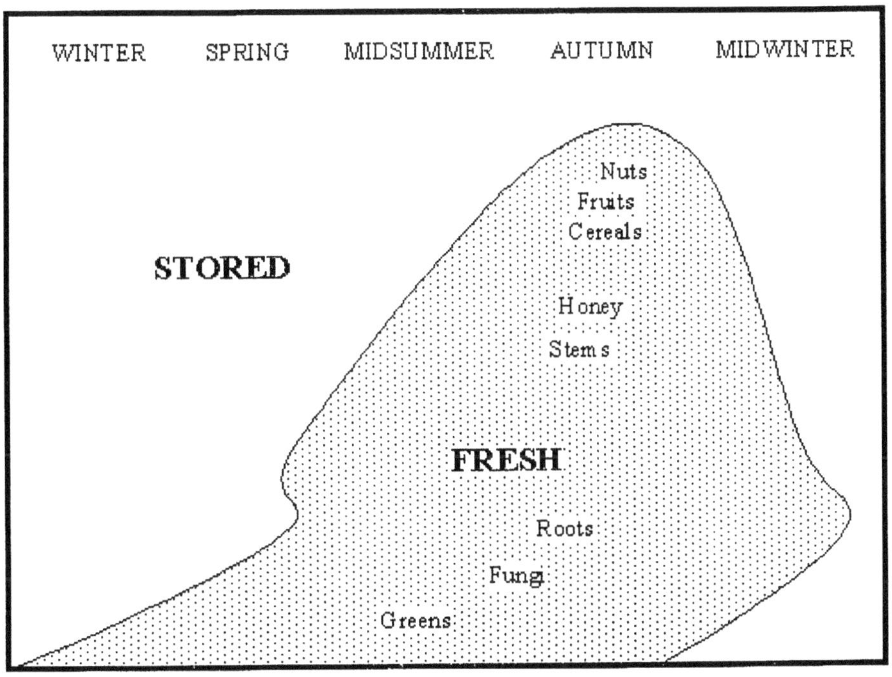

Figure 1.4. *A speculative reconstruction of the seasonality of Neolithic foods in southern Britain.*

milk to soft, mushy foods and thereafter, to use today's terminology, to solids. The use of cultivated cereals introduced greater elaboration into this process, with more flexibility in producing soft, boiled cereal-based meals. This process of hardening was one that was manifested not only through ingestion during maturation but also externally in the use of wood, soil and stone for ceremonial monuments (Parker Pearson and Ramilisonina 1998).

The nature/culture opposition was consequently a minor aspect of daily life, if indeed it had any reality at all for Neolithic people in Britain, since the sources of so much of the carbohydrate consumed, whilst appearing as 'wild' to our eyes, were the foods of indigenous ancestral tradition. Of course, other experiences of the European Neolithic must have been profoundly different and the wild/domesticated distinction appears to have been far more marked amongst those Early Neolithic communities that seem to have retained little in the way of a legacy from the Mesolithic. In the case of Greece, for example (Valamoti this volume), the new cereals almost certainly had a more central economic and symbolic role than wild plants.

Fish

The extent to which marine fish might have been subject to cultural proscriptions or even taboos in Neolithic and Bronze Age Britain is beginning to emerge. Fish remains are present in the middens of Neolithic settlements in Orkney such as Skara Brae (Clarke and Sharples 1985) and we could predict that coastal and riverine communities would have exploited these forms of food where available and/or when necessary. Yet fish bones are surprisingly few on these Neolithic sites in comparison to quantities from Mesolithic shell middens in Scotland (Barrett *et al.* 1999; Mellars 1978). New evidence from carbon and nitrogen stable isotopes from human bones further indicates that coastal Mesolithic people might have eaten large quantities of seafood whereas their Neolithic counterparts seem to have enjoyed a largely terrestrial diet (Richards and Hedges 1999; Richards 2000 and this volume; see Mays 1998: 182-90 for an introduction). Similar patterns have also emerged for the Mesolithic-Neolithic transitions in Sweden, Denmark and Portugal (Lidén 1995; Tauber 1981; Lubell *et al.* 1994).[11] Further work on sulphur isotopes from bone collagen may also help in identifying the contribution of freshwater foodstuffs (Richards *et al.* 2001: 189).

Fish - especially sea fish - seems to have formed only a minor part of the diet into the Bronze Age in Britain. It seems that the eating of fish might not have been an absolute taboo in northern Scotland but the scant remains hint at some form of proscription. That the middens of a coastal site such as Skara Brae should yield mainly trout and eels, with cod species and various types of wrasse as minor proportions of the assemblage (Barrett *et al.* 1999: tab. 2), makes it likely that sea fish were rarely eaten. The fruits of the sea need not be subject to formal taboo to be avoided: many coastal agricultural communities ignore the copious seafood on their doorsteps but cannot be said to proscribe its consumption. Some have explained the Mesolithic-Neolithic shift to terrestrial resources as an economically efficient transformation: women seem to have had higher intakes of fish and marine food (Meiklejohn and Zvelebil 1991; Bonsall *et al.* 1997; Lillie 1997; Schulting and Richards 2000) and if it was they who shifted their activities to cultivation, then fishing might have been abandoned. Additionally, sea fishing could have been seasonally incompatible with mixed farming (Zvelebil and Rowley-Conwy 1984). Yet such explanations fail to take into account the cultural and ideological power of foodstuffs in defining social values and lifestyles.

Those cultural and ideological categories were perhaps somewhat more complex than often envisaged. Julian Thomas (1991) has suggested that the British

[11] In eastern Scandinavia the exception is the Pitted Ware culture of the third millennium BC which appears to have been a seafood lifestyle co-existing with the terrestrial economy of the TRB.

Neolithic embodied a strong dichotomy between the wild and the tame although he concedes that there was extensive use of many wild food resources. As discussed above, the principle of a culture-nature dichotomy may not be the most appropriate structuring concept through which to understand Neolithic food moralities. Whatever the culinary status of wild animals, sea fish might well have been treated as a special category. A better understanding of their consumption and deposition has to be a key feature in future research.

Meat, grain and milk

In 1989 the standard view of the British Neolithic as a thoroughly agrarian and sedentary society received something of a battering as a result of two papers which emphasised the degree of reliance on wild resources and the mobile, non-intensive nature of farming and husbandry practices (Moffet *et al.* 1989; Entwhistle and Grant 1989). Entirely new interpretations of the British Neolithic as a largely pastoralist and mobile society (or at least a regionalised mosaic of different lifestyles, with the majority of the population largely transient) emerged in a series of syntheses which placed large-scale cereal consumption firmly in the Bronze Age (Edmonds 1999; Thomas 1991; 1999). Stable isotope results on human bones later appeared to confirm suspicions that cereals were not always a substantial part of the diet for certain Neolithic populations (Richards 2000 and this volume). Individuals buried in the south Welsh chambered tomb of Parc le Breos Cwm had ^{15}N values indicative of high levels of ingested animal protein - higher than those of individuals from Hazelton and West Kennet chambered tombs in southern England and probably higher than modern-day mixed diets (Whittle and Wysocki 1998; Wysocki and Whittle 2000).

Mark Robinson also points out that hazel nuts, crab apples, sloes, hawthorn, blackberries and raspberries are common occurrences on Neolithic sites and are virtually absent on Bronze Age settlements, suggesting that they played at least a small role in Neolithic food preparation (Robinson 2000). Glynis Jones (2000) has reassessed Neolithic cereal production, based especially on the site of Lismore Fields, Derbyshire, and highlights the extent to which carbonised cereal grains are massively under-represented in relation to carbonised wild foods such as hazel nuts. Cereal grains may be under-represented and yet assemblages like that at Lismore Fields do indicate intensive production and storage which does not sit easily with interpretations of mobile pastoralism and small-scale consumption (see also Rowley-Conwy 2000; Cooney 2000: 44-5; Monk 2000).

Entwhistle and Grant (1989) also raised doubts about Tony Legge's interpretation of cattle bones from the Hambledon Hill causewayed enclosure (*c.* 3400-3000 BC) as indicative of a milking herd (Legge 1981; 1989). Legge had argued that the high neonatal mortality and the small number of adult females represented in the bone assemblage matched a pattern of milk yield in which most calves were killed and most of those that grew to maturity were the mothers of the next generation. Entwhistle and Grant pointed out that Hambledon Hill was not a Neolithic dairy farm but a ceremonial centre - these were unlikely to be animals from a single herd but animals brought from a wide catchment area by people coming for the festivities. The debate remains unresolved but both could be right if the animals were brought from a number of different milking herds for the purposes of feasting (see Ray and Thomas this volume).

One solution to this apparent dilemma is to consider the possibility of distinct communities, each practising either sedentary cultivation or mobile pastoralism. Another is that both strategies were practised by the same communities, with labour divided accordingly (*cf* Parker Pearson 2000b). Thirdly, the isotopic evidence suggests a degree of regionalism in which communities in south Wales were more reliant on animal proteins than those across the Severn estuary. The problem, to some extent, may be one of funerary representativeness. Just as carbonised grains represent only the tiniest proportion of the whole, so the human remains recovered from long barrows, chambered tombs and causewayed enclosures are likely to have been only a tiny proportion of the total population, preserved because of their special treatment.

Mike Richards' comparison of isotope values of individuals buried in the long barrow at Hambledon Hill with those of individuals from the adjacent causewayed enclosure shows the former to have a higher average intake of animal proteins (2000). Individuals such as these might well have enjoyed a special diet which differed from others' in being high in animal protein, and this could have been achieved through regular attendance at feasts where meat was available in huge quantities. A diet high in animal proteins might have been attained by those who 'worked' the feasting network and achieved renown for themselves and their lineages, thus meriting burial in these special places. Of course, blood and milk are also animal proteins and these might have been consumed in quantity, not only fresh but also as conserved products of congealed blood (black pudding is our version) and cheese or curds.

Identification of products such as cows' milk (Craig *et al.* 2000; Dudd *et al.* 1999; Evershed and Dudd 1998), honey/beeswax, dripping (Needham and Evans 1987) and various kinds of meat from lipid and protein residue analyses of ceramics has opened up a new perspective on prehistoric foods. Yet interpretive issues remain. To what extent may the detection of beeswax, animal fats or milk actually relate to the water-proofing of the pots prior to cooking rather than the foodstuffs stored or prepared within them?

Comparisons of Neolithic dietary residues with those in the body of the Chalcolithic Alpine 'iceman' (*c.* 3300 BC; see Spindler 1994) have also helped to signpost some of the complexities of the evidence and the needs for interpretive subtleties. In contrast to the British Neolithic diets high in animal proteins, the stable isotopic analysis of the iceman's hair indicates that he had been living on a diet primarily of plant and cereal foods in the months prior to his death (Macko *et al.* 1999). Yet his colon contained meat in the form of muscle fibres, together with wheat and barley bran (Dickson *et al.* 2000). Whilst the analysts of his colon were keen to point out that he was not a vegan, as the stable isotope analysts had suggested, perhaps the main point to be grasped is that diet is not necessarily a daily, weekly or even monthly constant. In certain cultures people may go for months without eating meat in those periods when festivals or feasts are not programmed (Parker Pearson 2000b).

The iceman offers a remarkable opportunity to examine the temporality and seasonality of dietary variation in one individual over the short term, medium term and, if stable isotopic analysis of his bones is carried out, in the long term. New developments in the analysis of cholesterol in bone may make similar short-term dietary distinctions possible in skeletal material (Stott *et al.* 1999). Such temporality within individual dietary biographies may also be addressed by stable isotopic analysis, although the recent work has concentrated on cattle rather than on human bones and teeth (Balasse *et al.* 1997; 1999).

Nutrition and malnutrition

Food is most graphically 'written' on the body in terms of people's level of nutrition. Today the over-indulgence in the 'wrong' foods that leads to obesity has created a whole universe of conflicting emotional states and discourses (Lupton 1996). In prehistory it is far more likely that ample bodily proportions were considered in many circumstances as a sign of well-being, with thinness being a marker of poverty, hard times, ill health and perhaps disfavour in relations with the spirit world.

Osteological analyses of malnutrition have been almost entirely conducted within the evolutionary adaptive paradigm of processual archaeology, focusing on the estimation of levels of population 'stress' and health as abstracted and dubious markers of the human species' biological fitness at different times in prehistory. In part this is due to the difficulties of separating different factors - such as malnutrition, disease, or hereditary conditions - responsible for particular osteological and dental conditions. Even so, such approaches tend to downplay the cultural aspects of food and to treat it as no more than a necessary fuel, an amalgam of calories, proteins, carbohydrates and minerals. For example, the period of transition from food gathering to food production in both the Old and New Worlds has been characterised in a variety of case studies as a period showing a progressive decline in overall health and adaptive fitness (Cohen and Armelagos 1984) but there is little consideration of how this decline - attributed in no small measure to the dietary deficiencies associated with an increased reliance on cereal consumption - might have been considered an acceptable price to pay for embracing the ideological trappings associated with these new foodstuffs.

Evidence for the decline in health and nutrition at the Mesolithic-Neolithic transition comes from the eastern Mediterranean and the Near East, as measured by declining stature and increases in deprivation markers such as porotic hyperostosis (including cribra orbitalia) and enamel hypoplasia (Angel 1984; Smith *et al.* 1984; Agelarakis and Waddell 1994). Conversely, the transition in Scandinavia appears to have been marked by no decline in average stature but an increase in tooth diseases (perhaps caused by the new carbohydrate-rich, terrestrial diet; Bennike 1999: 28) and in western Europe by increases in average stature and no significant trends in deprivational markers (Meiklejohn *et al.* 1984; Meiklejohn and Zvelebil 1991).

Although the lack of Mesolithic skeletal material in Britain prevents any accurate assessment in this country, there are no apparent reasons as to why it should have been different from the coastal regions of the continent. The Danish evidence (Bennike 1999: fig. 6.3) shows a gradual increase in average male stature from the Mesolithic (166cm or 5' 6½") and Early Neolithic (165cm or 5' 6") to the Middle and Late Neolithic (171cm or 5' 8½") rising to 173cm/5' 9" in the Middle Bronze Age; this closely matches the British sequence of Early Neolithic (5' 9") to Beaker-period (5' 10½") average male height.[12] Women's average heights in Denmark appear to have risen from the Mesolithic (154cm or 5' 1½") and Early Neolithic (153cm or 5' 1") to the Middle and Late Neolithic (156cm or 5' 2½") before increasing to 164cm/5' 5½" in the Middle Bronze Age. Stature is also affected by genetic factors such as gene pool size and inbreeding (Larsen 1997: 13-19; Roberts and Manchester 1995: 25-7) so height increases could have been caused by immigration rather than by nutritional improvement.

Low incidences of porotic hyperostosis (caused mostly by acquired iron deficiency anaemia; Larsen 1997: 29-40) and enamel hypoplasia (caused by nutritional deficiency or disease; Larsen 1997: 43-56) amongst northwest European Neolithic samples also provide no support for considering malnutrition and starvation as spectres that regularly haunted the land. However there are two caveats; one is that famines can kill formerly well-nourished people without leaving any mark on the bones: 'In theory... periods of malnutrition can lead to decreased evidence of skeletal disease in archaeological populations, creating the impression of better health when, in fact, individuals were very sick and died quickly' (Ortner 1998: 82). The other caveat is that our samples come from the tiny minority of the Neolithic and Early Bronze Age population of a sufficient standing that their bones were laid to rest in protective monuments and who thus might well have led their lives cushioned from the worst of famine and scarcity.

It is not until the Mycenaean Middle/Late Bronze Age that there is any evidence for class-based dietary differences in Europe (Angel 1984: 66). Nor are there any signs from Neolithic or Early Bronze Age skeletons of any unusual sexual dimorphism in stature hinting at gender-based nutritional discrimination against women.[13]

[12] Table 2.1 in Roberts and Manchester (1995) is misleading because it implies that Danish Neolithic men were taller than British Neolithic men, whereas the former are actually contemporary with the British Bronze Age sample which is drawn from Beaker-period material.

[13] Bennike (1985: 51-2) suggested that this was a possibility for the Danish Early and Middle Neolithic but her analysis of a larger sample has now shown this not to be the case (1999: fig. 6.3).

Yet we should be prepared to look at the stable isotope data in particular for gender inequalities and differences in the British Neolithic which might well have had a nutritional dimension.

Malnutrition has a cultural dimension as much as a physical one. Famines are often treated as entirely natural disasters and yet humans are often implicated in their arrival and impact, by depredation of natural resources, unwillingness to transport food to affected areas, restrictions on migration and warfare. Biblical accounts give some idea of the extent and severity of famines in the Bronze Age Near East but human bones have so far not proved to be particularly useful guides to prehistoric malnutrition. It is estimated that only one individual skeleton in a hundred in a typical sample is likely to provide osteological evidence for malnutrition and correct identification of the type of malnutrition, whether mineral-, vitamin-, trace element- or protein-deficient, is not always easy (Ortner and Theobald 2000: 35). Future work on prehistoric malnutrition will require the integration of standard osteological methods with stable isotopic analysis, bone histology, dental pathology and infant skeletal studies within more holistic studies of the full range of evidence.

Cannibalism: 'we are *who* we eat'

Survival cannibalism is one strategy for coping with famine and food shortage. In a modern world in which gourmets continually push the boundaries of taste experience, the absolute food taboo remains that against eating human flesh. Human bone, blood and mummified tissue were consumed as medicine in Britain only 300 years ago (Parker Pearson 1999: 52) but today there is a deep cultural and psychological abhorrence of eating parts of other people. Consequently the revulsion with which we react to those who eat human flesh (Arens 1979; Lewis 1986: 63-77) has partly conditioned the, in some cases, hyper-scepticism which greets any archaeological claims for human cannibalism (Bahn 1990).

Restricting this brief survey to Late Mesolithic and Neolithic practices in western Europe, the Neolithic deposits from a cave at Fontbregoua in southeastern France contain identically butchered animal and human bones in which the longbones have also been broken as if to extract marrow (Villa *et al.* 1986; Villa and Mahieu 1991). Defleshing cutmarks on human bones at Dyrholmen in the Danish Late Mesolithic have been interpreted as cannibalism (Larsson 1990). Both of these examples could conceivably be instances of survival cannibalism since the human bones were found amongst food waste. Ritualised eating of the dead cannot be ruled out but is more convincing in contexts with ceremonial dimensions. The recognition of cutmarks on human bones from Neolithic funerary contexts at Haddenham and Hambledon Hill (Chris Evans pers. comm.; McKinley forthcoming) has raised issues of whether bodies were merely defleshed or symbolically or actually eaten as part of the mortuary practices.

Feasting

When considering culinary waste in the Neolithic and Bronze Age of the British Isles, as opposed to southeastern Europe and the Near East, it is perhaps surprising to find it best represented not in houses and their environs but in tombs, burial sites and ceremonial monuments. Even where bone is preserved, the quantities of waste recovered from in and around house sites have been surprisingly small (Barclay 1996; Darvill 1996: 98; Grogan 1996: 57-9). The exceptions are the settlements of Orkney where deep artefact- and bone-rich middens surround these sites (Childe 1931; Clarke and Sharples 1985). The reason for this difference may be due partly to post-depositional survival and partly to the peculiarly 'house-centred' lifestyle of the Orcadian Neolithic. Even so, John Hedges (1984: 216) has this to say about meat-eating in that setting: 'The slaughtering of stock would not have been a frequent occurrence and judging from comparatively recent practices among the rural population of Orkney meat would have been shared out and much of it pickled, such fare only really being eaten on festive and other special occasions.'

When we compare the evidence for feasting in Neolithic Britain with that in the earliest agricultural communities of the Near East, there is surprisingly little from that region (Thorpe 1996: 19) despite pronouncements on its importance (Hodder 1990; Hayden 1990). Feasting is thus far more visible in the archaeological record of Neolithic and Early Bronze Age Britain and is regularly invoked in standard texts (Bradley 1984; Edmonds 1999; Parker Pearson 1993; Pryor 1998; Thomas 1999), having been recognised as a feature of this period for many years (see Smith 1965). Forecourts and chambers of Neolithic tombs, causewayed enclosure ditches, henge monument interiors and ditches, and a few Early Bronze Age burial mounds, have produced substantial collections of butchered animal bones and/or large ceramic assemblages.

The funerary or ancestral associations of most of these contexts has not escaped notice by prehistorians and references have been made to feasting being not so much an act in itself as an accompaniment to rites of passage in which the giving and sharing of food were part of the funerary practices in which both living and dead participated. In some cases, such as causewayed enclosure ditches, the depositions appear to have been episodic and sequential. In others, such as certain burial mounds and chambered tombs or the West Kennet timber enclosures, the feasting debris appears to have been deposited during or prior to construction rather than during later use. There has been some discussion of the extent to which these feasting events were employed for community integration or as competitive strategies for dominance between lineages and clans (Whittle and Pollard 1998: 244) but little attempt to develop Dietler's agency-centred political economy of feasting.

Cattle and pigs

One of the classic contexts for feasting is that of the earthen henges and their associated timber monuments of the Late Neolithic, especially in the period 2500-2200 BC. These monuments produce few human remains and therefore seem not to have been the final destinations of human bones but their landscape contexts suggest that they might have been involved in ancestor-directed ceremonials (Parker Pearson and Ramilisonina 1998; Thomas 1999). The assemblages from these sites often include large quantities of a type of pottery known as Grooved Ware and bones of pigs in much larger numbers than those of sheep or cattle. Albarella and Serjeantson's reanalysis of the pig bones from Durrington Walls (2002) indicates that, even though the pigs were domesticated, some or all of them were shot with arrows and the carcasses were then burnt, roasted or barbecued, as indicated by burning of the extremities of the limbs. The large quantities of pig bones from ceremonial sites of this period have led some to consider it a 'pig culture' in which cattle, the more symbolically and economically important animal of the Neolithic (Ray and Thomas this volume) had temporarily been ousted from pole position. Yet Beaker burials from the same period indicate that cattle do dominate the funerary contexts, of which the most dramatic is a collection of 180 skulls and other body parts on top of a round barrow at Irthlingborough (Davis and Payne 1993).

Previous interpretations of a Grooved Ware pig society and a Beaker cattle culture as two separate cultural, ethnic, chronological or status groups in the Late Neolithic/Early Bronze Age are no longer the most likely possibility: these might have been complementary and contemporary culinary strategies and scenarios practiced among the same communities. Beaker folk were also Grooved Ware people, the same communities using different animals and pots in different contexts. By the Late Neolithic, Grooved Ware was used for storage, cooking and serving but, as noted above, the references embodied in its shapes and appearances might have had resonances with pre-cooking activities of collecting, preparation and storage whereas the shapes of Peterborough Ware and Beakers suggest styles of dispensing, sharing and consumption. This process from preparation to consumption was perhaps more metaphorical than actual in terms of the Late Neolithic relationship between Grooved Ware and Beaker pots and it might have embodied more spiritual transformations of raw/cooked by which the living may be transformed into ancestral dead (Parker Pearson 2000a).

The ritualised distinction between places for the living and places for the ancestors was one that had been maintained from the Early to the Late Neolithic. In the Early Neolithic this relationship was played out at a small scale, with a distinction between the tomb and its forecourt and exterior. Pork was consumed and the bones deposited in the ditches and features outside the tomb whereas cattle bones were placed within the tomb itself (Edmonds 1999: 64; Thomas 1999: 27-8). What was different in the Late Neolithic was the spatial scale of this relationship. Ceremonial enclosures such as Durrington Walls with its pig feasts may be as much as three miles away from their contemporary counterparts (in this case, Stonehenge where cattle bones and Beaker sherds predominate).

Preliminary research on the residues within Grooved Ware pots has produced intriguing results. The flat bottoms of these vessels appear to make them more suitable for storage and display than the round-bottomed pots of the Early and Middle Neolithic which seem only suited to sitting in the embers of the fire or resting in sand. That said, Grooved Ware pots were used for cooking and five pots from Upper Ninepence in Wales have produced lipid residues which indicate that pork was boiled within two of them (Dudd *et al.* 1999). Slightly earlier Peterborough Ware from the same site was used for processing ruminant milk, as were the three other Grooved Ware pots. A similar pattern of pork and ruminant milk has been recognised from different pots within a sample of Grooved Ware pots from Durrington Walls but the situation is more complex since most vessels show an indeterminate mix of pork with ruminant milk (or possibly also with ruminant flesh). The situation is similarly complex for Peterborough and Grooved Ware pots at Yarnton in Wiltshire (Muhkerjee *et al.* forthcoming; Hey *et al.* this volume). The Late Neolithic complex at Yarnton is of a much smaller order than, say, the Stonehenge ceremonial complex and indicates that certain ceremonials in this period were conducted at a much smaller scale than those staged at the monument complexes.

Botanical analysis of a burnt residue inside a Grooved Ware pot from the ceremonial timber structure at Balfarg in Scotland identified its former contents as a barley and oat porridge flavoured with honey (beeswax was present), meadowsweet, henbane (a plant with narcotic properties), fat hen and other plants (Moffatt in Barclay and Russell-White 1993). The latter discovery is of interest because Sherratt had previously predicted that the Grooved Ware used in these ceremonial sites must have had 'some sacramental significance, and might have involved more than everyday ingredients: something more sugary, spicy, or exciting (narcotic) than was encountered in secular existence' (1991; 1997: 413, 426-7). However, there are caveats to these claims: Grooved Ware seems not, for example, to have been used exclusively in ceremonial settings, as Sherratt believed, since it does occur in everyday contexts in settlements.[14] Furthermore, there is now some doubt about the presence of hallucinogens in the Balfarg pot in the wake of a re-analysis of its contents (Long *et al.* 1999) and it has subsequently been interpreted as a fermentation vessel (Dineley and Dineley 2000).

[14] There may, however, have been ways in which Grooved Ware in ceremonial settings was differentiated from the everyday; for example, the Grooved Ware from the ceremonial site at Durrington Walls can be distinguished from Grooved Ware elsewhere in the Wessex region by its fabric inclusions of marine shell.

Drinking

Alcohol consumption as cultural practice has attracted much interest from archaeologists as well as social anthropologists.[15] Andrew Sherratt's seminal papers on the social uses of alcohol and other drugs in prehistory provide a useful framework for exploring the adoption, spread and social contexts of drugs. Sherratt's association of the development of alcohol with his secondary products revolution (1981; 1983; 1987) in the fourth millennium BC may seem slightly conservative but he gives good reasons why it appears to have developed so long after the agrarian revolution. One is that very few undomesticated fruits would have contained enough sugars for fermentation to be possible, and that the use of appropriate yeasts and long-term preparation in closed pots would have been complex. Another is that there is simply no empirical evidence for alcohol before that era. Sherratt's model is one of substitution in which he speculates that palm fruit in the Near East was the first to be made into alcohol and thereafter the method spread using the regional ingredients of grapes in the Mediterranean and grain further north and west (Sherratt 1987).

Yet it may be premature to rule out the production of alcohol in the context of earlier Neolithic grain consumption. Suitable yeasts are found on elderberries (Wood 2001: 155) and wild honey would have provided ample sugar for fermentation (Crane 1983). The large storage vessels of the Early Neolithic might well have doubled up as fermentation bins. One of the houses at Late Neolithic Skara Brae (*c.* 2500 BC) has been interpreted as a beer-making area (Dineley and Dineley 2000a; 2000b). Dineley and Dineley make the point that the success of cereals as a Neolithic domesticate may have been due neither to their calorific value nor to their use in bread or 'cakes' but to their role in making alcohol. To extend their argument further, the Neolithic domestication 'package' of cereals, cattle, pigs and sheep may be viewed from the consumption perspective, providing its users with a mobile feasting network in which the timing, scale and certainty of feasts were of a higher order than amongst those communities relying entirely on wild resources.

As we move into the Early Bronze Age, so the evidence from Britain is more conclusive. The well-known Beaker in a burial at Ashgrove in Scotland contained honey pollen consonant with mead - lime (not local to the region), meadowsweet, heather, ribwort plantain and mint (Dickson and Dickson 2000: 79-80). The presence of meadowsweet pollen in other Scottish Bronze Age burials has been interpreted variously as evidence for floral tributes, ale or sweetened porridge and the matter is certainly not clear-cut (Bohncke 1983; Dickson 1978; Dickson and Dickson 2000; Moore 1978; Tipping 1994). The well-preserved Middle Bronze Age burial at Egtved in Denmark contained a birch-bark bucket with pollen of lime, meadowsweet and white clover (presumably from honey), grains of wheat, leaves of bog myrtle and fruits of cranberry/cowberry (Dickson and Dickson 2000: 81; see Koch this volume). Other likely alcoholic brews are known from Germany and Greece in this period (Vencl 1994) and Eva Koch has suggested that there may have been a gender division in the allocation of strong mead for men and less alcoholic mixed drinks for women among the grave goods (Koch this volume).

A carbonised grape pip and burnt vinewood from Hambledon Hill - all the more extraordinary given the rarity with which such things become carbonised - raise the possibility that if grapes or even wine were imported into, or even cultivated in, Britain as early as 3400 BC then wine may have been produced (Jones and Legge 1987).

... and drugs

Magic mushrooms (*Psilocybe semilanceata*) are indigenous to Britain and there is every reason to expect that their hallucinogenic properties were known by the Neolithic, if not long before. The availability of opium poppies in Early Neolithic LBK society is well documented from the carbonised seed evidence. Bakels points to the symbolic and ritual significance of a small LBK pot from Belgium tempered with opium poppy seeds, and points out that this poppy's wild ancestors derive from the western Mediterranean, a region not otherwise linked with the LBK (Bakels 2000: 101-4). Whether crop or weed, the opium poppy appears to have reached northern Europe not only by human agency but also from a direction different to the classic domesticates of cereals and animals. Less direct evidence of the presence of opium poppies in northern Europe comes from the collared flasks of the Danish Late Neolithic TRB culture which are, when inverted, strongly reminiscent of dried poppy heads (Sherratt 1991). Surprisingly, no chemical analysis has been carried out on their contents.

Cannabis was the only crop grown by hunter-fisher-gatherers in third millennium BC Lithuania (Rimantienė 1992; Zvelebil 1994) and there is a claim for *Cannabis sativa* in Bronze Age Scotland (Ryder 1999). Sherratt (1991) makes a good case for the Corded Ware and Beakers of the Late Neolithic and Early Bronze Age as drinking sets for infusions of *Cannabis sativa* mixed with alcohol. We do not, however, have the 'smoking gun' evidence present in the Iron Age burials at Pazyryk, where burnt cannabis seeds lay on a censer beneath a tepee-shaped inhaler (Rudenko 1970). Small pottery censers are known from the French Middle Neolithic and are interpreted as opium-burners by Sherratt. Similar ceramics are known from the British Early Bronze Age, including one from Stonehenge (Cleal *et al.* 1995).

A number of authors have pointed out the importance of mind-altering substances for the imprinting of ritual experience on the minds of participants. Michael Dietler has documented the role of alcohol as one of these psychoactive substances in European Iron Age feasting

[15] For social anthropological texts see Douglas 1987; Mandelbaum 1964; Spindler 1964; for archaeological papers see Dietler 1990; Dineley and Dineley 2000a and b; Hamilakis 1999; 2000; Hastorf 1991; Joffe 1998; McGovern *et al.* 1992; Sherratt 1987; 1991; Sherratt and Taylor 1989; Vencl 1994.

and it should be noted that the initial impact of alcohol on physiologies unused to it can be dramatic and entirely different to the stylised expressions of contemporary Western drinking (see Douglas 1987). Narcotics and hallucinogens may be classed as 'non-foods' and lie the other side of the 'edibility' category. They may even be poisons, as in the case of *Psilocybe* mushrooms, and thereby provide an element of risk and danger in their ingestion.

Medicines

A large number of plants native to Britain have medicinal properties (Grieve 1931; Launert 1981) which might have been known as far back as the Neolithic. Iris rhizomes, for example, can be used to treat minor ailments such as diarrhoea, toothache and bleeding and carbonised fragments of such rhizomes were found at Skara Brae (Dickson and Dickson 2000: 55). Birch-tar chewing gum could have been used for sore throats and has been found in European Mesolithic contexts (Aveling 1997). Skull-cap, dead-nettle, sheep's sorrel, lesser celandine, purging buckthorn and juniper all have medicinal properties and have all been found in Iron Age contexts (Dickson and Dickson 2000: 101-2; Taylor in Parker Pearson and Sharples 1999: 190). Meadowsweet is an astringent, antacid and antirheumatic and its consistent appearance in Scottish Early Bronze Age burials (Tipping 1994) may hint at prehistoric awareness of its remedial properties as well as its value for making mead (Dickson and Dickson 2000: 83).

There is a temptation to assume that prehistoric people had the same level of knowledge of medicinal plants as those recorded in the folk ethnopharmacopoeias of the last few centuries, as if rural communities regardless of time and space should have some shared absolute knowledge of their surroundings. Such medicinal knowledge cannot be assumed and must be demonstrated, if only to a degree of likelihood as in the case of the iris rhizomes at Skara Brae. Secondly, contemporary and folk-historical understandings of medicine are likely to have been entirely different to prehistoric concepts of knowledge in which curative properties might have been inherently bound up with magic, the supernatural and ancestral powers (see Edmonds 1999: 21).

Towards an agenda for culinary archaeology

In the last twenty years the scientific identification of food residues in all their manifestations has come a very long way and we now stand on the edge of a revolution in our knowledge. Some of those residues were ingested into the bodies of prehistoric people and are recoverable through bone chemistry, such as isotopic analyses of carbon, nitrogen, strontium, oxygen, lead and sulphur, to provide insights into diet and migration (Ambrosa and Katzenberg 2000; Richards 2000 and this volume; Richards *et al.* 2001; Budd *et al.* 2000 and this volume). Toothwear analysis, palaeopathology, stature and analysis of dental caries and calculus all provide further information on food practices.

Applications of the analysis of pollen and carbonised plant remains to contexts such as residues in pots are providing fine-grained insights into the compositions of meals and drinks. Analyses of lipid and protein residues from the interior fabrics of pots are also producing evidence of their former contents, information which archaeologists hardly dared dream of twenty years ago. Studies of animal bones are also changing: research priorities are no longer exclusively zoological or palaeoeconomic but now include more holistic appreciations of both production and consumption. Yet archaeology is still an interpretive business that can ill-afford to maintain a world of atheoretical technicians providing undirected results on the one hand and over-theoretical armchair interpreters criticising this blind empiricism on the other.

The archaeology of food is still in its infancy and perhaps some of the following guidelines may be of use in its future development:

1. Food is a cultural artefact as well as a natural product and merits investigation of its consumption as much as its production.

2. Culinary elaboration and the development towards a cuisine in Goody's terms may shed further light on the relationship between cooking, cuisine and social inequality.

3. The uses of different cooking methods - boiling, roasting, baking *etc.* - can be shown to have strong cultural associations in many cultures and should be a focus for prehistoric studies.

4. The introduction of psychoactive substances and other culinary innovations might have had an impact as dramatic as that of metals, pottery and other tangible products, and their contexts of introduction and use require close investigation.

5. Feasting might have been a major arena for political and social change and deserves detailed study, especially in its relationship with the household. Are the middens of later prehistoric settlements, for example, created by everyday consumption or by more sporadic feasting events?

6. Lévi-Strauss' famous nature-culture divide may say more about us than about societies which live within nature, as Descola has described them, and we need to investigate the changing axes of edibility/inedibility in prehistory and the various contexts in which these principles were mobilised and transformed.

7. Food is part of a much broader sphere of social relationships, and not merely a thing in itself; it is 'good to think with' and 'good to think about' as much as 'good to eat'.

8. Stable isotope studies may help to unlock social and geographical patterns of consumption and mobility, yet the food consumed by the animals consumed by people should be as much a subject of study as human food (yes, cows and sheep eat grass but they can also eat seaweed, hay and other fodder, and variations in their diet will affect isotopic values for human bones).

9. Integration, integration, integration: the archaeology of food requires specialists in many different fields of artefactual, ecofactual, architectural and formation process studies to work together.

10. Excavators need to remember that 'boring' artefact categories such as fire-heated rocks or unusual opportunities for particular kinds of sampling all present potential insights into past food practices. New techniques must become standard practices - we now accept that animal bones should be recovered and reported but how long must we wait before chemical residue analyses become standard requirements for any project?

Acknowledgements

The initial impetus for this paper came from teaching a course in contemporary material culture during the 1990s and I am indebted to the many students on this course who made me read and think about the subject. The volume would not, of course, have been possible without the conference itself and I am grateful for the help of Alison Blanchard and her team of fellow students who ensured its smooth running and success. Thanks to Marek Zvelebil for comments on the manuscript and to Jenny Moore, Ol Craig and Frances Healy for providing some of the references. Karen Godden copy-edited this and all the contributors' papers.

References

Adams, C. 1990. *The Sexual Politics of Meat: a feminist-vegetarian critical theory*. New York: Continuum.

Agelarakis, A. and Waddell, B. 1994. Analysis of non-specific indicators of stress in subadult skeletons during the agricultural transition from southwestern Asia. *Homo* 45: 9.

Albarella, U. and Serjeantson, D. 2002. A passion for pork: meat consumption at the British Late Neolithic site of Durrington Walls. In P. Miracle and N. Milner (eds) *Consuming Passions and Patterns of Consumption*. Cambridge: McDonald Institute. 33-49.

Ambrosa, S.H. and Katzenberg, M.A. (eds) 2000. *Biogeochemical Approaches to Palaeodietary Analysis*. New York: Kluwer.

Angel, J.L. 1984. Health as a crucial factor in the changes from hunting to developed farming in the eastern Mediterranean. In M.N. Cohen and G.J. Armelagos (eds) *Paleopathology at the Origins of Agriculture*. New York: Academic Press. 51-73.

Appadurai, A. 1981. Gastro-politics in Hindu south Asia. *American Ethnologist* 8: 494-511.

Appadurai, A. 1993. Consumption, duration and history. *Stanford Literature Review* 10: 11-33.

Arens, W. 1979. *The Man-eating Myth: anthropology and anthropophagy*. New York: Oxford University Press.

Arnott, M.L. (ed.) 1975. *Gastronomy: the anthropology of food and food habits*. The Hague: Mouton.

Aveling, E. 1997. Birch tar chewing gum. *British Archaeology* 21: 6.

Bahn, P.G. 1990. Eating people is wrong. *Nature* 348: 395.

Bakels, C. 2000. The Neolithicization of the Netherlands: two ways, one result. In A.S. Fairbairn (ed.) *Plants in Neolithic Britain and Beyond*. Neolithic Studies Group Seminar Papers 5. Oxford: Oxbow. 101-6.

Balasse, M., Bocherens, H. and Mariotti, A. 1999. Intra-bone variability of collagen and apatite isotopic composition used as evidence of a change of diet. *Journal of Archaeological Science* 26: 593-8.

Balasse, M., Bocherens, H., Tresset, A., Mariotti, A. and Vigne, J.D. 1997. Emergence of dairy production in the Neolithic? Contribution of isotopic analysis of cattle archaeological bones. *Comptes Rendus de l'Académie des Sciences Série 2 Fascicule A: Sciences de la Terre et des Planètes* 325: 1005-10.

Barclay, G.J. 1996. Neolithic buildings in Scotland. In T.C. Darvill and J. Thomas (eds) *Neolithic Houses in Northwest Europe and Beyond*. Oxford: Oxbow. 61-75.

Barclay, G.J. and Russell-White, C.J. 1993. Excavations in the ceremonial complex of the 4th to 2nd millennium BC at Balfarg/Balbirnie. *Proceedings of the Society of Antiquaries of Scotland* 123: 42-210.

Barrett, J. 1980. The pottery of the Later Bronze Age in lowland England. *Proceedings of the Prehistoric Society* 46: 297-330.

Barrett, J. 1989. Food, gender and metal: questions of social reproduction. In M.-L. Sørensen and R. Thomas (eds) *The Bronze-Iron Age Transition*. Oxford: British Archaeological Reports (International Series) S483. 304-20.

Barrett, J.H., Nicholson, R.A. and Cerón-Carrasco, R. 1999. Archaeo-ichthyological evidence for long-term socioeconomic trends in northern Scotland: 3500 BC to AD 1500. *Journal of Archaeological Science* 26: 353-88.

Barth, F. 1987. *Cosmologies in the Making*. Cambridge: Cambridge University Press.

Barthes, R. 1972. *Mythologies*. London: Jonathan Cape.

Battaglia, D. 1990. *On the Bones of the Serpent: person, memory, and mortality in Sabarl Island society*. Chicago: University of Chicago Press.

Belasco, W. 1993. *Appetites for Change: how the counterculture took on the food industry*. 2nd edition. Ithaca, NY: Cornell University Press.

Bell, D. and Valentine, G. 1997. *Consuming Geographies: we are where we eat*. London: Routledge.

Bender, B. 1978. Gatherer-hunter to farmer: a social perspective. *World Archaeology* 10: 204-22.

Bennike, P. 1985. *Palaeopathology of Danish Skeletons: a comparative study of demography, disease and injury*. Copenhagen: Akademisk Forlag.

Bennike, P. 1999. The Early Neolithic Danish bog finds: a strange group of people! In B. Coles, J.M. Coles and M. Schou Jørgensen (eds) *Bog Bodies, Sacred Sites and Wetland Archaeology*. Exeter: WARP Occasional Paper 12. 27-32.

Bloch, M. 1999. Commensality and poisoning. *Social Research* 66: 133-49.

Bohncke, S. 1983. The pollen analysis of deposits within a Food Vessel from the henge monument at North Mains. 178-80. In G.J. Barclay, Sites of the third millennium BC to the first millennium AD at North Mains, Strathallan, Perthshire. *Proceedings of the Society of Antiquaries of Scotland* 113: 122-281.

Bonsall, C., Lennon, R., McSweeney, K., Stewart, C., Harkness, D., Boroneant, V., Bartosiewicz, L., Payton, R. and Chapman, J. 1997. The Mesolithic and Early Neolithic in the Iron Gates: a palaeodietary perspective. *Journal of European Archaeology* 5: 50-92.

Bourdieu, P. 1977. *Outline of a Theory of Practice*. Cambridge: Cambridge University Press.

Bourdieu, P. 1984. *Distinction: a social critique of the judgement of taste*. London: Routledge & Kegan Paul.

Bourdieu, P. 1990. *The Logic of Practice*. Cambridge: Polity Press.

Bowden, R. 1983. *Yena: art and ceremony in a Sepik society*. Oxford: Pitt Rivers Museum.

Bradley, R. 1984. *The Social Foundations of Prehistoric Britain*. London: Longmans.

Braithwaite, M. 1982. Decoration as ritual symbol: a theoretical proposal and an ethnographic study in southern Sudan. In I. Hodder (ed.) *Symbolic and Structural Archaeology*. Cambridge: Cambridge University Press. 80-88.

Braudel, F. 1981. *The Structures of Everyday Life: the limits of the possible*. London: Collins.

Brück, J. 2001. Body metaphors and technologies of transformation in the English Middle and Late Bronze Age. In J. Brück (ed.) *Bronze Age Landscapes: tradition and transformation*. Oxford: Oxbow. 149-60.

Budd, P., Montgomery, J., Evans, J. and Barreiro, B. 2000. Human tooth enamel as a record of the comparative lead exposure of prehistoric and modern people. *The Science of the Total Environment* 263: 1-10.

Bulmer, R. 1967. Why the cassowary is not a bird: a problem of zoological taxonomy among the Karam of the New Guinea Highlands. *Man* 2: 5-25.

Burgess, C. 1980. *The Age of Stonehenge*. London: Dent.

Campbell, E. 2000. The raw, the cooked and the burnt: interpretations of food and animals in the Hebridean Iron Age. *Archaeological Dialogues* 7: 184-98.

Charles, N. and Kerr, M. 1988. *Women, Food and Families*. Manchester: Manchester University Press.

Chaudhuri, K.N. 1990. *Asia Before Europe: economy and civilisation of the Indian Ocean from the rise of Islam to 1750*. Cambridge: Cambridge University Press.

Childe, V.G. 1928. *The Most Ancient East*. London: Routledge and Kegan Paul.

Childe, V.G. 1931. *Skara Brae: a Pictish village in Orkney*. London: Routledge and Kegan Paul.

Clark, J.G.D. 1954. *Star Carr*. Cambridge: Cambridge University Press.

Clarke, D.V. and Sharples, N. 1985. Settlements and subsistence in the third millennium BC. In C. Renfrew (ed.) *The Prehistory of Orkney*. Edinburgh: Edinburgh University Press. 54-82.

Clarke, M.J. 2001. Akha feasting: an ethnoarchaeological perspective. In M. Dietler and B. Hayden (eds) *Feasts: archaeological and ethnographic perspectives on food, politics and power*. Washington DC: Smithsonian Institution. 144-67.

Classen, C. 1993. *Worlds of Sense:exploring the senses in history and across cultures*. London: Routledge.

Classen, C., Howes, D. and Synnott, A. 1994. *Aroma: the cultural history of smell*. London: Routledge.

Cleal, R.M.J., Walker, K.E. and Montagu, R. 1995. *Stonehenge in its Landscape: twentieth century excavations*. London: English Heritage.

Cohen, M.N. and Armelagos, G.J. 1984. Paleopathology at the origins of agriculture: editors' summation. In M.N. Cohen and G.J. Armelagos (eds) *Paleopathology at the Origins of Agriculture*. New York: Academic Press. 585-601.

Connah, G. 2001. *African Civilisations: an archaeological perspective*. Cambridge: Cambridge University Press.

Cook, I., Crang, P. and Thorpe, M. 1999. Eating into Britishness: multicultural imaginaries and the identity politics of food. In S. Roseneil and J. Seymour (eds) *Practicing Identities: power and resistance*. Basingstoke: Macmillan. 39-51.

Cooney, G. 2000. *Landscapes of Neolithic Ireland*. Routledge: London.

Counihan, C. 1999. *The Anthropology of Food and Body: gender, meaning and power*. London: Routledge.

Counihan, C. and Kaplan, S. 1998. *Food and Gender: identity and power*. Amsterdam: Harwood.

Counihan, C. and Van Esterik, P. (eds) 1997. *Food and Culture: a reader*. New York: Routledge.

Craig, O.E., Mulville, J., Parker Pearson, M., Sokol, R.J., Gelsthorpe, K., Stacey, R. and Collins, M.J. 2000. Detecting milk proteins in ancient pots. *Nature* 408: 312.

Crane, E. 1983. *The Archaeology of Beekeeping*. London: Duckworth.

Crowe, I. 2000. *The Quest for Food: its role in human evolution and migration*. Stroud: Tempus.

Cumberpatch, C.G. and Blinkhorn, P.W. (eds) 1997. *Not so Much a Pot, More a Way of Life: current approaches to artefact analysis in archaeology*. Oxford: Oxbow.

Curtis, R.I. 2001. *Ancient Food Technology*. Leiden: Brill.

Danforth, L.P. 1982. *The Death Rituals of Rural Greece*. Princeton: Princeton University Press.

Darvill, T.C. 1996. Neolithic buildings in England, Wales and the Isle of Man. In T.C. Darvill and J. Thomas (eds) *Neolithic Houses in Northwest Europe and Beyond*. Oxford: Oxbow. 77-111.

Davidson, J. 1997. *Courtesans and Fishcakes: the consuming passions of Classical Athens*. London: Fontana.

Davies, G. 2002. The 7,700-year-old woman who ate like a wolf. *British Archaeology* 66: 7.

Davis, S. and Payne, S. 1993. A barrow full of skulls. *Antiquity* 67: 12-22.

Deetz, J. 1977. *In Small Things Forgotten: the archaeology of early American life*. New York: Anchor Press/Doubleday.

Dennell, R. 1985. *European Economic Prehistory: a new approach*. London: Academic Press.

Descola, P. 1994. *In the Society of Nature: a native ecology in Amazonia*. Cambridge: Cambridge University Press.

Dickson, C. and Dickson, J.H. 2000. *Plants and People in Ancient Scotland*. Stroud: Tempus.

Dickson, J.H. 1978. Bronze Age mead. *Antiquity* 52: 108-13.

Dickson, J.H., Oeggl, K., Holden, T.G., Handley, L.L., O'Connell, T. and Preston, T. 2000. The omnivorous Tyrolean iceman: colon contents (meat, cereals, pollen, moss and whipworm) and stable isotope analyses. *Philosophical Transactions of the Royal Society B* 355: 1843-9.

Diener, P. and Robkin, E. 1978. Ecology, evolution and the search for cultural origins: the question of Islamic pig prohibition. *Current Anthropology* 19: 493-540.

Dietler, M. 1990. Driven by drink: the role of drinking in the political economy and the case of Early Iron Age France. *Journal of Anthropological Archaeology* 9: 352-406.

Dietler, M. 2001. Theorising the feast: rituals of consumption, commensal politics, and power in African contexts. In M. Dietler and B. Hayden (eds) *Feasts: archaeological and ethnographic perspectives on food, politics and power*. Washington DC: Smithsonian Institution. 65-114.

Dietler, M. and Hayden, B. 2001. Digesting the feast - good to eat, good to drink, good to think: an introduction. In M. Dietler and B. Hayden (eds) *Feasts: archaeological and ethnographic perspectives on food, politics and power*. Washington DC: Smithsonian Institution. 1-20.

Dietler, M. and Hayden, B. (eds) 2001. *Feasts: archaeological and ethnographic perspectives on food, politics and power*. Washington DC: Smithsonian Institution.

Dineley, M. and Dineley, G. 2000a. From grain to ale: Skara Brae, a case study. In A. Ritchie (ed.) *Neolithic Orkney in its European Context*. Cambridge: McDonald Institute. 196-200.

Dineley, M. and Dineley, G. 2000b. Neolithic ale: barley as a source of malt sugars for fermentation. In A.S. Fairbairn (ed.) *Plants in Neolithic Britain and Beyond*. Neolithic Studies Group Seminar Papers 5. Oxford: Oxbow. 137-53.

Douglas, M. 1966. *Purity and Danger: an analysis of the concepts of pollution and taboo*. London: Routledge and Kegan Paul.

Douglas, M. 1972. Deciphering a meal. *Daedalus* 101: 61-82.

Douglas, M. 1975. *Implicit Meanings: essays in anthropology*. London: Routledge and Kegan Paul.

Douglas, M. (ed.) 1987. *Constructive Drinking: perspectives on drink from social anthropology*. Cambridge: Cambridge University Press.

Douglas, M. and Nicod, M. 1974. Taking the biscuit. The structure of British meals. *New Society* 19: 744-7.

Dudd, S.N., Evershed, R.P. and Gibson, A.M. 1999. Evidence for varying patterns of exploitation of animal products in different prehistoric pottery traditions based on lipids preserved in surface and absorbed residues. *Journal of Archaeological Science* 26: 1473-82.

Eder, K. 1996. *The Social Construction of Nature: a sociology of ecological enlightenment*. London: Sage.

Edmonds, M. 1999. *Ancestral Geographies of the Neolithic: landscape, monuments and memory*. London: Routledge.

Elias, N. 1978 [1939]. *The Civilising Process. Volume 1: the history of manners*. New York: Urizen.

Entwhistle, R. and Grant, A. 1989. The evidence for cereal cultivation and animal husbandry in the southern British Neolithic and Bronze Age. In A. Milles, D. Williams and N. Gardner (eds) *The Beginnings of Agriculture*. Oxford: British Archaeological Reports (International Series) 496. 203-15.

Evershed, R. and Dudd, S.M. 1998. Direct demonstration of milk as an element of archaeological economics. *Science* 282: 1478-81.

Fairbairn, A. (ed.) 2000. *Plants in Neolithic Britain and Beyond*. Neolithic Studies Group Seminar Papers 5. Oxford: Oxbow.

Falk, P. 1994. *The Consuming Body*. London: Sage.

Farb, P. and Armelagos, G. 1980. *Consuming Passions: the anthropology of eating*. Boston: Houghton Mifflin.

Featherstone, M. 1987. Leisure, symbolic power and the life course. In J. Horne, D. Jary and A. Tomlinson (eds) *Sport, Leisure and Social Relations*. Sociological Review monograph 33. London: Routledge and Kegan Paul. 113-38.

Fenton, A. and Kisban, E. (eds) 1986. *Food in Change: eating habits from the Middle Ages to the present day*. Edinburgh: John Donald and National Museums of Scotland.

Fenton, A. and Myrdal, J. (eds) 1988. *Food and Drink and Travelling Accessories: essays in honour of Gosta Berg*. Edinburgh: John Donald and National Museums of Scotland.

Fenton, A. and Owen, T.M. (eds) 1981. *Food in Perspective: proceedings of the third international conference on ethnological food research*. Edinburgh: John Donald.

Fiddes, N. 1991. *Meat: a natural symbol.* London: Routledge.
Fieldhouse, P. 1986. *Food and Nutrition, Customs and Culture.* London: Chapman and Hall.
Fischler, C. 1988. Food, self and identity. *Social Science Information* 27: 275-92.
Forster, R. and Ranum, O. (eds) 1979. *Food and Drink in History.* Baltimore: Johns Hopkins University Press.
Fortune, R. 1932. *Sorcerors of Dobu: the social anthropology of the Dobu islanders of the western Pacific.* London: Routledge.
Friedman, J. 1975. Tribes, states and transformations. In M. Bloch (ed.) *Marxist Analyses and Social Anthropology.* London: Duckworth. 161-202.
Friedman, J. and Rowlands, M.J. 1978. Notes toward an epigenetic model of the evolution of 'civilisation'. In J. Friedman and M.J. Rowlands (eds) *The Evolution of Social Systems.* London: Duckworth. 201-76.
Garnsey, P. 1999. *Food and Society in Classical Antiquity.* Cambridge: Cambridge University Press.
Gerritsen, F. 2000. Of calories and culture. *Archaeological Dialogues* 7: 169-72.
Gibson, A.M. and Woods, A. 1990. *Prehistoric Pottery for the Archaeologist.* Leicester: Leicester University Press.
Giles, M. and Parker Pearson, M. 1999. Learning to live in the Iron Age: dwelling and praxis. In B. Bevan (ed.) *Northern Exposure: interpretative devolution and the Iron Ages in Britain.* Leicester: Leicester Archaeology Monograph 4. 217-31.
Gillison, G. 1980. Images of nature in Gimi thought. In C. MacCormack and M. Strathern (eds) *Nature, Culture and Gender.* Cambridge: Cambridge University Press. 143-73.
Glob, P.V. 1974. *The Mound People: Danish Bronze-Age man preserved.* London: Faber and Faber.
Godelier, M. 1999. *The Enigma of the Gift.* Oxford: Polity.
Goodman, J., Lovejoy, E. and Sherratt, A. (eds) 1995. *Consuming Habits: drugs in history and anthropology.* London: Routledge.
Goody, J. 1982. *Cooking, Cuisine and Class: a study in comparative sociology.* Cambridge: Cambridge University Press.
Gosden, C. 1999. Introduction. In C. Gosden and J. Hather (eds) *The Prehistory of Food: appetites for change.* London: Routledge. 1-9.
Gosden, C. and Hather, J. (eds) 1999. *The Prehistory of Food: appetites for change.* London: Routledge.
Grieve, M. 1931. *A Modern Herbal.* London: Jonathan Cape.
Grogan, E. 1996. Neolithic houses in Ireland. In T.C. Darvill and J. Thomas (eds) *Neolithic Houses in Northwest Europe and Beyond.* Oxford: Oxbow. 41-60.
Hage, P. 1979. Symbolic culinary mediation: a group model. *Man* 14: 81-92.
Halstead, P.J. 1989. Like rising damp? An ecological approach to the spread of farming in south east and central Europe. In A. Milles, D. Williams and N. Gardner (eds) *The Beginnings of Agriculture.* Oxford: British Archaeological Reports (International Series) 496. 23-53.
Halstead, P.J. 2001. Neighbours from hell? The household in Neolithic Greece. In P.J. Halstead (ed.) *Neolithic Society in Greece.* Sheffield: Sheffield Academic Press. 77-95.
Hamilakis, Y. 1999. Food technologies/technologies of the body: the social context of wine and oil production and consumption in Bronze Age Crete. *World Archaeology* 31: 38-54.
Hamilakis, Y. 2000. The anthropology of food and drink consumption and Aegean archaeology. In S.J. Vaughan and W.D.E. Coulson (eds) *Palaeodiet in the Aegean.* Oxford: Oxbow. 13-27.
Harding, A.F. 2000. *European Societies in the Bronze Age.* Cambridge: Cambridge University Press.
Harris, D., Hooper, S. and Trudgian, P. 1984. Excavation of three cairns on Stannon Downs, St Breward, Cornwall. *Cornish Archaeology* 23: 141-54.
Harris, M. 1968. *The Rise of Anthropological Theory: a history of theories of culture.* London: Routledge and Kegan Paul.
Harris, M. 1975. *Cows, Pigs, Wars and Witches: the riddles of culture.* London: Hutchinson.
Harris, M. 1978. *Cannibals and Kings: the origins of cultures.* London: Collins.
Harris M. 1985. *Good to Eat: riddles of food and culture.* New York: Simon Schuster.
Harris, M. and Ross, E. 1987. *Food and Evolution.* Philadelphia: Temple University Press.
Hastorf, C. 1991. Gender, space and food in prehistory. In J. Gero and M. Conkey (eds) *Engendering Archaeology: women and prehistory.* Oxford: Blackwell. 132-62.
Hastorf, C. 1998. The cultural life of domestic plant use. *Antiquity* 72: 773-82.
Hayden, B. 1990. Nimrods, piscators, pluckers and planters: the emergence of food production. *Journal of Anthropological Archaeology* 9: 31-69.
Hayden, B. 2001. Fabulous feasts: a prolegomenon to the importance of feasting. In M. Dietler and B. Hayden (eds) *Feasts: archaeological and ethnographic perspectives on food, politics and power.* Washington DC: Smithsonian Institution. 23-64.
Hedges, J.W. 1984. *Tomb of the Eagles: a window on Stone Age tribal Britain.* London: John Murray.
Hill, J.D. 1989. Rethinking the Iron Age. *Scottish Archaeological Review* 6: 16-24.
Hill, J.D. 1995. *Ritual and Rubbish in the Iron Age of Wessex: a study in the formation of a specific archaeological record.* Oxford: British Archaeological Reports (British Series) 242.
Hodder, I. 1982. *Symbols in Action: ethnoarchaeological studies of material culture.* Cambridge: Cambridge University Press.
Hodder, I. 1990. *The Domestication of Europe.* Oxford: Blackwell.
Hodder, I. 1999. *The Archaeological Process: an introduction.* Oxford: Blackwell.

Howes, D. 1987. Olfaction and transition: an essay on the ritual uses of smell. *Canadian Review of Sociology and Anthropology* 24: 398-416.

Howes, D. (ed.) 1991. *The Varieties of Sensory Experience: a sourcebook in the anthropology of the senses.* Toronto: University of Toronto Press.

Huffman, T.N. 1996. *Snakes and Crocodiles: power and symbolism in ancient Zimbabwe.* Johannesburg: Witwatersrand University Press.

Hugh-Jones, C. 1979. *From the Milk River: spatial and temporal processes in Amazonia.* Cambridge: Cambridge University Press.

Ingold, T. (ed.) 1989. *What is an Animal?* London: Harper Collins.

James, A. 1996. Cooking the books: global or local identities in contemporary British food cultures? In D. Howes (ed.) *Cross-Cultural Consumption: global markets, local realities.* London: Routledge. 77-92.

Jerome, N.W., Kandel, R.F. and Pelto, G.H. (eds) 1980. *Nutritional Anthropology: contemporary approaches to diet and culture.* Pleasantville, NY: Redgrave Publishing.

Joffe, A. 1998. Alcohol and social complexity in ancient western Asia. *Current Anthropology* 39: 297-322.

Jones, A. 1996. Food for thought: material culture and the transformation in food use from the Mesolithic to Neolithic. In T. Pollard and A. Morrison (eds) *The Early Prehistory of Scotland.* Edinburgh: Edinburgh University Press. 291-301.

Jones, A. 1998. Where eagles dare: landscape, animals and the Neolithic of Orkney. *Journal of Material Culture* 3: 301-24.

Jones, A. 1999. The world on a plate: ceramics, food technology and cosmology in Neolithic Orkney. *World Archaeology* 31: 55-78.

Jones, A. 2000. Life after death: monuments, material culture and social change in Neolithic Orkney. In A. Ritchie (ed.) *Neolithic Orkney in its European Context.* Cambridge: McDonald Institute. 127-38.

Jones, A. 2002. *Archaeological Theory and Scientific Practice.* Cambridge: Cambridge University Press.

Jones, G. 2000. Evaluating the importance of cultivation and collecting in Neolithic Britain. In A.S. Fairbairn (ed.) *Plants in Neolithic Britain and Beyond.* Neolithic Studies Group Seminar Papers 5. Oxford: Oxbow. 79-84.

Jones, G. and Legge, A.J. 1987. The grape (Vitis vinifera L.) in the Neolithic of Britain. *Antiquity* 61: 452-5.

Jones, M. 2002. Eating for calories or company? Concluding remarks on consuming passions. In P. Miracle and N. Milner (eds) *Consuming Passions and Patterns of Consumption.* Cambridge: McDonald Institute. 131-6.

Kahn, M. 1986. *Always Hungry, Never Greedy: food and the expression of gender in a Melanesian society.* Cambridge: Cambridge University Press.

Keller, F. 1878. *The Lake Dwellings of Switzerland and other parts of Europe.* London: Longman..

Kiple, K.F. and Ornelas K.C. (eds) 2000. *The Cambridge World History of Food.* Cambridge: Cambridge University Press.

Korsmeyer, C. 1999. *Making Sense of Taste: food and philosophy.* Ithaca, NY: Cornell University Press.

Kuper, J. (ed.) 1997. *The Anthropologists' Cookbook.* 2nd edition. London: T.J. International.

Kus, S. 1992. Towards an archaeology of body and soul. In J.-C. Gardin and C.S. Peebles (eds) *Representations in Archaeology.* Bloomington: Indiana University Press. 168-77.

Larsen, C.S. 1997. *Bioarchaeology: interpreting behavior from the human skeleton.* Cambridge: Cambridge University Press.

Larsson, L. 1990. The Mesolithic of southern Scandinavia. *Journal of World Prehistory* 4: 257-309.

Launert, E. 1981. *Edible and Medicinal Plants of Britain and Northwest Europe.* London: Hamlyn.

Leach, E. 1964. Anthropological aspects of language: animal categories and verbal abuse. In E.H. Lennenberg (ed.) *New Directions in the Study of Language.* Cambridge MA: MIT Press. 23-64.

Leach, E. 1976. *Culture and Communication: the logic by which symbols are connected.* Cambridge: Cambridge University Press.

Legge, A.J. 1981. Aspects of cattle husbandry. In R. Mercer (ed.) *Farming Practice in British Prehistory.* Edinburgh: Edinburgh University Press. 161-81.

Legge, A.J. 1989. Milking the evidence: a reply to Entwhistle and Grant. In A. Milles, D. Williams, and N. Gardner (eds) *The Beginnings of Agriculture.* Oxford: British Archaeological Reports (International Series) 496. 217-42.

Legge, A.J., Payne, S. and Rowley-Conwy, P. 1998. The study of food evidence in prehistoric Britain. In J. Bayley (ed.) *Science in Archaeology.* London: English Heritage. 89-94.

Lentz, C. (ed.) 1999. *Changing Food Habits: case studies from Africa, South America and Europe.* Amsterdam: Harwood.

Lévi Strauss, C. 1966a. The culinary triangle. *New Society* 166: 937-40.

Lévi-Strauss, C. 1966b. *The Savage Mind.* London: Weidenfeld and Nicolson.

Lévi-Strauss, C. 1969. *The Raw and the Cooked. Introduction to a science of mythology: 1.* London: Jonathan Cape.

Lévi-Strauss, C. 1974. *From Honey to Ashes. Introduction to a science of mythology: 2.* London: Jonathan Cape.

Lévi-Strauss, C. 1978. *The Origin of Table Manners. Introduction to a science of mythology: 3.* London: Jonathan Cape.

Lewis, I.M. 1986. *Religion in Context: cults and charisma.* Cambridge: Cambridge University Press.

Lidén, K. 1995. Megaliths, agriculture, and social complexity: a diet study of two Swedish megalith populations. *Journal of Anthropological Archaeology* 14: 404-17.

Lillie, M.C. 1997. Women and children in prehistory: resource sharing and social stratification at the Mesolithic-Neolithic transition in Ukraine. In J.

Moore and E. Scott (eds) *Invisible People and Processes: writing gender and childhood into European archaeology*. Leicester: Leicester University Press. 213-28.

Long, D., Milburn, P., Bunting, J.M., Tipping, R. and Holden, T.G. 1999. Black henbane (*Hyoscyamus niger* L.) in the Scottish Neolithic: a re-evaluation of palynological findings from Grooved Ware pottery at Balfarg riding school and henge, Fife. *Journal of Archaeological Science* 26: 45-52.

Lubell, D., Jackes, M., Schwarcz, H., Knyf, M. and Meiklejohn, C. 1994. The Mesolithic-Neolithic transition in Portugal: isotopic and dental evidence of diet. *Journal of Archaeological Science* 21: 201-16.

Lupton, D. 1994. Food, memory and meaning: the symbolic and social nature of food events. *Sociological Review* 42: 665-85.

Lupton, D. 1996. *Food, the Body and the Self*. London: Sage.

Mabey, R. 1972. *Food for Free*. London: Collins.

MacClancy, J. 1992. *Consuming Culture*. London: Chapmans.

McCracken, R. 1982. Cultural differences in food preferences and meanings. *Human Organization* 41: 161-7.

McGovern, P.E., Glusker, D.L., Exner, I.J. and Voigt, M.M. 1992. Neolithic resinated wine. *Nature* 360: 24-5.

Mack, J. 1986. *Madagascar: island of the ancestors*. London: British Museum.

McKinley, J.I. Forthcoming. The human remains. In R. Mercer and F. Healy, *Hambledon Hill, Dorset, England: excavation and survey of a Neolithic monument complex and its surrounding landscape*. London: English Heritage.

Macko, S.A., Engel, M.H., Andrusevich, V., Lubec, G., O'Connell, T.C. and Hedges, R.E.M. 1999. Documenting the diet of ancient human populations through stable isotope analysis of hair. *Philosophical Transactions of the Royal Society of London B* 354: 65-76.

Mandelbaum, D.G. 1964. Alcohol and culture. *Current Anthropology* 6: 281-93.

Mays, S. 1998. *The Archaeology of Human Bones*. London: Routledge.

Meadows, K.I. 1997. Much ado about nothing: the social context of eating and drinking in early Roman Britain. In C.G. Cumberpatch and P.W. Blinkhorn (eds.) *Not so Much a Pot, More a Way of Life: current approaches to artefact analysis in archaeology*. Oxford: Oxbow. 21-35.

Meigs, A.S. 1984. *Food, Sex, and Pollution: a New Guinea religion*. New Brunswick, NJ: Rutgers University Press.

Meigs, A.S. 1997. Food as a cultural construction. In C. Counihan and P. Van Esterik (eds) *Food and Culture: a reader*. New York: Routledge. 95-106.

Meiklejohn, C. and Zvelebil, M. 1991. Health status of European populations at the agricultural transition and the implications for the adoption of farming. In H. Bush and M. Zvelebil (eds) *Health in Past Societies: biocultural interpretations of human skeletal remains in archaeological contexts*. Oxford: British Archaeological Reports (International Series) 567. 129-45.

Meiklejohn, C., Schentag, C., Venema, A. and Key, P. 1984. Socioeconomic change and patterns of pathology and variation in the Mesolithic and Neolithic of western Europe: some suggestions. In M.N. Cohen and G.J. Armelagos (eds) *Paleopathology at the Origins of Agriculture*. New York: Academic Press. 75-100.

Mellars, P.A. 1978. Excavation and economic analysis of Mesolithic shell middens on the island of Oronsay (Inner Hebrides). In P.A. Mellars (ed.) *The Early Post-Glacial Settlement of Northern Europe*. London: Duckworth. 371-96.

Mennell, S. 1985. *All Manners of Food: eating and taste in England and France from the Middle Ages to the present*. Oxford: Blackwell.

Mennell, S., Murcott, A. and van Otterloo, A. 1992. *The Sociology of Food: eating, diet and culture*. London: Sage.

Messer, E. 1984. Anthropological perspectives on diet. *Annual Review of Anthropology* 13: 205-49.

Miller, D. 1985. *Artefacts as Categories: a study of ceramic variability in central India*. Cambridge: Cambridge University Press.

Milner, N. and Miracle, P. 2002. Introduction: patterning data and consuming theory. In P. Miracle and N. Milner (eds) *Consuming Passions and Patterns of Consumption*. Cambridge: McDonald Institute. 1-5.

Miracle, P. and Milner, N. (eds) 2002. *Consuming Passions and Patterns of Consumption*. Cambridge: McDonald Institute.

Moffett, L. 1991. Pignut tubers from a Bronze Age cremation at Barrow Hills, Oxfordshire, and the importance of vegetable tubers in the prehistoric period. *Journal of Archaeological Science* 18: 187-91.

Moffett, L., Robinson, M.A. and Straker, V. 1989. Cereals, fruit and nuts: charred plant remains from Neolithic sites in England and Wales and the Neolithic economy. In A. Milles, D. Williams, and N. Gardner (eds) *The Beginnings of Agriculture*. Oxford: British Archaeological Reports (International Series) 496. 243-61.

Monk, M. 2000. Seeds and soils of discontent: an environmental archaeological contribution to the nature of the Early Neolithic. In A. Desmond, G. Johnson, M. McCarthy, J. Sheehan and E. Shee Twohig (eds) *New Agendas in Irish Prehistory: papers in commemoration of Liz Anderson*. Bray, Co. Wicklow: Wordwell. 67-87.

Montón Subías, S. 2001. Cooking in zooarchaeology: is this issue still raw? In P. Miracle and N. Milner (eds) *Consuming Passions and Patterns of Consumption*. Cambridge: McDonald Institute. 7-15.

Moore, H. 1982. The interpretation of spatial patterning in settlement residues. In I. Hodder (ed.) *Symbolic and Structural Archaeology*. Cambridge: Cambridge University Press. 74-9.

Moore, H. 1986. *Space, Text and Gender: an anthropological study of the Marakwet of Kenya*. Cambridge: Cambridge University Press.

Moore, P.D. 1978. Bronze Age honey. *Nature* 276: 443.

Morris, B. 1998. *The Power of Animals: an ethnography*. Oxford: Berg.

Muhkerjee, A., Evershed, R.P. and Gibson, A.M. Forthcoming. The significance of the Grooved Ware pottery tradition in Neolithic Britain in relation to human diet, animal husbandry and ritual practices. In A.M. Gibson (ed.) *Prehistoric Pottery: people, pattern and purpose*. Oxford: British Archaeological Reports (International Series).

Needham, S. and Evans, J. 1987. Honey and dripping: Neolithic food residues from Runnymede Bridge. *Oxford Journal of Archaeology* 6: 21-8.

Ohnuki-Tierney, E. 1993. *Rice as Self: Japanese identities through time*. Princeton, NJ: Princeton University Press.

O'Kelly, M. 1954. *Excavations and Experiments in Ancient Irish Cooking-Places*. Dublin: Royal Society of Antiquaries of Ireland.

Okely, J. 1983. *The Traveller-Gypsies*. Cambridge: Cambridge University Press.

Ortner, D.J. 1998. Male-female immune reactivity and its implications for interpreting evidence in human skeletal paleopathology. In A.L. Grauer and P. Stuart-Macadam (eds) *Sex and Gender in Paleopathological Perspective*. Cambridge: Cambridge University Press. 79-92.

Ortner, D.J. and Theobald, G. 2000. Palaeopathological evidence of malnutrition. In K.F. Kiple and K.C. Ornelas (eds) *The Cambridge World History of Food*. Cambridge: Cambridge University Press. 34-44.

Parker Pearson, M. 1993. *Bronze Age Britain*. London: Batsford and English Heritage.

Parker Pearson, M. 1996. Food, fertility and front doors in the first millennium BC. In T.C. Champion and J.R. Collis (eds) *The Iron Age in Britain and Ireland: recent trends*. Sheffield: J.R. Collis Publications. 117-32.

Parker Pearson, M. 1999. *The Archaeology of Death and Burial*. Stroud: Sutton.

Parker Pearson, M. 2000a. Ancestors, bones and stones in Neolithic and Early Bronze Age Britain and Ireland. In A. Ritchie (ed.) *Neolithic Orkney in its European Context*. Cambridge: McDonald Institute. 203-14.

Parker Pearson, M. 2000b. Eating money: a study in the ethnoarchaeology of food. *Archaeological Dialogues* 7: 217-32.

Parker Pearson, M. and Ramilisonina. 1998. Stonehenge for the ancestors: the stones pass on the message. *Antiquity* 72: 308-26.

Parker Pearson, M. and Sharples, N. 1999. *Between Land and Sea: excavation of a broch at Dun Vulan, South Uist*. Sheffield: Sheffield Academic Press.

Paston-Williams, S. 1993. *The Art of Dining: a history of cooking and eating*. London: National Trust.

Pearce, J. and Luff, R. 1994. The taphonomy of cooked bone. In R. Luff and P. Rowley-Conwy (eds) *Whither Environmental Archaeology?* Oxford: Oxbow. 51-6.

Pullar, P. 1970. *Consuming Passions: a history of English food and appetite*. London: Hamish Hamilton.

Pryor, F. 1998. *Farmers in Prehistoric Britain*. Stroud: Tempus

Rappaport, R. 1968. *Pigs for the Ancestors: ritual in the ecology of a New Guinea people*. New Haven: Yale University Press.

Renfrew, J.M. 1985. *Food and Cooking in Prehistoric Britain: history and recipes*. London: English Heritage.

Revel, J.F. 1982. *Culture and Cuisine: a journey through the history of food*. New York: Da Capo Press.

Richards, A. 1939. *Land, Labour and Diet in Northern Rhodesia*. Oxford: Oxford University Press.

Richards, C. 2002. *Dwelling Amongst the Monuments: excavations at Barnhouse and Maeshowe, Orkney*. Cambridge: McDonald Institute.

Richards, M.P. 2000. Human consumption of plant foods in the British Neolithic: direct evidence from bone stable isotopes. In A.S. Fairbairn (ed.) *Plants in Neolithic Britain and Beyond*. Neolithic Studies Group Seminar Papers 5. Oxford: Oxbow. 123-35.

Richards, M.P., Fuller, B.T. and Hedges, R.E.M. 2001. Sulphur isotopic variation in ancient bone collagen from Europe: implications for human palaeodiet, residence mobility, and modern pollutant studies. *Earth and Planetary Science Letters* 191: 185-90.

Richards, M.P. and Hedges, R.E.M. 1999. A Neolithic revolution? New evidence of diet in the British Neolithic. *Antiquity* 73: 891-97.

Richmond, A. 1999. *Preferred Economies: the nature of the subsistence base throughout mainland Britain during prehistory*. Oxford: British Archaeological Reports (British Series) 290.

Rimantienė, R. 1992. Neolithic hunter-gatherers at Sventoji in Lithuania. *Antiquity* 66: 367-76.

Roberts, C. and Manchester, K. 1995. *The Archaeology of Disease*. 2nd edition. Stroud: Alan Sutton.

Robinson, M.A. 2000. Further considerations of Neolithic charred cereals, fruit and nuts. In A.S. Fairbairn (ed.) *Plants in Neolithic Britain and Beyond*. Neolithic Studies Group Seminar Papers 5. Oxford: Oxbow. 85-90.

Rowley-Conwy, P. 2000. Through a taphonomic glass, darkly: the importance of cereal cultivation in Neolithic Britain. In J.P. Huntley and S. Stallibrass (eds) *Taphonomy and Interpretation*. Oxford: Oxbow. 43-53.

Rowley-Conwy, P. and Zvelebil, M. 1989. Saving it for later: storage by prehistoric hunter-gatherers in Europe. In P. Halstead and J. O'Shea (eds) *Bad Year Economics*. Cambridge: Cambridge University Press. 40-56.

Rudenko, S.I. 1970. *Frozen Tombs of Siberia: the Pazyryk burials of Iron Age horsemen*. London: Dent.

Ryder, M.L. 1999. Probable fibres of hemp (*Cannabis sativa* L.) in Bronze Age Scotland. *Environmental*

Archaeology 4: 93-5.

Sahlins, M. 1976. *Culture and Practical Reason.* Chicago: University of Chicago Press.

Sahlins, M. 1985. *Islands of History.* Chicago: University of Chicago Press.

Scholliers, P. (ed.) 2001. *Food, Drink and Identity: cooking, eating and drinking in Europe since the Middle Ages.* Oxford: Berg.

Schulting, R.J. and Richards, M.P. 2000. The use of stable isotopes in studies of subsistence and seasonality in the British Mesolithic. In R. Young (ed.) *Mesolithic Lifeways: current research from Britain and Ireland.* Leicester: University of Leicester. 55-66.

Shanklin, E. 1985. Sustenance and symbol: anthropological studies of domesticated animals. *Annual Review of Anthropology* 14: 375-403.

Sherratt, A. 1981. Plough and pastoralism: aspects of the secondary products revolution. In N. Hammond, I. Hodder and G. Isaac (eds) *Pattern of the Past: studies in honour of David Clarke.* Cambridge: Cambridge University Press. 261-305.

Sherratt, A. 1983. The secondary exploitation of animals in the Old World. *World Archaeology* 15: 90-104.

Sherratt, A. 1987. Cups that cheered: the introduction of alcohol to prehistoric Europe. In W. Waldren and R. Kennard (eds) *Bell Beakers of the Western Mediterranean: the Oxford international conference 1986.* Oxford: British Archaeological Reports (British Series) 331. 81-106.

Sherratt, A. 1991. Sacred and profane substances: the ritual use of narcotics in later Neolithic Europe. In P. Garwood, D. Jennings, R. Skeates and J. Toms (eds) *Sacred and Profane: proceedings of a conference on archaeology, ritual and religion.* Oxford: Oxford University Committee for Archaeology monograph 32. 50-64.

Sherratt, A. 1997. *Economy and Society in Prehistoric Europe: changing perspectives.* Edinburgh: Edinburgh University Press.

Sherratt, A. 1999. Cash-crops before cash: organic consumables and trade. In C. Gosden and J. Hather (eds) *The Prehistory of Food: appetites for change.* London: Routledge. 13-34.

Sherratt, A. and Taylor, T. 1989. Metal vessels in Bronze Age Europe and the context of Vulchetrun. In J. Best and N. de Vries (eds) *Thracians and Myceneans.* Leiden: Brill. 106-34.

Simoons, F.J. 1979. Questions in the sacred cow controversy. *Current Anthropology* 20: 467-93.

Simoons, F.J. 1994. *Eat not this Flesh.* 2nd edition. Milwaukee: University of Wisconsin Press.

Smith, I.F. 1965. *Windmill Hill and Avebury.* Oxford: Clarendon Press.

Smith, P., Bar-Yosef, O. and Sillen, A. 1984. Archaeological and skeletal evidence for dietary change during the Late Pleistocene/Early Holocene in the Levant. In M.N. Cohen and G.J. Armelagos (eds) *Paleopathology at the Origins of Agriculture.* New York: Academic Press. 101-36.

Soler, J. 1979. The semiotics of food in the Bible. In R. Forster and O. Ranum (eds) *Food and Drink in History.* Baltimore: Johns Hopkins University Press. 126-38.

Sørensen, M.L.S. 2000. *Gender Archaeology.* Cambridge: Polity Press.

Spindler, G.D. (ed.) 1964. Alcohol symposium. *American Anthropologist* 66: 341-84.

Spindler, K. 1994. *The Man in the Ice.* London: Weidenfeld & Nicolson.

Stott, A.W., Evershed, R.P., Jim, S., Jones, V., Rogers, J.M., Tuross, N. and Ambrose, S. 1999. Cholesterol as a new source of palaeodietary information: experimental approaches and archaeological applications. *Journal of Archaeological Science* 26: 705-16.

Sutton, D.E. 2001. *Remembrance of Repasts: an anthropology of food and memory.* London: Berg.

Synnott, A. 1991. A sociology of smell. *Canadian Review of Sociology and Anthropology* 28: 437-59.

Tambiah, S.J. 1969. Animals are good to think and good to prohibit. *Ethnology* 8: 424-59.

Tannahill, R. 1973. *Food in History.* Harmondsworth: Penguin.

Taube, K.A. 1996. The Olmec maize god: the face of corn in Formative Mesoamerica. *Res* 29/30: 39-81.

Taube, K.A. 2000. Lightning celts and corn fetishes: the Formative Olmec and the development of maize symbolism in Mesoamerica and the American Southwest. In J.E. Clark and M.E. Pye (eds) *Olmec Art and Archaeology in Mesoamerica.* New Haven: National Gallery of Art and Yale University Press. 297-337.

Tauber, H. 1981. ^{13}C evidence for dietary habits of prehistoric man in Denmark. *Nature* 292: 332-3.

Thomas, J. 1991. *Rethinking the Neolithic.* Cambridge: Cambridge University Press.

Thomas, J. 1999. *Understanding the Neolithic.* London: Routledge.

Thomas, J. 2000. Death, identity and the body in Neolithic Britain. *Journal of the Royal Anthropological Institute* 6: 653-68.

Thorpe, I.J. 1996. *The Origins of Agriculture in Europe.* London: Routledge.

Tilley, C. 1996. *An Ethnography of the Neolithic: early prehistoric societies in southern Scandinavia.* Cambridge: Cambridge University Press.

Tipping, R. 1994. 'Ritual' floral tributes in the Scottish Bronze Age - palynological evidence. *Journal of Archaeological Science* 21: 133-9.

Vaughan, S.J. and Coulson, W.D.E. (eds) 1999. *Palaeodiet in the Aegean.* Oxford: Oxbow.

Vencl, S. 1994. The archaeology of thirst. *Journal of European Archaeology* 2: 299-326.

Verpoorte, A. 2000. Grateful prey and animal sign: reflections on food, the Palaeolithic and the nature-culture dichotomy. *Archaeological Dialogues* 7: 173-83.

Villa, P. and Mahieu, E. 1991. Breakage patterns in human longbones. *Journal of Human Evolution* 21:

27-48.

Villa, P., Bouville, C., Courtin, J., Helmer, D., Mahieu, E., Shipman, P., Belluomini, G. and Branca, M. 1986. Cannibalism in the Neolithic. *Science* 233: 431-6.

Visser, M. 1986. *Much Depends on Dinner: the extraordinary history and mythology, allure and obsessions, perils and taboos of an ordinary meal.* Toronto: McClelland and Stewart.

Walcher, D.N., Kretchmer, N. and Barnett, H.L. (eds) 1976. *Food, Man and Society*. New York: Plenum.

Warde, A. 1997. *Consumption, Food and Taste: culinary antinomies and commodity culture*. London: Sage.

Wheeler, R.E.M. 1956. *Archaeology from the Earth.* Harmondsworth: Penguin.

Whittle, A. and Pollard, J. 1998. Windmill Hill causewayed enclosure: the harmony of symbols. In M. Edmonds and C. Richards (eds) *Understanding the Neolithic of North-western Europe.* Glasgow: Cruithne Press. 231-47.

Whittle, A. and Wysocki, M. 1998. Parc le Breos Cwm chambered long cairn, Gower, West Glamorgan: date, contents and context. *Proceedings of the Prehistoric Society* 64: 139-82.

Wiessner, P. and Schiefenhovel, W. (eds) 1996. *Food and the Status Quest: an interdisciplinary perspective.* Oxford: Berghahn.

Willis, R. (ed.) 1990. *Signifying Animals: human meaning in the natural world.* London: HarperCollins.

Wilson, H. 1988. *Egyptian Food and Drink.* Princes Risborough: Shire.

Wood, J. 2000. Food and drink in European prehistory. *European Journal of Archaeology* 3: 89-111.

Wood, J. 2001. *Prehistoric Cooking*. Stroud: Tempus.

Woodward, A. 1999. When did pots become domestic? Special pots and everyday pots in British prehistory. *Medieval Ceramics* 22-3: 3-10.

Wright, K. 2000. The social origins of cooking and dining in early villages of western Asia. *Proceedings of the Prehistoric Society* 66: 89-121.

Wysocki, M. and Whittle, A. 2000. Diversity, lifestyles and rites: new biological and archaeological evidence from British earlier Neolithic mortuary assemblages. *Antiquity* 74: 591-601.

Zvelebil, M. 1994. Plant use in the Mesolithic and its role in the transition to farming. *Proceedings of the Prehistoric Society* 60: 35-74.

Zvelebil, M. and Rowley-Conwy, P. 1984. Transition to farming in northern Europe: a hunter-gatherer perspective. *Norwegian Archaeological Review* 17: 104-28.

Explaining the dietary isotope evidence for the rapid adoption of the Neolithic in Britain

Mike P. Richards

Department of Archaeological Sciences, University of Bradford

Introduction

There are significant material differences between the Mesolithic and Neolithic periods in Britain, and new evidence from isotope studies strongly suggests that the dietary transition between these two distinct periods was abrupt. The rapid change in diet indicated by the isotope analyses may seem counter-intuitive, and has met opposition, as many models used by archaeologists to explain changes in material culture through time are based on analogies which assume a gradual, piecemeal adoption of new technologies and ideas. In this paper I contrast the British and Danish isotope data with two analogies that seem appropriate to our understanding of the processes involved in the Mesolithic/Neolithic transition: the colonisation of Greenland by Danish settlers and the spread of maize agriculture into northeastern North America. These analogies were chosen as they have stable isotope data that can be directly compared and contrasted with the Danish and British evidence. From these analogies I suggest possible interpretations of why there was such a rapid adoption/transition to Neolithic material culture in Britain, and the important role that food, and perhaps cattle in particular, played.

Summary of the evidence for dietary and cultural changes at the Mesolithic/Neolithic transition

The main characteristics of the British Neolithic 'package' - pottery, domesticated animals and burial monuments - appear throughout the British Isles at about 4000 BC (between 5300 and 5200 radiocarbon years BP). In terms of the precision of radiocarbon dating, the spread is instantaneous, with radiocarbon dates from Neolithic sites in Scotland being indistinguishable from dates for sites in southern England. Similar pottery styles appear throughout the British Isles in this early period and we also see the presence of domestic animals, particularly cattle, throughout Britain at these Early Neolithic sites.

Perhaps the most striking new innovation of this period are the burial monuments, the chambered tombs and long barrows that appeared throughout Britain, thousands of which still survive today. Of course, we also see similar types of chambered tombs in coastal Europe, dating to about the same period. In Denmark the change in material culture associated with the appearance of TRB-style pottery, domesticated plants and animals and chambered tombs is remarkably similar to the phenomenon that occurred in Britain at about the same time.

As the theme of this volume and this paper is food, what about diet at this time? Was there a change from the Mesolithic mode of subsistence to a new Neolithic economy, largely based on the production, rather than the collection, of foods? Most of the evidence for a change in diet is anecdotal. We do get the appearance of domesticated species at Neolithic sites, but as we have so few settlement sites from this period the domesticates come from contexts that clearly had a strong ritual component to them, specifically causewayed enclosures as well as chambered tombs. Therefore it has been argued that the economy of Neolithic peoples was a continuation of Mesolithic economy with the gradual addition of domesticates. Specifically, the Neolithic economy has been considered as one based on hunting and gathering of wild foods, in which the domesticates were only important as ritual commodities, contributing only marginally to diet, at least in the Early Neolithic (*e.g.* Dennell 1983; Barker 1985). Others have argued that the significant change in the archaeological record from the Late Mesolithic to Early Mesolithic was associated with a change in diet, but that wild foods were still important, and the transition was fairly gradual (*e.g.* Whittle 1996; Thomas 1991; 1999).

Recently there has been a fair amount of work on the application of direct measures of diet, particularly stable isotope analysis, to Mesolithic and Neolithic humans in Britain (Richards and Hedges 1999a; 1999b; Richards and Mellars 1998; Schulting and Richards 2000) building on existing and newer work for northwest Europe (Tauber 1981; 1983; 1986; Lubell *et al.* 1994; Schulting and Richards 2001). Details of the stable isotope method are reviewed extensively elsewhere (*e.g.* Schwarcz and Schoeninger 1991) and will not be summarised here. However, I do want to look at the results of the application of this method specifically in Britain and Denmark, as it can provide direct evidence of diets and changes in diets and economy associated with the appearance of Neolithic material culture in these two regions.

Compared to other coastal countries in western Europe, Britain has a relatively small amount of Mesolithic human remains, so it is difficult to contrast the isotope values of Mesolithic and Neolithic humans. This is not an issue in Denmark, however, where there are a number of Late Mesolithic (Ertebølle) cemeteries. In a key early paper, Henrik Tauber (1981) measured the carbon isotopes of human bone collagen from various coastal Mesolithic and Neolithic humans. Each individual was radiocarbon dated, and he found that there was a remarkable change in the isotope values from a

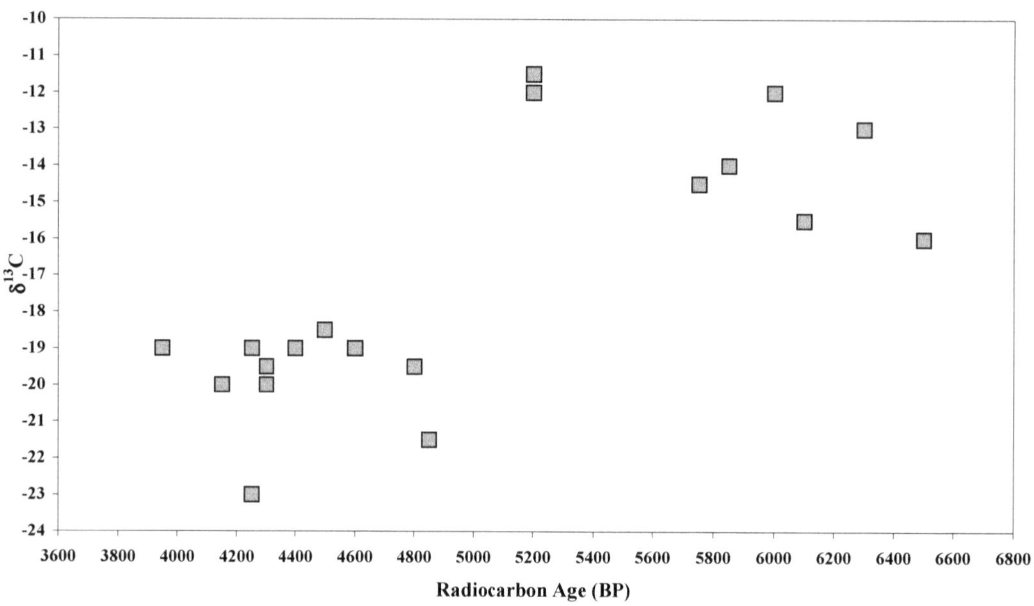

Figure 2.1. *Bone collagen $\delta^{13}C$ and radiocarbon ages from Mesolithic and Neolithic humans from Denmark. There is a sharp change in the isotope data at approximately 5200 BP (c. 4000 cal BC), from a diet where the main source of protein was marine food to one in which terrestrial protein dominated. Taken from Richards et al. in press.*

Mesolithic signal which indicated the importance of marine foods in diets, to a more terrestrial isotope signal in the Neolithic humans. This finding was further supported by more data Tauber published subsequently (Tauber 1983; 1986) as well as more recent tabulations of isotope data (Richards *et al.* in press b) (Figure 2.1). Tauber concluded that there was a rapid change in diet associated with the introduction of the Neolithic into Denmark, about 4000 cal BC (approximately the same time as Neolithic material culture appears in Britain).

Richards and Hedges (1999a) attempted to carry out a similar study to Tauber's using British material. There are very few Late Mesolithic human remains in Britain, and a number of coastal sites, where it is likely that further Mesolithic human remains and shell middens may be found, are now underwater due to sea level changes. Limited work on some of the few human bone samples from Mesolithic contexts, from the shell middens on Oronsay, Scotland, confirmed that Late Mesolithic humans in this context had a primarily marine-based diet (Richards and Mellars 1998). Therefore, Richards and Hedges (1999a) concentrated on measuring human remains dating from the Early Neolithic, after 4000 BC, to see if there were any traces of marine food diets. Individuals were taken from a range of contexts, including caves as well as Neolithic monuments. The evidence is compelling: there is not a single individual with any evidence of any significant consumption of marine foods after the introduction of the Neolithic into Britain. Subsequent detailed and targeted research (Richards and Hedges in prep.) has confirmed this pattern by producing isotope data on over 300 Early and Middle Neolithic humans, none of which have isotope values indicating marine foods in their diets.

It is difficult not to interpret the isotope evidence as indicative of a very clear and dramatic change in subsistence associated with the appearance of Neolithic material culture into Britain at approximately 4000 BC. This dietary evidence does link in with the other lines of evidence at this time, especially with the first appearance in Britain of classic characteristics of the northwest European Neolithic 'package' such as domesticated plants and animals, chambered tombs, causewayed enclosures and pottery. These innovations appear in the archaeological record throughout Britain within a few hundred years and, matching this evidence with the compelling dietary evidence from stable isotope analysis, it is difficult not to conclude that there must also have been a rapid and dramatic change in society associated with this appearance of Early Neolithic material culture throughout the British Isles. The isotope evidence, and indeed many other lines of archaeological evidence, do not support the idea of a slow, gradual and 'piecemeal' adoption of Neolithic material culture and subsistence over centuries.

Perhaps the most basic model to explain this evidence for a rapid transition is one that sees limited migration into Britain by people with an already established Neolithic culture (including subsistence practices), and then the very rapid and wholesale adoption of this Neolithic package by indigenous Mesolithic peoples. However, this model, which requires people who have had a successful Mesolithic lifestyle for millennia to completely and rapidly change to something very different, seems almost incomprehensible and counter-intuitive. A difficulty in accepting this model is that there are few modern analogies that we can draw upon to explain these dramatic changes in lifestyle and society. There are, however, more recent analogies of colonisation and the spread of new practices of food production and consumption that we can draw upon to perhaps explain what the Neolithic transition was not.

Can we look at recent colonisation events for parallels?

As I am suggesting that there was likely to be some limited migration into Britain in the Neolithic, can we look at more recent colonisation events for parallels with the British evidence? The colonisation of Greenland by Danish settlers about 1000 AD may be an appropriate analogy. This colonisation event is well documented, and here I will only focus on the role of food and diet in these events using new evidence from stable isotope analysis.

At first sight, this seems an interesting analogy for the Mesolithic-Neolithic transition in Britain, as in this example we have immigrants who bring with them their established culture with pottery, metalworking and domesticated plants and animals, and they encounter a nomadic indigenous population which does not have any domesticated foods.

The evidence of diet from stable isotope evidence comes from a recent paper by Arneborg *et al.* (1999). They measured the isotope values of humans from the Norse settlement graveyard dating from the late tenth to the mid-fifteenth centuries and contrasted the data with isotope measurements of contemporary indigenous peoples. As expected, the native peoples had a largely marine-food diet, based on fish and especially marine mammals, and this does not seem to have changed after the Danish settlers arrived. When the Danish settlers first arrived in the late tenth century AD, they had a completely terrestrial diet, based on the consumption of cattle and sheep meat and milk as well as grains. What the stable isotope evidence eloquently shows is that, century by century, there was an increasing input of more and more marine foods into their diets, until they approached the same proportion of marine foods in their diets as that consumed by the native peoples.

Therefore, instead of seeing the rapid spread of European domesticates into the existing indigenous population, we are seeing the opposite, as the Norse settlers start consuming the same foods as the native peoples and the indigenous peoples are largely unaffected by the colonisation. This is, unfortunately, not an adequate analogy for the Mesolithic and Neolithic in Britain. The foods the Danes brought with them did not survive well in the harsh Greenland climate, so there were significant practical reasons why these new foods were not adopted by the indigenous inhabitants of this region and were supplemented by local, marine-based, foods.

How do these changes in Britain and Denmark compare to the spread of agriculture in North America?

As the Neolithic stable isotope evidence strongly suggests the abandonment of wild marine resources in favour of terrestrial foods, likely to be domesticates, it is useful to compare the British and Danish isotope evidence with the isotope data for the spread of a domesticate in North America. Specifically, this is the spread of maize agriculture into northeastern North America from Mesoamerica.

The area with the best stable isotope evidence of dietary changes associated with the introduction of maize agriculture is the eastern Woodlands area, although isotope and archaeological data from other areas of North America show similar patterns. In an early and classic paper on stable isotopes, van der Merwe and Vogel (1978; Vogel and van der Merwe 1977) measured the radiocarbon dates and associated carbon isotope values of male and female adults from a number of archaeological sites in Illinois, Ohio and West Virginia dating from 4000 cal BC to 1200 cal AD (Figure 2.2). Maize is a C_4 photosynthetic pathway plant which has a stable carbon isotope value that is very different from C_3 pathway plants, which is the main pathway that plants in this region (and indeed in Europe) follow. The earliest humans in this study have stable carbon isotope ($\delta^{13}C$) values that indicate that their foods, whether plants or the animals that consumed those plants, were C_3 based. This correlates with the archaeological evidence and indicates that maize did not appear in this region until about 200 cal AD. After the introduction of maize at this time there is a gradual change in the $\delta^{13}C$ values, indicating the slow incorporation, over hundreds of years, of this new domesticate. By the end of this time period the $\delta^{13}C$ value indicates that the vast majority of dietary protein was from maize, and terrestrial wild foods were no longer important in the diet. The isotope data from 1000-1200 cal AD matches well with the archaeological evidence of changes in settlement patterns, including increased sedentism, and increasing population.

What these isotope data indicate very clearly is the gradual adoption through time of maize agriculture in this region. It takes approximately a millennium from the first introduction of the new domesticate into this region for it to become the main dietary staple. This is a clear example of the spread of an idea that is gradually adopted and incorporated into an existing society through time. This pattern of gradual adoption of a domesticate is very similar to models suggested for the gradual introduction of Neolithic domesticates into Mesolithic societies in the Early Neolithic, which are then only fully adopted to the exclusion of wild foods after a millennium or so. Yet, these North American isotope data showing a gradual uptake of a new food source over a millennium are in sharp contrast with the Danish and British isotope evidence from the Mesolithic-Neolithic transition. The

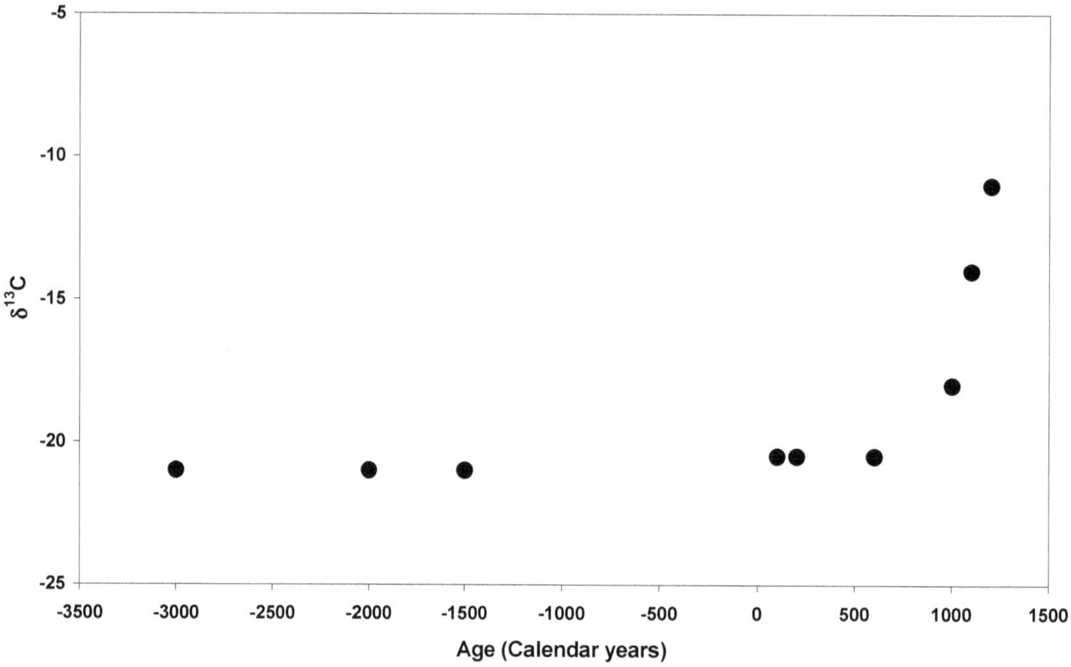

Figure 2.2. *Bone collagen $\delta^{13}C$ and radiocarbon ages for humans from various sites in the northeastern U.S. The data show that before maize agriculture was introduced to this region, human $\delta^{13}C$ indicates that dietary protein came from C_3 plant sources. After the introduction of maize (a C_4 plant), at approximately 200 cal AD there is a gradual change in the $\delta^{13}C$ value indicating the increasing importance of this crop in the diet, until it is the dominant protein source by about 1200 cal AD. Redrawn from van der Merwe 1982.*

European data reveal a very rapid change in diets, and are not consistent with either the gradual uptake of a new (terrestrial based) food source or the concomitant gradual decrease in the use of marine foods through time.

I argue that we see such a difference in the isotope patterns between these two regions because in North America we are talking about the spread of a domestic *plant*, whereas I believe that in Europe, at least in Britain and Denmark, the spread is of domesticated *animals*, perhaps cattle in particular. The differing evidence from these two regions is reflecting the different foodstuffs and the different requirements for producing the different foods.

Discussion

To summarise the stable isotope evidence from these two more recent contexts, the Greenland case is an example of the small-scale initial colonisation of a new area by people with agriculture and animal husbandry. Yet instead of seeing the rapid spread of these supposedly superior foodstuffs into the indigenous populations, we actually see the opposite, as, for various reasons, the colonists adopted the indigenous foods instead. When we look at the spread of a domesticate in North America, we do not see the rapid adoption of this new foodstuff but instead, and perhaps more satisfying intuitively, we see its gradual uptake over many centuries.

So, in comparison, I cannot see how Neolithic material culture and the associated change in diets could spread through Britain so rapidly from a small-scale colonisation, and it is not consistent with the gradual adoption of new types of food production. So, an alternative needs to be presented, focusing on the less definable aspects of this new Neolithic culture. Perhaps there was something more to the new ways of life that the Neolithic represented other than just increased productivity. Perhaps those foods also had something else associated with them, a profound symbolic meaning bound up with their practical values.

Was the Neolithic package spread throughout this region with a strong spiritual and religious association? And if so, was a vital part of this package the adoption of Neolithic foodstuffs? I think that this proposition is a fruitful starting point, and becomes yet more useful if we treat the spread of cattle domestication as central to the Neolithic package.

We have evidence for the symbolic importance of cattle. In Britain we see cattle remains interred in ritually charged sites such as causewayed enclosures and chambered tombs (see Ray and Thomas this volume; Jones and Richards this volume). There are purposeful, ritual, cattle burials in central Europe, at sites of the so-called Globular Amphora culture in Poland. Pollex (1999) has recently reviewed the evidence for these central European cattle burials and also believes that these ritual treatments of cattle are indicative of the religious and symbolic importance of cattle in Neolithic society. Evidence for the ritual treatment of cattle remains is not unique to the European Neolithic, as attested by the numerous depictions of cattle at sites like Catalhöyük (although, interestingly, they do not seem to have been an important part of the diet at this site [Richards *et al.* in press a]).

Are the cattle burials in central Europe and the deposits of cattle bones in Neolithic sites in Britain the last vestiges of the spread of something like a cattle cult that started so much earlier in the Near East? That is unlikely. It is more probable that people in the British Neolithic shared symbolic, and even spiritual, perceptions of cattle similar to those that had also occurred three millennia earlier at sites like Catalhöyük. And indeed the depiction of cattle and the ritual treatment of cattle remains is a phenomenon that persisted throughout European prehistory, and we can see elaborations of it in the Aegean Bronze Age.

Why were cattle so ritually charged and symbolically important? I could imagine the impact that these domestic cattle might have had on Mesolithic foragers, who were well acquainted with the huge and forbidding wild cattle (*Bos primigenius*). Were these wild cattle highly prized in Mesolithic societies? Were these Neolithic colonists a source of wonder to the Mesolithic peoples, as they controlled and produced their own cattle?

Maize in pre-contact North America often had strong religious associations and it is very likely that its spread into North America was due to its symbolic, as well as nutritional, attributes. However, in the North American example discussed above, it took a very long time for maize to become the main dietary staple, while I am arguing that in Neolithic Britain and Denmark the very different isotope pattern indicative of a rapid change in diet was also due to the adoption of new domestic foods with strong symbolic/religious associations. How then can we explain the very different isotope values in these two contexts? There may be good, practical reasons for the differences. Maize agriculture, as well as the domestication of other plants, required significant changes - including increased sedentism - in the lifestyles of nomadic hunter-gatherers who subsisted on a diet of wild plants and animals,. Alternatively, the adoption of a domestic animal by hunter-gatherers who were already subsisting on mainly animal protein - from marine sources in the case of Britain and Denmark - might have required less structural change in society. Perhaps it was easier to substitute domesticated animals into an existing animal-based economy that previously relied on gathered *marine* foods. I need to stress here that the isotope evidence presented here for Neolithic peoples indicates *terrestrial* food, not the presence of cattle in the diet, and I am inferring that cattle were the main adopted domesticate from the contextual data.

The notion of the spread of the Neolithic as being closely related to the spread of a religious belief system is not my original idea, and the sentiments in this paper echo work by Alasdair Whittle and Richard Bradley:

> '...another possibility to consider is whether, as the frame of explanation is shifted to the conceptual and the cultural, some kind of spiritual conversion was involved. Enculturation involved increased attention to the past, to honour ancestors and to frame lines of descent. The model of 'megalithic missionaries' used by an earlier generation of scholars has been much ridiculed, and I am not proposing to revive it in unaltered form.' (Whittle 1996: 210).

> 'The transmission of these new ideas has been compared with the spread of religious beliefs ... Such beliefs may even have extended to new ways of thinking about fertility and the natural world. Thus the rapid expansion of farming in many different parts of Europe may have been facilitated by ideological change.' (Bradley 1998: 161)

Conclusions

I argue that we now have good and compelling evidence for the rapid spread of Neolithic material culture, and especially a dramatic change in diet, in Britain at the Mesolithic-Neolithic transition *c.* 4000 BC. However, we have difficulty explaining the mechanisms behind the rapid spread of Neolithic material culture throughout Britain because we have almost no modern, historical or archaeological analogies for similar phenomena to draw upon. I argue that the rapid adoption of the Neolithic package by indigenous inhabitants of Britain makes more sense if it included powerful religious elements. I think this Neolithic spirituality must have been strongly associated with food, and cattle must have been a vital part of this. The idea of religion as a key factor in the rapid spread of the Neolithic has only rarely been considered in Neolithic studies, and I hope that this paper will at least provide a point of departure for future debate.

Acknowledgements

I would like to thank Mike Parker Pearson for inviting me to attend the conference at which this paper was presented, as well as for very helpful comments on the paper. The ideas expressed also benefited greatly from discussions with Alasdair Whittle, Rick Schulting and Richard Bradley.

References

Arneborg, J., Heinemeir, J., Lynnerup, N., Nielsen, H.I., Rud, N. and Sveinbjörnsdóttir, A.E. 1999. Change of diet of the Greenland Vikings determined from stable carbon isotope analysis and ^{14}C dating of their bones. *Radiocarbon* 41: 157-68.

Barker, G. 1985. *Prehistoric Farming in Europe.* Cambridge: Cambridge University Press.

Bradley, R. 1998. *The Significance of Monuments.* London: Routledge.

Dennell, R.W. 1983. *European Economic Prehistory.* London: Academic Press.

Lubell, D., Jackes, M., Schwarcz, H., Knyf, M. and Meiklejohn, C. 1994. The Mesolithic-Neolithic

transition in Portugal: isotopic and dental evidence of diet. *Journal of Archaeological Science* 21: 201-16.
Pollex, A. 1999. Comments on the interpretation of the so-called cattle burials of Neolithic central Europe. *Antiquity* 73: 542-50.
Richards, M.P. and Hedges, R.E.M. 1999a. A Neolithic revolution? New evidence of diet in the British Neolithic. *Antiquity* 73: 891-7.
Richards, M.P. and Hedges, R.E.M. 1999b. Stable isotope evidence for similarities in the types of marine foods used by Late Mesolithic humans at sites along the Atlantic coast of Europe. *Journal of Archaeological Science* 26: 717-22.
Richards, M.P. and Mellars, P. 1998. Stable isotopes and the seasonality of the Oronsay middens. *Antiquity* 72:178-84.
Richards, M.P., Pearson, J.A., Molleson, T.I., Russell, N. and Martin, L. In press a. Palaeodietary evidence from Neolithic Çatalhöyük, Turkey. *Journal of Archaeological Science.*
Richards, M.P., Price, T.D. and Koch, E. In press b. The Mesolithic/Neolithic transition in Denmark: new stable isotope data. *Current Anthropology.*
Schulting, R.J. and Richards, M.P. 2000. Mesolithic subsistence and seasonality: the use of stable isotopes. In R. Young (ed.) *Current Research on the Mesolithic of Britain and Ireland.* Leicester: Leicester University Press. 55-65.
Schulting, R.J. and Richards, M.P. 2001. New palaeodietary and AMS dating evidence from the Breton Mesolithic cemeteries of Téviec and Höedic. *Journal of Anthropological Archaeology* 20: 314-44.
Schwarcz, H. and Schoeninger, M. 1991. Stable isotope analyses in human nutritional ecology. *Yearbook of Physical Anthropology* 34: 283-321.
Tauber, H. 1981. ^{13}C evidence for dietary habits of prehistoric man in Denmark. *Nature* 292: 332-3.
Tauber, H. 1983. ^{13}C dating of human beings in relation to dietary habits. *PACT* 8: 365-75.
Tauber, H. 1986. Analysis of stable isotopes in prehistoric populations. In B. Herrmann (ed.) *Innovative Trends in Prehistoric Archaeology.* Berlin: Mitteilungen der Berliner Gesellschaft für Anthropologie, Ethnologie und Urgeschichte 7.
Thomas, J. 1991. *Rethinking the Neolithic.* Cambridge: Cambridge University Press.
Thomas, J. 1999. *Understanding the Neolithic.* London: Routledge.
van der Merwe, N.J. 1982. Carbon isotopes, photosynthesis, and archaeology. *American Scientist* 70: 209-15.
van der Merwe, N.J. and Vogel, J.C. 1978. ^{13}C content of human collagen as a measure of prehistoric diet in Woodland North America. *Nature* 276: 815-16.
Vogel, J.C. and van der Merwe, N.J. 1977. Isotopic evidence for early maize cultivation in New York State. *American Antiquity* 42: 238-42.
Whittle, A.W.R. 1996. *Europe in the Neolithic: the creation of new worlds.* Cambridge: Cambridge University Press.

In the kinship of cows: the social centrality of cattle in the earlier Neolithic of southern Britain

Keith Ray[1] *and Julian Thomas*[2]

[1]Herefordshire Archaeology, Hereford
[2]School of Art History and Archaeology, University of Manchester

Introduction

'At a subsequent period Mr Cunnington made a second attempt on this tumulus by opening more ground on the east as well as the west end; at the former he found the heads and horns of seven or more oxen; also a large cist close to the skeletons; but owing to the great height of the barrow, and the large stones continually rolling down upon the labourers, he was obliged to stop his operations.' (Cunnington 1889: 105).

The mortuary area at Boles Barrow in Wiltshire, described here by the descendant of the original excavator, was located within an earthen long barrow. The mortuary area was floored with a compact spread of flint nodules, among which were found bones from some 16 adults and children, in association with a pit and with a group of the crania and metatarsals of seven cattle. The whole group of deposits was covered by a ridged heap of large sarsens, further flint nodules, and a bluestone block (Kinnes 1992: 25).

For more than a hundred years, the close association between human and cattle bones in a variety of contexts in the earlier Neolithic of Southern Britain has been noted and remarked upon. In this paper we examine how it is possible to theorise a social centrality of cattle to Neolithic communities. We shall do so by considering the pattern of such associations, and of the deliberate structured location of cattle bones in a variety of focal depositional contexts, in reference to the very nature of human-animal relations engendered in this period.

We argue that the scale and intricacy of deliberate deployment of cattle remains, and of their direct association with human remains, must derive from a series of relations (both conceptual and practical) that existed between the two communities, human and bovine. These phenomena are not explicable by extrapolating the existence of a fixed symbolic meaning for cattle or by simply recognising a key economic status for them. We infer, rather, that cattle were a means of social reckoning that denoted wealth and status, but which also evoked tradition, custom, intercession, and descent.

Establishing some patterns

Since the early 1980s, prehistorians have begun to pay greater attention to the structure as well as the fact of depositional associations. We have realised that the frequently complex nature of such associations has resulted from acts of deliberate structuring through selection among and arrangement of the object-world of past communities. Such structuring has consequently come to be appreciated as part of a deliberate manipulation of materials in order to further a variety of social purposes, simultaneously emblematic and strategic.

What, then, is this pattern of associations between the interment of human and of cattle remains? Firstly, we should note the degree to which the circumstance pervades the structured contexts of the period. Also, it is evident that the pattern is established from the inception of the Neolithic in southern Britain. For example, one of the Handley Hill pits near Wor Barrow on Cranborne Chase contained a disarticulated human burial, a deliberate arrangement of cattle bones, and a large Grimston Ware vessel (Barrett *et al.* 1991: 34). An upright post in this pit presumably served as an above-ground marker. This feature links the site into a local tradition of marking Early Neolithic pits that either contained burials, as at Launceston Down (Piggott and Piggott 1944: 74-5) or had mortuary associations, as at Hambledon Hill (Mercer 1980: 23).

Cattle bones are prominent among finds from the ditches of long barrows, as, for example, at Thickthorn Down and Wor Barrow on Cranborne Chase (Drew and Piggott 1936; Pitt Rivers 1898), at Kingston Deverill on Salisbury Plain (Thomas 1999b: 142), at Fussell's Lodge east of Salisbury (Grigson 1966), at Horslip on the Marlborough Downs (Ashbee *et al.* 1979) and at Lambourn on the Berkshire Downs. Direct association with human mortuary deposits at long barrows is at least equally as common. Indeed, it has been noted that bones of domesticated cattle have been recorded in long barrows, whether chambered or not, wherever accurate identifications of animal bones have been made (Grigson 1966: 65).

In the Salisbury Plain area alone, besides Heytesbury 1 (Boles Barrow), cattle bones are documented from mortuary contexts and directly among or close to human remains at Amesbury 42, Corton, Fussell's Lodge, Sherrington 1, Tilshead Lodge, and Norton Bavant (Grigson 1966). Meanwhile, cattle bones predominate in similar animal bone deposits at Nutbane, and in the mortuary enclosure at Normanton Down, and were apparently present in cremated deposits at Knook Barrow and at King Barrow (Grigson 1966; Kinnes 1992). Nor is such direct association limited to the chalklands of Wessex. Further afield, cattle bones have been recorded from earthen long barrows at Badshot, Surrey (Jackson 1939), and the Giant's Hills, Skendleby, in Lincolnshire (Phillips 1935). Moreover, cattle bones have been recovered among the contents of the chambers of at least 14 Cotswold-Severn long barrows. Entire calf skeletons have been found in two others, an articulated example from Bown Hill, and scattered remains of one at Notgrove (Clifford 1936; Thomas 1988).

The other major series of sites in which cattle bones are found in direct or in looser association with human remains are causewayed enclosures, and the evidence from these sites is dispersed widely across the landscapes of southern Britain. Purposeful deposition of cattle bones is attested at the Maiden Castle enclosure (Dorset; Grigson in Smith *et al.* 1981: 199), at Whitesheet Hill (Wiltshire; Piggott 1952), and at Staines near London (Serjeantson 1996). Deposits featuring the placing of bovid bones in direct association with human bones are attested at the enclosures at Hambledon Hill (Dorset; Thomas 199b: 28), at Windmill Hill (Wiltshire; Smith 1965; Whittle and Pollard 1998), and at Etton near Peterborough (Armour-Chelu 1998).

Finally, towards the end of the fourth millennium BC, cattle bones are again found in association with major monuments with mortuary referents, as at the bank barrow at Maiden Castle. More striking still is their presence 'in place of' human remains at the Beckhampton Road barrow near Avebury (Ashbee *et al.* 1979: 245).

Conceptualising human - animal relations in the Neolithic

This repeated close association between the remains of people and cattle is therefore both striking and of long duration throughout the fourth millennium BC in southern Britain. In the tradition of generalisation that has characterised much recent writing on the Neolithic, we could deduce here the operation of profound symbolic systems represented or exemplified by these traces. From such a perspective, these systems might have privileged the idea of domestication itself (*cf* Hodder 1990), or of the cosmological ordering of inner and outer worlds (*cf* Whittle and Pollard 1998). The problem, as we see it, with both these symbolist views, as with more overtly economistic explanations, is the tendency to think of cattle emblematically, either as objects or as a food resource.

Such a limited view derives ultimately from the way in which western thought has increasingly segregated humans from non-humans. This is a process that Bruno Latour refers to as 'conceptual purification', in which the capacity for reason is believed to separate people from other beings, whether animate or inanimate (Latour 1993: 13). In practice, of course, we are ever more closely tied into networks of relationships with non-humans, principally machines and material goods, through which we pursue our everyday projects. But the understanding remains pervasive, that culture and the mind represent a sphere of human activity that is quite distinct from 'nature'. The West has, as a consequence, been able to promote and sustain a vision of humanity as 'the rational animal': a biological substrate to which a soul and a rational mind have been added (Heidegger 1993). The corollary of this logic is that, the more we exercise our reason, the more we transcend the animal in us. It is perhaps instructive to recall in this context that René Descartes, the originator of many similar aspects of Western dualistic thinking, considered animals to be biological automata (Descartes 1912: 116).

In this pattern of thought, most clearly articulated in the Enlightenment, it is our capacity for reason that enables us to act decisively to improve our own material conditions (Cassirer 1951: 199). As this modern age of reason has progressed, therefore, humanity has increasingly been seen to supplant God as the being for whose benefit the world exists. The prospect of 'the coming of the Kingdom of Heaven' has been replaced in the Western imagination by the immanence of the processes of self-realisation and human emancipation. Our ever-greater mastery over Nature is understood as the means by which these ends are to be achieved. As a result, we talk of an 'environment' that is defined as somehow external to humanity, and that therefore exists as a vast storehouse of resources that we are at liberty to plunder.

'Environmentalists' of course recoil from the destruction of natural habitats that occurs as a consequence of such views. Nonetheless, they sometimes do so on the understanding that 'humans are animals too': that despite our technological and cultural achievements, we still have all the main attributes of biological organisms. We would rather address the issue not in terms of our common biology, but by contradicting the idea of conceptual dislocation in the first place. We would therefore suggest that, for people in the distant past at least, the notion of 'a mere animal', a creature that existed without social relations, cultural significance, and spiritual values, would have been all but incomprehensible.

As such, we should not see cattle as having been 'bracketed off' by Neolithic people from a wider world of animate being. The dialogues that existed for these people are likely to have been maintained with a complex variety of beings, both corporeal and ethereal. We would propose, therefore, that a fictive kinship with cattle, both genealogised and metaphorical, was created by these people in the intimacy of their relationships with their herds. In turn, the creation of such 'kinship' enabled the dead of both communities to intercede on behalf of humans in the unseen world that influenced the destiny of both hearths and herds.

The Neolithic is traditionally 'the period of domestication'. Despite the very different forms of critical evaluation that the term 'domestication' has been subjected to by Higgs and Jarman (1969) and by Ian Hodder (1990), we still routinely understand this process in terms of the way that human beings have acquired their means of subsistence. However, if domestication is understood instead as implying primarily a social relationship with animals, our appreciation of what the Neolithic was *experienced as* could also change profoundly (Ingold 1986: 252-5). For instance, it may be perceived as a period characterised not so much by a novel exploitation of animals, as much as by a first entry into an intimate partnership with them.

Of human groups and herds of cattle: some novel dimensions of mobility

In considering the earliest development of Neolithic communities in the British Isles, it has become widely accepted that, for southern England at least, domesticated animals had more of an initial impact than cereals (Entwhistle and Grant 1989; Richmond 1999). Cereals were certainly grown, but their use might have been sporadic or episodic, and their dietary role was supplemented with wild plants (Robinson 2000). Woodland clearance in the earlier Neolithic no doubt therefore took place largely to provide pasture and must, to some extent, have been generated and maintained by the grazing of herds. It follows from this that movement between seasonal pastures will have provided one of the defining rhythms of life for Early Neolithic communities.

It has recently been suggested that the large clearings that appear to have existed in the surroundings of at least some major ceremonial monuments will have served as pasture for cattle (Barclay and Hey 1999: 71). In this respect, monument complexes might have been seen to mark the seasonal gathering points for herding communities. Given the role that some causewayed enclosures at least also performed in respect of the processing and circulation of the remains of humans, it is not entirely surprising to find deposits of cattle bones alongside human remains at these sites.

Cattle bones are repeatedly found in bundles that also include human bones placed in causewayed ditch segments. Both cattle skulls and human skulls are frequently found in locations marked by particular densities of deposition, such as the butt-ends of such ditch segments. Consideration of depositional contexts from two geographically widely separated but recently published sites can be used to illustrate the complexities involved.

At Etton on the edge of the Cambridgeshire fens near Peterborough, a causewayed camp featured the remarkable survival of woodworking debris and placed deposits including birch-bark containers (Pryor 1998). The faunal specialist noted that the numbers of cattle bones and large ungulate fragments found at the site meant that 'this taxon far exceeded all others in importance' at the site (Armour-Chelu 1998: 284). Since 'the number of cattle bones from the structured deposits was small', it was concluded that 'in most respects the assemblage represented domestic refuse' (*ibid.*). This interpretation is difficult to sustain, however, given the nature and pattern of the occurrence of placed deposits across the site. For example, 'cattle bones were not deposited as partial bodies or skeletons, but more groups of vertebrae and/or ribs, skulls and neonatal remains of this taxon were found' (*ibid.*). Moreover, 'there was a disproportionately high number of meat-yielding elements, which were placed as long linear deposits in several ditch segments. These deposits were similar to those described from Windmill Hill and could represent feasting debris' (*ibid.*).

It is of particular interest how the site at Etton has been interpreted as having been organised, with a deliberate distinction in depositional practice detected between the western and eastern ditch segment arcs. While structured deposits existed in both arcs, it was only in the eastern arc segments that the cattle bones were found in long linear spreads (Pryor 1998: 361). It was in these segments that the more overtly structured deposits occurred, and where finds of human remains were concentrated. It was also in just those later phases here, in which the cattle bone 'stacks' become more frequent, that the human bones (mostly skull fragments and all showing signs of excarnation) are present: located in the terminals of segments 6, 8, 10, 12, 13 and 14. In the northern terminal of segment 6, a skull fragment had been placed in a pile of cattle bones with a red deer antler 'baton', next to a 'pyre' deposit (*ibid.*: 29-31, fig. 25). Cattle skulls had deliberately been placed near the southern end of segment 10, close to a placed fragment of human skull (*ibid.*: 44-5).

Both the selection and the association of animal bones, pottery, flint, stone axes and other materials at Etton were complex, and clearly followed elaborate rules. What needs to be stressed here is not so much the exclusiveness of association specifically between cattle and human remains at the site, but the very pervasiveness of cattle remains in the context of feasting and the incorporation of selection of bones from excarnated human bodies. Pryor sees the eastern arc deposits as possibly the product of elaborate kin-group ceremonies (*ibid.*: 361), and this is of particular interest when we go on to consider why cattle should be expected to be in evidence in such social contexts.

From the recent excavations at Windmill Hill, there is evidence that cattle remains predominate in just those deposits of animal bones that are most obviously placed. The excavators have suggested, indeed, that some deposits represent the 'conspicuous non-consumption' of meat rather than the dumping of animal bones after butchery (Whittle *et al.* 2000: 357). It was noted that there appeared to be a direct equivalence in the way that human and animal remains were treated and that, moreover, in many contexts human and cattle bones were buried in significant juxtaposition (*ibid.*: 360). For example, in one ditch segment, a human child's femur was found pushed down into the marrow cavity of a distal ox humerus (*ibid.*: 110).

Faunal analysis from the causewayed enclosures, including Windmill Hill, Etton, Maiden Castle and Hambledon Hill, has revealed a characteristic age and sex structure which is dominated by old and moderately old cows, with many bullocks being killed off young rather than being kept as steers. As Tony Legge has argued, this suggests that milk was of some importance, and that females were being slaughtered only when they had finished lactating (Legge 1981). The remains at the enclosures may sometimes derive from sporadic events of deposition rather than significant feasts. Nonetheless, it may be that beef was not so much a part of everyday life as a prized delicacy associated with ceremonial events. Even then, access to meat might well have been socially divided.

At the time that Legge was writing it seemed natural to assume that, being 'farmers like us', Neolithic people were operating an efficient dairying economy, the surplus of which could be used to provide the calorific base for Wessex chiefdoms. Of course, post-Medieval Britain is not the only exemplar for a cattle-keeping economy. East African cattle herders both milk and bleed their animals. It is interesting to ponder what pattern of stable isotopes could we expect to find in the bones of people who had, on a daily basis, consumed bowls of mixed blood and milk (*cf* Richards 2000). Of course, we do not know to what degree the cattle-keeping populations of earlier Neolithic Britain were lactose-tolerant, although secondary products such as butter and cheese do contain reduced quantities. Meanwhile, equally special relationships with cattle obtain elsewhere, with Indian Hindu cattle-keepers abstaining from slaughtering their charges except under the most ritualised of circumstances, and then never partaking of their flesh.

Why a place for cattle with the human dead?

The fact and the manner of association of cattle bones with human remains in causewayed enclosures, then, evokes a series of relations concerning food, circulation, transition, mobility, transit and exchange. While other animals are bound up in complex ways with such relations, the depth and frequency of the link to domesticated cattle, their centrality in the economy, and their impact within the landscape and upon the pattern of human movement marks them out as a special category.

It is time to consider more explicitly the association between cattle, death and the ancestors, by focusing upon the contexts of mortuary structures. Here, again, it is the striking associations and arrangement of cattle bones with human bones in the mortuary structures and sequences of the Salisbury Plain area that provide a starting point for discussion. The deliberate selection of particular parts of cattle skeletons in this area has been noted. At Tilshead Lodge, Thurnham found two skulls, six or seven cervical vertebrae, and the metatarsus and phalanges still articulated nearby (Thurnham 1869). Similar associations and selection were noted at Boles Barrow, Corton, King's Barrow, Knook 2, and Sherrington 1 (Grigson 1966: 65).

At Fussell's Lodge, in south Wiltshire, an ox skull was recovered from one end of the linear timber chamber, which also contained four discrete groups of human remains. The bones of three *Bos* feet and a tail were found in the blocking of the same chamber. The excavator's suggestion was that all of these bones may come from a single animal, and that they may represent an ox hide, with head and feet intact, that had been draped over the timber chamber prior to the construction of the long mound (Ashbee 1966). This might or might not have been the case, but it certainly demonstrates an intimate association between the skeletal remains of human beings and those of cattle. Yet, as we have already mentioned, the sense of there being some kind of equivalence between humans and cattle arises given not only from their association, but also from their apparent inter-changeability. For instance, at the long barrow Amesbury 42, near the Stonehenge Cursus, Thurnham failed to find any human remains as a primary burial deposit. In their place were at least one ox skull, and the foot bones of four or five cattle (Thurnham 1869: 182).

Curation and tradition in the deposition of cattle bones

In most cases, it has been assumed that cattle bones have been placed as deposits soon after the killing of the individual animal(s) concerned. However, there are instances where this could be questioned: where, for instance, mandibles are found individually, and lacking the teeth that were once located in them. In such circumstances, the possibility of curation should, but rarely has, been considered. At Stonehenge, the dating of two mandibles separately and apparently deliberately deposited near to two ditch segment terminals of Stonehenge I has raised this question to some prominence.

It has been suggested (Cleal *et al.* 1995: 529-31) that the date of the deliberate deposits of animal bone (the two right jaws of cattle, a crushed ox skull, and a red deer tibia) differs significantly from the preferred dating of the digging of the enclosure ditch. The inference is that these bones must have been kept for between 50 and 850 years before being placed in their 'final' contexts. The jawbones had lost their teeth, and yet the surface of the bone was in good condition, indicating that they had not been exposed to the elements for any lengthy period of time. So, either they must have been retrieved from an ancient midden, or they had been deliberately stored for some generations by the time that eventual deposition was deemed appropriate for them.

At Etton, the two bones with scored marks deposited near the butt-ends of segment 9 were among four bones so marked from that segment, and are complemented by a rib with scored striations from segment 3 (Armour-Chelu 1998: 286-8). Although these marks could have been made immediately after defleshing, they were not made as part of that process. The excavator and specialist appear to have regarded these bones as artefacts, the segment 3 example being referred to as a possible 'tally stick'. Whether or not used as tools, it seems likely that, once scored, the bones concerned were curated, at least until appropriate circumstances for deposition obtained.

At Fussell's Lodge long barrow, the ox skull found sealed by a flint stack, between the human Bone Group B and the entrance to the enclosure (see above), was 'very badly preserved, dirty and broken into many small pieces' (Grigson 1966: 65). It was inferred that the cattle skull had therefore been subjected to the same previous defleshing treatment as the human bones, either through prior burial and exhumation, or through excarnation (*ibid.*). It seems at least a possibility here that, like the skull at Stonehenge, it might have been curated.

This practice of storage of animal bones is widely attested in the ethnographic record. In myriad contexts in West Africa, for instance, shrine-priests, chiefs, and secret societies store bones of sacrificed animals (KR's personal observation: for early first-hand descriptions see

Partridge 1905; for an impression of the wider ritual context see Cole and Aniakor 1984). These bones are often carefully wrapped in cloths and concealed in lofts or storehouses. They are brought out on special occasions, and either prominently or covertly displayed, sometimes only to initiates. Such bones are passed from office-holder to office-holder down the generations, and they often accumulate a rich patina of age. The keeping of cattle bones in Early Neolithic southern Britain echoes such practices, and perhaps served as a particular mnemonic also of the herds to which the individual animals belonged.

Another fascinating aspect of 'cattle orientation' in the earlier Neolithic is the degree of continuity of practices evident through to the end of the third millennium BC. These practices are specifically those that closely associate the communities of the living and the dead through funerary display. The most prominent features of such practices have been established through recognition of instances of what has come to be known as 'head and hooves' burial, initially from the recurrent phenomenon noted among the Salisbury Plain long barrows, noted above. In at least one instance, it has also been noted through evidence of mass deposition of cattle remains in acts of closure of funerary ceremonies.

In the first case, there are burials such as that at Hemp Knoll, where the head and feet of an individual cow were placed in a grave pit containing a Beaker burial in a coffin (Robertson-Mackay 1980). Stuart Piggott has noted the frequency of occurrence of this circumstance in Beaker burials, and its antecedents in the earlier Neolithic, as well as its more widespread occurrence throughout prehistoric Europe and Eurasia, and its persistence into the Roman period in Britain (Piggott 1962). One interpretation of this recurrence of heads and hooves is that it represents the enactment of a performative mortuary ritual, shamanic or otherwise, involving the clothing of an individual, or the deceased, in the skin of an ox/cow. Whether or not this was literally the case, the phenomenon could suggest at least a metaphoric 'cloaking' of the human dead in the body of an animal closely associated with the sustaining, movement, and 'transit' of members of human communities. Alternatively, the head and hooves might have formed that part of the carcass given to the dead.

The second case is best represented by the capping of the complex mortuary deposit at one of four round barrows excavated at Irthlingborough in Northamptonshire in 1986-87 (Parker Pearson 1993: 78-81). This deposit comprised a limestone rubble cairn in among the stones of which were recovered 184 cattle skulls, 38 mandibles, 33 shoulder blades and 15 pelvises. The cairn covered a collapsed timber mortuary structure housing a Beaker burial.

Tracing kinship, in parallel

We have implied above, then, that it is not only possible to trace a close connection between cattle and humans through the earlier Neolithic, but that the intricacy of this association demands further explanation. We would assert that the links are sufficiently complex to transcend any simple iconic or symbolic reference, whether or not this formed part of the matrix of allusion at play in such contexts.

The gatherings of people and animals at significant locations such as the causewayed enclosures would have provided premium opportunities for the display of herds, so that the quantity and quality of cattle that a community possessed might easily have amounted to a source of prestige. Cattle might have represented wealth, but not necessarily in the modern sense of goods to be disposed at will. Instead, we may think of them as inalienable wealth, intimately bound up with the identity of the community (*sensu* Weiner 1992). We have suggested that the slaughter and eating of cows was a special event, and that such events were often enacted at mortuary sites as well as at causewayed enclosures, in concert with other often highly ritualised activities. If so, then the sacrifice or formal consumption of cattle could be seen as necessary for any intercession with the dead or with deities. Similarly, cattle are very likely to have been used as gifts in procuring alliances, as bridewealth, or as death payments. If for Karl Marx commodities in modern society are congealed lumps of labour, then it is possible to envisage that, for people in the Neolithic, cattle were congealed lumps of social relations. They represented a store of potential ways in which people could extend their social presence across space and time.

Cattle bones, then, might routinely have complemented the intimate remains of the dead. As we have just suggested, they could moreover have provided a means whereby people extended their social presence across time and space. Such extension could be referenced materially: in time, for instance through curation of their bones, and in space, for instance through their substitution for human remains. It is reasonable to infer, therefore, that there existed among these communities a close connection between cattle and kinship. The social relationships of the human communities were so deeply interwoven with the dynamics of their herds that the latter could have provided both metaphor and mnemonic for the very social constitution of those human groups.

In this way, the composition of each herd itself would have 'recorded' past human alliances, marriages, gifts, cattle raids and arrangements with other groups, because at least some people would have been well aware which cattle had come from where, and which had mated with which. This record would have made each herd literally a living history, the population structure of the herd providing a parallel for that of the human group, each herd/group characterised by genealogies and lines of descent.

The corollary of such a situation is that cattle, whether living or dead, might have been used as powerful resource for the articulation of human social relations. For instance, they could been both a means of reckoning descent, and a means of constructing the very fabric of human genealogy (understood, *cf* Vansina 1985, as a supremely 'elastic' medium in societies without writing).

One 'cattle community' or more?

So far we have been talking of 'cattle' bones as if it were obvious that we were referring to domesticated cattle (*Bos taurus*), but now it is time to turn our attention, if briefly, to *Bos primigenius* (wild cattle, or aurochs). The bones of these sometimes massive animals frequently found their way on to earlier Neolithic sites. For instance, at Stonehenge I, they were retrieved from no fewer than five locations in different ditch segments (Cleal *et al.* 1995). However, primary associations often exist also with human remains, and with human mortuary contexts. Most striking, perhaps, is the purposeful deposition of a complete skull in one of the flanking ditches of the Thickthorn Down long barrow in Dorset (Drew and Piggott 1936). The continuity of the presence of aurochs is widely attested through the Neolithic, and it comes as no surprise to find that, at Irthlingborough, teeth and a horn from one of these animals found their way into the sealing deposit with all the young cattle skulls.

Despite this frequency of occurrence, the 'history' of *Bos primigenius* cattle in Britain through to their extinction in the Late Bronze Age is poorly understood. If domesticated cattle had a parallel 'kin' relationship with humans, why do their wild cousins appear so often in the same, or similar, contexts? One explanation is provided by Pollard and Whittle's nature/culture interpretation of the spatial ordering of acts of deposition in the circuits of sites like Windmill Hill (Whittle *et al.* 2000). This proposes that wild cattle provided a symbolic opposite of domesticated cattle, and therefore served to point a metaphoric contrast between the human 'civilisation' of farmers, and the un-ordered world beyond. While such concepts are attractive, not least for their elegance, we believe that something more complex was going on.

It seems at least probable that people in the earlier Neolithic (and later) recognised the existence of parallel (as well as contrasting) cattle communities. If their intent had been to 'promote' farming, then presumably not only could herds of *Bos primigenius* have been captured, but presumably they could also have been cross-bred with *Bos taurus* herds. That this did not happen, at least while prehistoric communities were mobile, can perhaps be inferred from the continued existence of wild cattle beyond the Neolithic. In his musings on the Irthlingborough discovery, Parker Pearson (1993: 81) suggested that wild cattle were 'presumably becoming rare and might well have been ranched as a semi-domesticate'. Alternatively they were 'a prized trophy, hunted in the wild'.

This latter is an interesting suggestion, not least because it presupposes a deliberate maintenance of communities of wild cattle for hunting, a practice attested elsewhere in the world into recent times. An example is instructive in this context. In eastern Nigeria, the Igbo communities of the Nsukka area have long kept small domesticated cattle in small herds dispersed among the compounds of their villages. At the same time, in order to progress through 'title grades', it was necessary to hunt and kill an *atu* (bush-cow, or buffalo) to obtain its tail for use as a fly-whisk (Afigbo 1981: 85). For this purpose it was necessary to travel into a large tract of forest between the Anambra and Niger rivers, near their confluence, where wild buffalo were deliberately left to roam.

A not entirely dissimilar situation might have existed in Neolithic Britain, where there are likely to have been extensive areas of forest that were deliberately retained as wilderness areas. An example claimed for southern Britain is a tract of land on the western Hampshire/Berkshire border, where the pollen record for Snelsmore shows that a wide area at the foot of the downs only ceased being forested in the Iron Age (Waton 1982).

We can therefore perhaps re-cast this 'separate existence' of wild cattle populations from domestic herds in an alternative way. While it may be that *Bos primigenius* groups were *either* reserved for hunting *or* managed as semi-domesticates, in either case they were presumably deliberately segregated from *Bos taurus* herds, at least until the point of slaughter. Perhaps this, too, can be taken to suggest that there was a concern among these communities to protect the blood-stock of their domestic cattle: even, again, that the 'kinship' they possessed was a resource in its own right.

Earlier Neolithic people and cattle: recasting a relationship

The dominance of cattle remains in so many earlier Neolithic contexts in southern Britain indicates that it is realistic to infer that the cattle-holding communities concerned moved to the tempo of their cattle: their grazing, watering, lactation and reproduction. People and cattle moved together around the landscape, periodically fissioning into smaller groups, or gathering together for ceremonial activities at the great monument clusters. Just as the living circulated, so too did the remains of the dead, as we now appreciate comprising the bones of both people and cattle (Thomas 1999a). Also bound up in these patterns of circulation were certain artefacts such as stone axes, which passed from hand to hand, and from group to group. It seems possible then, that shared metaphors or understandings linked these dissimilar things: people, bones, artefacts, cattle. Axes might have belonged to herds, and pots that routinely contained the bodily fluids of cattle were smashed in funerary practice, perhaps because they were like human bodies.

More certainly, the bones of cattle 'tracked' the destinies of human remains: buried whole when people were buried whole, placed in selected bone groups together when selection of body parts for deposition was deemed appropriate. When heads (or skulls) were to be placed (sometimes dramatically, in mortuary houses, in ditch bottoms or terminals, or in mounds) cattle or people could be selected, apparently interchangeably (or replaced by substitute pots or axes). Where human remains were cremated, cattle too had been burned on the pyre and were ready for placement. Such repeated associations are intimate and personal, as well as articulate about social relations: they reflect ties of emotion as well as lineage.

If this seems far-fetched to us, we perhaps need to

realise that our modernity is not as universally shared as we have often imagined. For instance, the attachment to domesticated cattle even in Britain has survived more strongly than widely appreciated. In a curious way, the bloodstock ties might also have survived longer than we may expect, as reflected in the experience of the reforming landlords in their introduction of farming innovations in the west and north of Britain. So it was that Sir Henry Trelawny of Pelynt in southeast Cornwall (d. 1834) tried to introduce the Suffolk horse-drawn plough onto his Trelawne estate lands. The innovative plough was said to perform double the work at much lower expense than the traditional systems in use. However, Trelawny's tenants could not by any means be persuaded to adopt the Suffolk plough, preferring instead to use oxen to pull their old, inefficient, Cornish ploughs. He also introduced Leicestershire longhorn cattle bred by the improver Robert Bakewell, 'but neither his own people, nor the farmers of the neighbourhood, could be adduced to attend to them' since they so preferred their small black cattle (Grigson 1954: 60).

In earlier Neolithic Britain, if herds of cattle were understood as communities that mirrored those of human beings, then it is not surprising that their bones were treated in similar ways. Cattle were not the same as human beings, but their 'society' might have been understood in terms of equivalence. The ancestry of each herd was therefore of as much consequence as that of the human group that tended it, because the destiny of each was so finely interwoven. Of course, the people killed and ate individual cattle, and also drank their milk and (perhaps occasionally) their blood, while the cattle didn't kill or eat any people. At the same time, however, the people walked with the cattle, defended them from predators, tended them and at times watered and fed them. Those herding people could easily have envisaged such an arrangement as being characterised by reciprocity. Moreover, if a creature possesses a kinship that parallels your own, you may choose to kill and eat it in the kind of ritualised circumstances for such slaughter that appear to have obtained in earlier Neolithic southern Britain.

Since Gordon Childe at least, the Neolithic has been seen as one of the great 'revolutions' of Old World prehistory, marking the point at which human beings assumed control over the reproduction of other species. As such, this transition has come to form an important element in a larger narrative about how humanity has managed to prevail over Nature. Such views are challenged in a present whose narratives are dominated by the consequences of such arrogance. In this light, it would be singularly ironic if, for those who lived through a small part of this 'revolutionary' process of domestication, it was understood not in terms of domination, but of partnership.

References

Afigbo, A.E. 1981. *Ropes of Sand: studies in Igbo history and culture*. Ibadan: University Press Limited/Oxford University Press.

Armour-Chelu, M., 1998. The animal bone. In F. Pryor, *Etton: excavations at a Neolithic causewayed enclosure near Maxey, Cambridgeshire, 1982-7*. London: English Heritage Archaeological Report 18. 273-88.

Ashbee, P. 1966. The Fussell's Lodge long barrow excavations 1957. *Archaeologia* 100: 1-80.

Ashbee, P., Smith, I.F. and Evans, J.G. 1979. Excavation of three long barrows near Avebury. *Proceedings of the Prehistoric Society* 45: 207-300.

Barclay, A. and Hey, G. 1999. Cattle, cursus monuments and the river: the development of ritual and domestic landscapes in the Upper Thames Valley. In A. Barclay and J. Harding (eds) *Pathways and Ceremonies: the cursus monuments of Britain and Ireland*. Oxford: Oxbow. 67-76.

Barrett, J.C., Bradley, R. and Green, M. 1991. *Landscape, Monuments and Society: the prehistory of Cranborne Chase*. Cambridge: Cambridge University Press.

Cassirer, E. 1951. *The Philosophy of the Enlightenment*. Princeton: Princeton University Press.

Cleal, R.M.J., Walker, K.E. and Montague, R. 1995. *Stonehenge in its Landscape: twentieth-century excavations*. London: English Heritage Archaeological Report 10.

Clifford, E.M. 1936. Notgrove long barrow, Gloucestershire. *Archaeologia* 86: 119-62.

Cole, H.C. and Aniakor, C. 1984. *Igbo Arts: community and cosmos*. Los Angeles: UCLA Museum of Cultural History.

Cunnington, W. 1889. Notes on Boles Barrow. *Wiltshire Archaeological Magazine* 24: 104-25.

Descartes, R. 1912 [1637]. *Discourse on Method.* New York: E.P. Dutton.

Drew, C.D. and Piggott, S. 1936. The excavation of long barrow 163a on Thickthorn Down, Dorset. *Proceedings of the Prehistoric Society* 2: 77-96.

Entwhistle, R. and Grant, A. 1989. The evidence for cereal cultivation and animal husbandry in the southern British Neolithic and Bronze Age. In A. Milles, D. Williams and N. Gardner (eds) *The Beginnings of Agriculture*. Oxford: British Archaeological Reports (International Series) 496. 203-15.

Grigson, C. 1966. The animal remains from Fussell's Lodge long barrow. In P. Ashbee, The Fussell's Lodge long barrow excavations 1957. *Archaeologia* 100: 63-73.

Grigson, G. 1954. *The Freedom of the Parish*. London: Phoenix House.

Heidegger, M. 1993. Letter on humanism. In D.F. Krell (ed.) *Martin Heidegger: basic writings*. 2nd edition. London: Routledge. 213-65.

Higgs, E.S. and Jarman, M.R. 1969. The origins of agriculture: a reconsideration. *Antiquity* 43: 31-41.

Hodder, I. 1990. *The Domestication of Europe: structure and contingency in Neolithic societies*. Oxford: Blackwell.

Ingold, T. 1986. *The Appropriation of Nature: essays on*

human ecology and social relations. Manchester: Manchester University Press.

Jackson, J.W. 1939. 'Bone' from Badshot long barrow. In K.P. Oakley *et al.*, *A Survey of the Prehistory of the Farnham Region.* Guildford. 146-8.

Kinnes, I. 1992. *Non-Megalithic Long Barrows and Allied Structures in the British Neolithic.* London: British Museum Occasional Paper 52.

Latour, B. 1993. *We Have Never Been Modern.* London: Harvester Wheatsheaf.

Legge, A.J. 1981. Aspects of cattle husbandry. In R.J. Mercer (ed.) *Farming Practice in British Prehistory.* Edinburgh: Edinburgh University Press. 169-81.

Mercer, R., 1980. *Hambledon Hill: a Neolithic landscape.* Edinburgh: Edinburgh University Press.

Parker Pearson, M. 1993. *Bronze Age Britain.* London: Batsford and English Heritage.

Partridge, C. 1905. *Cross River Natives: being some notes on the primitive pagans of Obubura Hill district, southern Nigeria.* London: Hutchinson & Co.

Phillips, C.W. 1935. The excavation of the Giant's Hills long barrow, Skendlebury, Lincolnshire. *Archaeologia* 85: 37-106.

Piggott, S. 1952. The Neolithic camp on Whitesheet Hill, Kilmington Parish. *Wiltshire Archaeological Magazine* 54: 404-9.

Piggott, S. 1962. Heads and hoofs. *Antiquity* 36: 110-18.

Piggott, S. and Piggott, C.M. 1944. Excavations of barrows on Crichel and Launceston Downs, Dorset. *Archaeologia* 90: 47-80.

Pitt Rivers, A.L.F. 1898. *Excavations in Cranborne Chase near Rushmore, on the Borders of Dorset and Wilts.* Volume IV. Privately printed.

Pryor, F. 1998. *Etton: excavations at a Neolithic causewayed enclosure near Maxey, Cambridgeshire, 1982-7.* London: English Heritage Archaeological Report 18.

Richards, M.P. 2000. Human consumption of plant foods in the British Neolithic: direct evidence from bone stable isotopes. In A.S. Fairbairn (ed.) *Plants in Neolithic Britain and Beyond.* Neolithic Studies Group Seminar Papers 5. Oxford: Oxbow. 123-35.

Richmond, A. 1999. *Preferred Economies: the nature of the subsistence base throughout mainland Britain during prehistory.* Oxford: British Archaeological Reports (British Series) 290.

Robertson-Mackay, M.E. 1980. A 'head and hoofs' burial beneath a round barrow, with other Neolithic and Bronze Age sites, on Hemp Knoll, near Avebury, Wiltshire. *Proceedings of the Prehistoric Society* 46: 123-76.

Robinson, M.A. 2000. Further considerations of Neolithic charred cereals, fruit and nuts. In A.S. Fairbairn (ed.) *Plants in Neolithic Britain and Beyond.* Neolithic Studies Group Seminar Papers 5. Oxford: Oxbow. 85-90.

Serjeantson, D. 1996. The animal bones. In S. Needham and T. Spence (eds) *Refuse and Disposal at Area 16 East, Runnymede.* London: British Museum. 194-223.

Smith, A.G., Grigson, C., Hillman, G. and Tooley, M. 1981. The Neolithic. In I. Simmons and M. Tooley (eds) *The Environment in British Prehistory.* London: Duckworth. 125-209.

Smith, I.F. 1965. *Windmill Hill and Avebury.* Oxford: Clarendon Press.

Thomas, J.S. 1988. The social significance of Cotswold-Severn burial practices. *Man* (N.S.) 23: 540-59.

Thomas, J.S. 1999a. An economy of substances in earlier Neolithic Britain. In J. Robb (ed.) *Material Symbols: culture and economy in prehistory.* Carbondale: Southern Illinois University Press. 70-89.

Thomas, J.S. 1999b. *Understanding the Neolithic.* London and New York: Routledge.

Thurnam, J. 1869. On ancient British barrows, especially those of Wiltshire, and the adjoining counties (part I, long barrows). *Archaeologia* 42: 161-244.

Vansina, J. 1985. *Oral Tradition as History.* Oxford: James Currey.

Waton, P.V. 1982. Man's impact on the chalklands: some new pollen evidence. In M. Bell and S. Limbrey (eds) *Archaeological Aspects of Woodland Ecology.* Oxford: British Archaeological Reports (International Series) 146. 75-92.

Weiner, A.B. 1992. *Inalienable Possessions: the paradox of keeping-while-giving.* Berkeley: California University Press.

Whittle, A.W.R. and Pollard, J. 1998. Windmill Hill causewayed enclosure: the harmony of symbols. In M. Edmonds and C. Richards (eds) *Understanding the Neolithic in North-western Europe.* Glasgow: Cruithne Press. 231-47.

Whittle, A.W.R., Pollard, J. and Grigson, C. 2000. *The Harmony of Symbols: the Windmill Hill causewayed enclosure.* Oxford: Oxbow.

Animals into ancestors: domestication, food and identity in Late Neolithic Orkney

Andrew Jones[1] and Colin Richards[2]

[1] Department of Archaeology, University of Southampton
[2] School of Art History and Archaeology, University of Manchester

Introduction

Traditionally the relationship between people and animals during the Neolithic is described in terms of domestication and we shall begin by considering a number of recent models for the process of domestication in order to consider what the notion entails with regard to the relations that pertain between people and animals.

Importantly a number of recent authors have stressed that domestication is part of a long-term process rather than a single event (Hodder 1990; Bradley 1993). This point is critical since it encompasses the notion that the Neolithic was a gradual and contingent process rather than a stable or fixed cultural package (Thomas 1996). Associated with this understanding of domestication as a contingent process is the realisation that, as a result, the *concept* of domestication does not occur after the economic process of domestication (Hodder 1990; Thomas 1991; Bradley 1993, 1998). It is the conceptual aspects of domestication that we intend to review here.

The notion of domestication as concept is perhaps exemplified by the work of Ian Hodder (1990), although it is also critical to the analyses of both Cauvin (2000) and Thomas (1991). Although Hodder describes the Neolithic in terms of process, he conceptualises this process as a series of structural oppositions. The focus of these oppositions is the house, described by the concept of the *domus* which is broadly linked with ideas surrounding the inhabitation of the house. In opposition to this Hodder draws in the idea of the *agrios* - the outside, the wild - to distinguish the region beyond the nurturing domain of the house. The transition between these two areas he describes as the *foris*, a term that comes into play especially in central Europe to define the boundary or doorway between inside and outside (Hodder 1990: 130). Broadly, then, we can summarise Hodder's ideas in terms of inside (*domus*) and outside (*agrios*) with the *foris* - or doorway - standing as a transition point between them.

Probably the most striking aspect of this notion of the conceptual apparatus of the Neolithic is that the Neolithic 'as concept' stands in opposition to nature. However we define the *domus* or *agrios*, we cannot escape the fact that - in each spatial and temporal instance - the *domus* exists as a bounded homogeneous entity in a fixed relation to the *agrios*. The intake of matter or things from the *agrios* affects their transformation to the *domus*. For example, cattle skulls or the human dead placed in the house are considered either as instances of attempts to control the wild (or the dead) or in terms of making the house wild. In each case the values associated with one bounded concept (*domus* or *agrios*) affect or bleed into the other concept and transform it. In this sense the Neolithic as idea and as a practice materialised in the form of the economy appears in the guise of an intervention on nature. According to this model the boundaries of individual entities -wild and domestic, animal and human - have to be carefully managed.

Rather than viewing domestication as a series of practices which intervened on the natural world, as an alternative we wish to conceptualise domestication as action *within* nature, action which involved the establishment of a series of dynamic relations between people and animals. Our aim then is to examine how particular kinds of economic relations and particular kinds of person emerged from this nexus of relations between people and animals. We need to rethink the prevailing model which presents domestication as the simple transformation of one entity (human society) with regard to another (animals). Rather we wish to consider the way in which transformation occurs as the result of the relationships established between entities in which both entities are transformed through the relationship. In this case the relations we seek to understand are those created between animals and people.

In this instance we want to problematise the discrete relationship perceived to pertain in this model between the bounded human body and the bounded and discrete animal body. Following Tim Ingold (1996) we would point out that both animals and humans are ontologically linked by their co-presence. As such animals and humans are related through their mutual embodied inhabitation of the world. The point, then, is to investigate how animal bodies and human bodies are treated as part of the process commonly described as domestication.

Accounts such as Hodder's tend to emphasise the mental aspects of domestication over the consideration of the ontological status of people and animals in this process. Despite this, Hodder's conceptualisation of domestication does shift the emphasis away from a characterisation of domestication as a simple mode of production (see also Thomas 1999: 11-12) and enables us to also consider domestication in a more active light. Recent accounts of domestication emphasise the Neolithic as a set of practices that were open to negotiation in differing contexts (Thomas 1996; 1999). In this sense we may think of domestication as specialised kind of consumption, or more properly realise that consumption is a critical aspect of the domestication process. If we are to consider the process of domestication as a form of consumption there are a number of additional points we need to consider, the most important of which is the creative potential of consumption practices. Recent studies of consumption in the social sciences highlight the relationship between consumption and the expression of identities (*e.g.* Miller 1995). Not only are objects used to define identities, but the act of consumption is creative of both social relations and identities. It is to these consumption practices that we shall now turn.

The cultural biography of animals

At the outset it is important to point out that animals are likely to be given cultural order according to the position of different animal species in cosmological schemes. The cosmogonic root of such schemes provides the apparatus by which differing ontological statuses are imputed to different animals. While these schemes will have existed prior to domestication we should also expect them to have been transformed through the process. We shall examine how these cultural schemes are materialised in specific practices and in turn how these practices serve to constitute and transform the set of relations between people and animals.

We have become familiar with the notion of the cultural biography of things. Objects are imbued with histories and personalities by virtue of the relationships they establish between people. Yet we have seen little discussion of the cultural lives of things that already have biographies, such as animals. We want to pose the question 'what would happen if we examine the cultural life of animals?' In order to undertake such an analysis of the cultural life of animals we want to consider a *chaine opératoire* approach as a viable means of looking at how a series of broad cultural choices are made in the treatment of animals over the duration of their lives. Such a scheme allows us to consider the ways in which animals are used to establish relations between people.

When we consider the *chaines opératoires* associated with animals we need to understand that, through the practical process of slaughter and dismemberment, a particular relationship is established (see Chapman 2000). What we have to consider here is the relationship between parts and wholes. Prior to slaughter all animal species are in a fully articulated state. However, owing to the anatomy of animals on dismemberment each element of the animal skeleton stands for the whole. Not only this, but the relationship between parts and wholes is enchained: each part articulates together to form the whole. So there is a metonymic relationship between part and whole, but metaphor also relates each skeletal fragment to another. So the anatomy of the animal articulates particular sets of relationships which we need to take into consideration as we examine the engagement of people with animals.

With these points in mind, we will describe the biographies of a number of animal species as they are transformed over their lives. Here it is important to consider both the contexts in which we find animals of different species and the components (parts, wholes *etc.*) of the animal found in each context.

The biography of animals in Late Neolithic Orkney

The case study examined here is the Late Neolithic of Orkney. The Orkney Isles are situated some 10 miles off the north of the Scottish mainland. In terms of archaeology the islands have some of the most spectacular upstanding monuments in Europe. Orkney has been drawn into Hodder's discussion of the *domus* owing to the prevalence of impressive settlement complexes such as Skara Brae, Rinyo and Barnhouse (Hodder 1991; 1998). While the discussion of the later Neolithic side-steps the question of the period of transition to the use of domesticates, we believe that domestication is a long-term process and the discussion of this issue is valid in this context. Furthermore it allows us to exploit the rich material evidence from this region.

In this paper we draw on evidence for faunal remains from three main sources: passage graves/chambered tombs, houses and henges. In addition to faunal remains, we also draw on evidence derived from gas chromatography analysis of lipid residues on Late Neolithic Grooved Ware from the settlement at Barnhouse (Jones 1997; Richards 1990).

Birds

One of the most celebrated sets of depositional practices concerns the white-tailed sea eagle. This large fish-eating bird lived on the dramatic sea cliffs that characterise Orkney and it is found in a number of unusual archaeological deposits, in settlements and passage graves. At Isbister, dubbed the Tomb of the Eagles (Hedges 1983; 1984), disarticulated eagle bones formed a foundation deposit along with the bones of sheep, cattle and black-backed gull, as well as human long bones. At least 13 sea eagles were placed articulated within the tomb itself. Eagle bones were also recovered from the chambered tombs of Midhowe and the Knowe of Ramsay, Rousay. A sea eagle was also placed spread-eagled as part of the closure deposits of the Links of Noltland settlement (Clarke and Sharples 1990: 68). When placed with human bones, eagle bones are often in an articulated form.

A variety of bird species are found in settlement and mortuary contexts. There is not enough space to discuss these on a species by species basis but, generally, we note a distinction between the kinds of bird bone in each type of context. In settlements we observe the deposition of disarticulated bones associated with the torso. At Skara Brae the humerus bones of gannets were fashioned into points (Clarke and Sharples 1990: 78). In the context of the passage grave we observe the deposition of disarticulated head, feet and wings. In the settlement, bird bones are transformed and worked; in the passage grave those elements of the bird's skeleton such as the head, foot and wing that metonymically refer to the bird are deposited.

Whales

Species of whale also found in many contexts. Whales only appear in an articulated state when washed ashore, since it would appear that the processing of whales occurred outside the settlement. The excavation of a butchery site some 90m south of Skara Brae (Richards forthcoming) revealed a whale mandible mixed with numerous red deer bones and Late Neolithic stone edge-tools known as 'Skaill knives'. After processing, the disarticulated bones of whales were included in settlement and passage grave deposits in a number of ways. The skull of a pilot whale was embedded over the central passageway (passage A) at Skara Brae (Childe 1931: plate XII2) while a whale bone slab was also found embedded in the wall in the junction between two passages at Skara Brae. On the excavation of house 1 at Skara Brae, whale mandibles and ribs were found as part of the collapsed roof structure (Petrie 1867). As structural components of the house, certain attributes of the whale such as strength are actively transferred to the house.

A number of houses at Skara Brae contained cups made from whale vertebrae. These were used to contain pigments, haematite and white clay (Childe 1931; Petrie 1867; Clarke and Sharples 1990). Arguably, it is the strength afforded by the backbone of the whale that provides it with the ability to swim which is transferred to the paint enabling it to flow. The carved and polished teeth of whales were also employed. A number of killer whale teeth were found in house contexts at Skara Brae and they were also found as components of a necklace at Isbister. A composite necklace which contained the teeth of sperm, killer and pilot whales was found in the inner chambers at the Point of Cott (Barber 1996). The very fragments of the whale that afford them their fearsome nature, the teeth, are therefore re-articulated together. Individual teeth, each tooth with its own place-specific identity and history, are brought together to make a whole in the form of a necklace.

Dogs

Dogs are found in both disarticulated and articulated states. A dog rib was found in the settlement midden at Skara Brae (Noddle unpublished); dog long bones and ribs are also found in chambered tombs, as at the Knowe of Yarso and in henge ditches as at Stenness (Clutton-Brock 1976). Dogs are found deposited in close proximity to people at Burray where a total of seven fully articulated dogs were deposited in the side cells and stalls of the chambered tomb with either two or three human skulls placed with them (Davidson and Henshall 1989). At Cuween Hill passage grave a total of 24 dog skulls were found in chamber and infill deposits with five human skulls (Charleson 1902). The relationship between the skull of the human and that of the dog is established through their deposition together. Occasionally, as at Isbister, dogs' teeth are found as components of necklaces.

Deer

On present evidence it seems that red deer were introduced to Orkney in the Early Neolithic. The 'butchery' deposit outside Skara Brae alluded to earlier was largely composed of the remains of a number of articulated red deer (Richards forthcoming); red deer are also noted as minor components of the midden assemblages at Skara Brae in the form of worked antler tines (Noddle unpublished). There is a dichotomy then between the deposition of articulated red deer outside Skara Brae and their absence within the faunal assemblages in Skara Brae. This distinction is mirrored at Links of Noltland where the articulated skeletons of 15 red deer were found in the midden and wall deposits beyond the settlement (Clarke and Sharples 1990; Sharples 2000).

At Isbister, a red deer skull was deposited on top of a pile of human bones in the central chamber and a number of rib, fore and hind limb skeletal elements from immature deer were deposited (Barker 1983). Immature red deer also predominate at the Knowe of Ramsay and Knowe of Yarso chambered tombs (Barker 1983). At both tombs the major components included were from ribs, fore and hind limbs. Again there is a distinction between part and whole: the articulated animal is associated with areas just beyond the settlements, whereas fragments of deer, especially the skull, are associated with human remains in mortuary contexts.

Sheep

Sheep were a major component of middens at Skara Brae (Clarke and Sharples 1990; Noddle unpublished) where, as at Barnhouse, the remains are dominated by immature individuals. Anatomical analysis of sheep bones from the Skara Brae middens suggests that, proportionally, the whole carcass was deposited in a disarticulated state (Noddle unpublished). However, Childe (1931: 115-27) observed that sheep metatarsals and metacarpals were selected for the production of piercing tools, while tibiae were used to make smoothing instruments. Outside the settlements, sheep ribs were found in the ditch of the henge at Stenness (Clutton-Brock 1976). Sheep also occur in mortuary contexts: a lamb's skull was found in a pit at Quoyness (Childe 1951), and sheep ribs, and fore and hind limbs were found at Isbister, Midhowe, the Knowe of Rowiegar, the Knowe of Ramsay and the Knowe of Yarso (see Barker 1983 for discussion). Sixteen sheep were deposited in the primary and infill deposits at Blackhammer tomb (Callander and Grant 1937). In all contexts in which they are deposited, sheep bones are disarticulated.

Cattle

Cattle are the predominant species in Late Neolithic Orkney. The proportion of immature cattle in the midden at Skara Brae was high (Noddle unpublished). Again the proportion of cattle remains in the midden suggests that the whole carcass was deposited in a disarticulated state (*ibid.*). Despite this, cattle metapodials were used as adzes, while jawbones were used to produce flint fabricators and scapulae were used as shovels (Childe 1931: 115-27). However, cattle remains are found in a number of striking contexts both within and outside the settlements. Cattle limb bones were used, like whale bones, as structural components of the outer wall at Skara Brae and the revetment of the passage grave at Pierowall Quarry (Sharples 1984). A cattle skull was embedded in the junction between passage and house at Skara Brae

(Childe 1931). Like whales, it would seem that the strength of the animal is transferred to the house and used to secure the boundary of the settlement. Cattle skulls formed part of the closure deposit at the Links of Noltland settlement (Clarke and Sharples 1990) and were deposited with piles of human long bones and skulls in the central chamber at Isbister (Barker 1983). Cattle ribs, fore and hind limbs also formed deposits in Isbister, Midhowe, the Knowe of Rowiegar, Ramsay and Yarso, Quoyness, Howe and Pierowall Quarry.

Cattle products form the major contents of Grooved Ware pottery from the Late Neolithic settlement at Barnhouse (Richards 1990). We need to examine this evidence in some detail in order to see how the *chaines opératoires* associated with cattle intersect with ceramic production and use. Analysis of the pottery indicates three different sizes of vessel: large, medium and small (Jones 1997; 2000). These were produced and decorated in different ways. The production of pottery appears to have been broadly divided between the inner ring of houses, which tempered their pottery with shell, and the outer ring of houses which employed rock temper. While the inner ring of houses appears to have practiced a more communal mode of production related to the main firing site in the centre of the settlement, the occupants of the outer ring of houses were each using distinct sources of temper (see Jones 2000; 2001).

Despite this, the decorative schemes and motifs on the pottery from all houses were identical, appearing to refer to a settlement-wide notion of identity. Decoration was distinguished by vessel size: large vessels were decorated with a simple cordon, and medium-sized vessels with an incised curvilinear or applied serpentine scheme. The decoration of small vessels also employs this scheme with the addition of passage-grave art 'rosette' motifs.

As well as being variably produced, and differentially decorated, vessels of different sizes were employed for different functions. In the earliest phase of the site large and small vessels both appear to have contained a plant substance, likely to be barley (Jones 1999).[1] Medium-sized vessels contained cows' milk.[2] A number of medium vessels from a single house, the unusual structure House 2, appear to have contained cattle meat.[3]

It is likely that large vessels were used as storage containers since the interior of one vessel was sealed with what might have been birch pitch (Heron 1996).[4] By the final phase of settlement at Barnhouse, however, large vessels were also used for cooking cows' milk on the platform of the monumental building, structure 8 (Jones 1997). The production of Grooved Ware pottery was therefore closely allied to the expression of settlement-specific identities and also associated with the containment, presentation and consumption of cattle products.

The components of cattle are used in a number of different ways. The use of cows' milk suggests a process of exchange between cattle and humans. By the very process of milking, milk metonymically refers to cattle. The use of cattle meat appears to have been much more restricted, and only occurred within the confines of House 2 (Jones 1997; 2000). Here fragments of cattle bones, meat from the stomach (indicated by the presence of bile), and other parts of the animal are brought together for cooking and consumption within Grooved Ware vessels.

A summary of animal biographies

To recap, human engagement with animals in Late Neolithic Orkney involved a process in which animals were articulated, disarticulated, and re-articulated again. A number of animals were slaughtered and deposited in a fully articulated state; these include sea eagles and dogs. Red deer were slaughtered and deposited fully articulated beyond the limits of habitation at Skara Brae and Links of Noltland. In most other cases we observe the deposition of fragments of animals. We often observe, however, the deposition of very particular fragments: in the case of dogs, red deer, sheep, cattle and whales, the skull appears to be a significant element of the skeleton. In the case of cattle, sheep, pig and red deer in the context of the passage grave, the partially articulated meat-bearing elements, of ribs, hind and forelimbs are deposited. In settlements the disarticulated fragments of the skeletons of cattle, sheep, birds and whales are transformed into tools. The milk of cows is disassociated from the animal in order to be consumed by humans.

While in many contexts we observe the disarticulation of animals through processes of slaughter and consumption, in other contexts we note particular elements of the animal being brought back together. For example, teeth - especially those of the whale and dog - are re-articulated together in the form of necklaces. In the case of the cooking/consumption of cattle meat, elements of the cattle carcass are brought together and contained within a Grooved Ware vessel. The disarticulated fragments of the carcasses of animals are brought together in the homogenised accumulation of deposits that make up the middens surrounding all Late Neolithic Orcadian settlements. Finally, in mortuary contexts significant parts of the skeleton are brought into relation with the human skeleton.

Animals and ancestorhood in Late Neolithic Orkney

What is the significance of the processes of articulation, disarticulation and re-articulation undergone by animals? Are these practices simply a consequence of the anatomy of animals and the functional processes of slaughter and dismemberment? How do these practices relate to the expression of cultural identities?

In order to answer these questions we need to consider the analysis of another species of animal: humans. We observe the deposition of fully articulated human remains in a number of contexts. Two adult women were buried beneath the right-hand box bed in House 7 at Skara Brae

[1] Indicated by the presence of unsaturated fatty acids such as hexadec-9-enoic acid and a triene acid 9, 12, 15-octadecanoic acid.

[2] Indicated by the presence of 2-methylhexadecanoic acid.

[3] Indicated by the presence of a bile acid, Cholan-24-oic acid.

[4] Indicated by the presence of 16-hydroxyhexadecanoic acid and 22-hydroxydocosanoic acid, both of which are specific to tree resin.

(Childe 1931: 140-2). At Blackhammer two individuals were deposited in the tomb, one in the entrance passage, the other in the extreme left-hand compartment (Callander and Grant 1937). At Midhowe (Callander and Grant 1934) a series of crouched human skeletons were found on raised stone shelves (Davidson and Henshall 1989). At the chambered tomb of Quanterness three stone cists cut into the floor held humans remains (Renfrew 1979); the two excavated examples contained crouched inhumations. We also observe a sequence of activity with a teenager and child inserted at a later stage into one of the cists. Finally, a further inhumation was inserted into a pit cut into the layer of disarticulated bone that covered the floor at Quanterness.

Evidence from Quanterness, and other mortuary sites such as Isbister, suggests a sequence of activity associated with human remains. These commenced with the deposition of fully articulated human individuals into the tomb, and the subsequent disarticulation and sorting of certain bone elements, followed by later deposition of elements such as long bones and skulls in the side chambers of passage graves, or at the rears of stalled tombs. The principles involved in articulation and fragmentation are therefore not arbitrary. The processes of articulation and fragmentation are meaningful and there are homologies between the treatment of human remains and animal remains.

How do the practices of articulation, disarticulation and re-articulation relate to the expression of identity? John Chapman (2000) has recently pointed out that the processes of fragmentation and re-articulation are critical to the creation of social relations since the act of breaking and sharing material culture establishes affiliation between actors. Similarly the act of creating composites out of distinct fragments harnesses the relations established through sharing by cementing and articulating together shared social bonds and thereby re-incorporates, or re-articulates, a new set of social relations. Importantly, he notes that the practices of butchery and dismemberment are a primary means by which social relations of this kind are both negotiated and established (*ibid.*: 33-5).

How are we to understand these processes in relation to animals in Late Neolithic Orkney? Firstly, we need to be aware that, owing to their position in cosmological schemes of order, different species of animals are classified by habitat, by their proximity to humans and their relationship to significant and meaningful places in the lived landscape (Jones 1998). Their identity is inscribed by their relationship to humans. The practices of disarticulation involve separating out elements of the animal. However, as we argued above, certain elements of the skeleton may stand metonymically for the whole. More importantly, this process of disarticulation and the subsequent circulation of components of animals' bodies mean that specific components of animals may be brought into metaphorical relation with each other.

Since cattle share their lives closely with humans the process of consumption of milk and meat in Grooved Ware vessels is socially and symbolically significant since it involves the incorporation of cattle into the human body. The sharing of cows' milk in Grooved Ware throughout the Barnhouse settlement and the Stones of Stenness henge indicates a communal sense of identity, while the restricted consumption of cattle meat in House 2 at Barnhouse suggests a more closely defined, intimate identity. In each case the distinction between these identities can be read from the distinctive decoration of the Grooved Ware employed in each of these contexts (Jones 1997: chapter 9). Identities are also expressed in the construction and adornment of the human body with animal-tooth necklaces. Here identities are expressed through a fusion between the place-specific significance of animals (Jones 1998) and the social relations involved in accumulating the components of the necklace.

One of the critical aspects of the process of fragmentation is that it enables elements of the world which were hitherto discrete to be brought into metaphorical relation. Here we need to consider the relationship established between elements of the human body and those of the animal. In the context of the passage grave at Burray the articulated skeletons of seven dogs were buried with two or three human skulls; in this context we would argue that skulls metonymically represent elements of humanity[5]. In the context of Cuween Hill passage grave, the skulls of human beings were deposited alongside 24 dog skulls (Charleson 1902). Similarly the articulated skeletons of 13 sea eagles were buried at Isbister along with the disarticulated skulls and long bones of humans (Hedges 1983). At the same tomb two cattle skulls and a red deer skull were placed on piles of human long bones. At Quoyness the only skull to be found in the central chamber was that of a lamb placed in a pit, not unlike those dug at Quanterness to contain human remains. How was the relationship between people and animals constituted? What kinds of identities were being expressed through the congruence between people and animals in thesechambered tombs and passage graves? It is obvious from the deposition of ribs, fore and hind limbs, that joints of meat were being incorporated with the dead. But the act of incorporation in mortuary contexts goes beyond the deposition of food. We observe that animals are treated in parallel to humans; in some cases, as at Burray and Isbister, dogs and eagles appear to have been treated more formally than human burials.

Conventionally when we think of the storage and reallocation of human remains we consider that these are the remains of the once living, those who toiled on the land prior to present generations. Given this, the curation of their remains is significant. They are considered to be the bones of the ancestors. However, we appear to observe precisely the same practices with regard to animals. They too are deposited in an articulated state; the bones of once-articulated animals are associated with the human dead and their bones often appear to substitute or stand in for those of humans, or are placed in relation

[5] See Chapman 2000, chapter 5 for discussion on the use of human bone as metonym or synecdoche.

with humans. These practices embody a metaphorical unity between people and animals. Just as human bones imply the trace of previous generations so, in precisely the same terms, animals are ancestors: they too once shared the world with people and, in the form of food, enabled humans to propagate. Animals are ancestors of, and are ancestral to, the human population.

Conventional views of ancestorhood are dependant upon the relationship established by the transference of substance between generations of people, a genealogical model (Ingold 2000: 132-51). However an alternative conceptualisation of ancestorhood is that described by Ingold (2000: 140-51) as a relational model. This model depends less upon the transference of substance from one generation of humans to another and more on the series of connections established through inhabitation between both human and non-human beings. As Ingold notes (2000: 140), this relationship can be varied: 'ancestors can be ordinary humans who lived in the past, or spirit inhabitants of the landscape, or mythic other-than-human characters, or original creator beings'.

Conventionally, the relationship between humans and animals is described in terms of totemism (Fraser 1983), a process in which specific groups of the human population ally themselves with ancestral animals. If we are to understand how these relationships may be constituted during the Neolithic we need to examine the relationship between people, animals and place (for details of this argument see Jones 1998). Different species of animals evoke different aspects of the landscape: eagles and other birds evoke the sky, whales the sea, red deer the distant hill top, cattle, sheep and dogs the settlement. Dogs have many close similarities with humans, being carnivorous and living within settlements in intimate proximity to people (*ibid.*: 314).

All animals are not equal; rather they evoke quite distinct qualities of place and existence. Animals presence the relationship between people and different places in the landscape (Jones 1998), and it is for this reason that animals are deposited in tombs and in significant parts of the landscape. Red deer, for instance, are deposited at the edge of settlement because their fully articulated bodies embody the qualities of a distant past and a distant place (see also Sharples 2000). Their bodies define the distinction between the human and the animal, the settlement and the world beyond. On the other hand the 'tomb of the eagles' is in a dramatic cliff-top location, while those tombs containing domestic species are situated on lower ground, closer to settlements. The species of animals deposited in tombs are a means of classifying the human dead. The primacy of animal bone deposits in tombs and their association with the human dead suggest not only the ontological congruence between animal and human, but that tombs were constructed as much for the commemoration of animal ancestors as for human ancestors. Just as ancestral animal bones presence the human relations with specific places, so the bones of whale and cattle embedded into the settlement may be considered as protective, providing security against the forces those animals represent.

Conclusion

Unlike the 'mentalist' models of domestication discussed at the beginning of this paper we have suggested that we need to think of domestication as a process that is defined not by its location within specific places - such as the house - but by its extension across and between different landscape locales. There is no single site of domestication. Rather there are multiple sites where the relationships between people and animals are negotiated. Domestication is therefore created out of the network of relationships that occur at these sites.

We need not draw a sharp distinction between humans and non-humans in the process of domestication. Furthermore, less division between hunting and gathering and farming may be in evidence than was once supposed. The distinction between people and animals amongst hunter-gatherer and farmer populations has often been overdrawn. For example, Ingold (2000: 61-76) has described the relations between people and animals in hunter-gatherer and farmer populations as a shift from relations of trust to those of domination. A closer examination of the evidence suggests that while animal/human relationships in the Neolithic were different from earlier periods, just as hunter-gatherer groups perceive animals and humans to be engaged in a mutual process of dwelling, equally animals and people were considered to be mutually related during the Neolithic. Ingold (2000: 77-88) notes that it may be more appropriate to think of people and animals being engaged in a mutual process of growth. As farmers and cultivators, people therefore provide the conditions for the growth of animals. It is within the conditions of the concept of growth that we may focus on the mutually entangled history of people and animals.

As Thomas (1999: 229) points out, 'Neolithic social life involved a series of equivalent or interlocked flows of substance'; these substances include both human bodies and animal bodies. Animals and people therefore share a mutual and relational history. Animal species are physically changed through their involvement with humans (they become 'domesticated'), and humans are physically changed through their involvement with animals (the patterns of the seasons are related to the movement of animals, people become lactose tolerant by drinking milk). This entangled history is made explicit through the congruence of people and animals in life and death. The incorporation of people into the lives of animals is enacted by the consumption of flesh and milk and other animal products on a daily basis. Animals are neither wholly ancestral, nor wholly economic entities - they are simply other forms of 'people' whose lives have an important impact upon humanity in the Neolithic.

References

Barber, J. 1996. *The Excavation of a Stalled Cairn at the Point of Cott, Westray, Orkney*. STAR Monograph 1. Edinburgh: AOC.

Barker, G. 1983. The animal bones. In J.W. Hedges (ed.) *Isbister: a chambered tomb in Orkney*. Oxford:

British Archaeological Reports 115. 133-40.

Bradley, R. 1993. *Altering the Earth: the origins of monuments in Britain and continental Europe*. Edinburgh: Society of Antiquaries of Scotland Monograph 8.

Bradley, R. 1998. *The Significance of Monuments*. London: Routledge.

Callander, J.G. and Grant, W.G. 1934. A long stalled chambered cairn or mausoleum (Rousay type) near Midhowe, Rousay, Orkney. *Proceedings of the Society of Antiquaries of Scotland* 68: 320-50.

Callander, J.G. and Grant, W.G. 1937. Long stalled cairn at Blackhammer, Rousay, Orkney. *Proceedings of the Society of Antiquaries of Scotland* 71: 297-308.

Cauvin, J. 2000. *The Birth of the Gods and the Origins of Agriculture*. Cambridge: Cambridge University Press.

Chapman, J. 2000. *Fragmentation in Archaeology*. London: Routledge.

Charleson, M.M. 1902. Notice of a chambered cairn in the parish of Firth, Orkney. *Proceedings of the Society of Antiquaries of Scotland* 36: 733-8.

Childe, V.G. 1931. *Skara Brae: a Pictish village in Orkney*. London: Kegan Paul.

Childe, V.G. 1951. Re-excavation of the chambered cairn of Quoyness, Sanday, on behalf of the Ministry of Works in 1951-52. *Proceedings of the Society of Antiquaries of Scotland* 86: 155-72.

Clarke, D.V. and Sharples, N. 1990. Settlements and subsistence in the third millennium BC. In C. Renfrew (ed.) *The Prehistory of Orkney*. Edinburgh: Edinburgh University Press. 54-82.

Clutton-Brock, J. 1976. Animal remains from the Stones of Stenness. In J.N.G. Ritchie, The Stones of Stenness, Orkney. *Proceedings of the Society of Antiquaries of Scotland* 107: 1-60.

Davidson, J.D. and Henshall, A.S. 1989. *The Chambered Cairns of Orkney*. Edinburgh: Edinburgh University Press.

Fraser, D. 1983. *Land and Society in Neolithic Orkney*. Oxford: British Archaeological Reports (British Series) 131.

Hedges, J.W. 1983. *Isbister: a chambered tomb in Orkney*. Oxford: British Archaeological Reports 115.

Hedges, J. 1984. *The Tomb of the Eagles*. London: John Murray.

Heron, C. 1996. The chemistry and use of resinous substances. In A.M Pollard and C Heron (eds) *Archaeological Chemistry*. London: RSC Paperbacks.

Hodder, I. 1990. *The Domestication of Europe*. Oxford: Blackwell.

Hodder, I. 1991. Burials, houses, women and men in the European Neolithic. In I. Hodder (ed.) *Theory and Practice in Archaeology*. London: Routledge. 45-83.

Hodder, I. 1998. The *domus*: some problems reconsidered. In M. Edmonds and C. Richards (eds) *Understanding the Neolithic of North-western Europe*. Glasgow: Cruithne Press. 84-101.

Ingold, T. 1996. Hunting and gathering as ways of perceiving the environment. In R. Ellen and K. Fukui (eds) *Redefining Nature*. Oxford and Washington DC: Berg. 117-57.

Ingold, T. 2000. *The Perception of the Environment: essays in livelihood, dwelling and skill*. London: Routledge.

Jones, A. 1997. A biography of ceramics: food and identity in Late Neolithic Orkney. Unpublished PhD thesis, Glasgow University.

Jones, A. 1998. Where eagles dare: landscape, animals and the Neolithic of Orkney. *Journal of Material Culture* 3: 301-24.

Jones, A. 1999. The world on a plate: ceramics, food technology and cosmology in Neolithic Orkney. *World Archaeology* 31: 55-78.

Jones, A. 2000. Life after death: monuments, material culture and social change. In A. Ritchie (ed.) *Neolithic Orkney in its European Context*. Cambridge: McDonald Institute Monograph. 127-38.

Miller, D. 1995. Introduction by way of a polemic. In D. Miller (ed.) *Acknowledging Consumption*. London: Routledge.

Noddle, B. unpublished. The animal bones from Skara Brae. Final draft. Unpublished manuscript.

Petrie, G. 1867. Notice of ruins of ancient dwellings at Skara, Bay of Skaill, in the parish of Sandwich, Orkney. *Proceedings of the Society of Antiquaries of Scotland* 1866-68: 201-22.

Renfrew, C. 1979. *Investigations in Orkney*. London: Society of Antiquaries.

Richards, C.C. 1990. Postscript: the late Neolithic settlement complex at Barnhouse Farm, Stenness. In C. Renfrew (ed.) *The Prehistory of Orkney*. Edinburgh: Edinburgh University Press. 305-16.

Richards, C.C. forthcoming. Report on the Neolithic butchery site at Skaill Bay, Mainland, Orkney.

Sharples, N. 1984. Excavations at Pierowall quarry, Westray, Orkney. *Proceedings of the Society of Antiquaries of Scotland* 114: 75-125.

Sharples, N. 2000. Antlers and Orcadian rituals: an ambiguous role for red deer in the Neolithic. In A. Ritchie (ed.) *Neolithic Orkney in its European Context*. Cambridge: McDonald Institute Monograph. 107-17.

Thomas, J. 1991. *Rethinking the Neolithic*. Cambridge: Cambridge University Press.

Thomas, J. 1996. *Time, Culture and Identity*. London: Routledge.

Thomas, J. 1999. *Understanding the Neolithic*. London: Routledge.

Early Neolithic diets: evidence from pathology and dental wear

Andrew Chamberlain[1] *and Annsofie Witkin*[2]

[1] Department of Archaeology, University of Sheffield
[2] Oxford Archaeology, Oxford

Introduction

Investigations in archaeology and biological anthropology furnish copious information concerning past human diets. Trace element and stable isotopic analysis of human remains can provide direct evidence for diet, as can the analysis of food residues preserved on or within bodies at the time of death or on artefacts associated with the consumption of food. Indirect evidence for diet can be obtained from skeletal remains through the analysis of nutritional deficiency diseases, patterns of dental wear, dental pathology and masticatory biomechanics. The dental evidence is particularly important because teeth have a high survival potential in the archaeological record and teeth are not remodelled, and therefore they carry a cumulative record of their growth and subsequent usage. Furthermore, there are good standards for the diagnosis of dental diseases that are indicative of dietary practices (Hillson 2000).

Many studies have demonstrated changes in dental morphology and pathology at the transition to agriculture in several world regions. From these studies some generalisations can be drawn concerning the changes that might have occurred in Britain (Table 5.1), while recognising the caveat that there is great variation in patterns of animal and plant husbandry at the origins of agriculture in different parts of the world. The increased use of food preparation technology and consequent reduction in masticatory load are thought to have influenced rates and patterns of dental wear and, over many generations, might have resulted in tooth size reduction. Changes in the physical and chemical composition of food constituents might also have affected the incidence of dental pathology and the microscopic wear patterns that can be seen on the occlusal surfaces of the teeth. Specific changes in wear patterns seen during the transition to agriculture include the tendencies for the dentitions of agriculturalists to develop oblique dental wear planes and relatively smaller interproximal wear facets, but it should be noted that the processing techniques used in the preparation of some agricultural products might have introduced abrasives to the diet that in turn might have influenced dental wear (Molleson and Jones 1991).

The purpose of this paper is to undertake a preliminary examination of the dental evidence for dietary change at the transition to agriculture in Britain. This task is constrained by the very limited availability of a comparative sample of human skeletal remains from the Mesolithic period in Britain. However, the probable high mobility of early Holocene hunter-gatherers, coupled with the availability of a broad terrestrial connection between Britain and mainland Europe well into the Late Mesolithic (Coles 1998), justifies the use of Mesolithic populations from northwestern Europe and southern Scandinavia as an appropriate basis for comparison with the British Neolithic. Stable isotope studies of Mesolithic skeletons from the Breton Mesolithic cemeteries of Téviec and Hoëdic have shown considerable diversity of diet, perhaps reflecting long-distance migration between inland and coastal communities (Schulting and Richards 2001). Similar dietary diversity is suggested by a recent stable isotope determination on one of the few Mesolithic human bones recovered from an inland site in Britain (Davies *et al.* 2001), which contrasts markedly with the isotopic composition of Mesolithic human remains recovered from coastal locations in Britain (Richards and Sheridan 2000; Bronk Ramsey *et al.* 2000).

Evidence for dental disease in Neolithic Britain

Two dental conditions that have been linked to agriculture are caries and calculus. Dental caries is a disease caused by bacterial fermentation of dietary carbohydrates. Consumption of cereals increases the amount of cariogenic carbohydrates in the diet, and an additional risk factor for dental caries is the reduction in dental occlusal wear that is sometimes a correlate of the transition to agriculture (Powell 1985). Dental calculus is a mineralised deposit that accumulates at the attachment surface of the plaque, a bacteria-rich organic material that coats the tooth surfaces. Calculus accumulation has been linked to alkaline oral conditions and by inference to high protein consumption, but the relationship of calculus to diet is complex and protein consumption is only one of many factors that can affect the rate and amount of calculus deposition (Lieverse 1999).

Both dental caries and calculus are reported as having a higher frequency in agriculturalists than among hunter-gatherers (Larsen 1997; Lukacs 1989). Data for the prevalence of dental calculus are not consistently recorded among available skeletal samples in Britain, but most published reports on human skeletal remains include estimates of the prevalence of dental caries. Figure 5.1 depicts the prevalence of dental caries (percentage of teeth affected) in 14 northwest European Mesolithic and British Early/Middle Neolithic skeletal samples. Estimates based on less than 100 teeth have been excluded and caries rates for the Neolithic samples do not include deciduous teeth (the milk teeth are more susceptible to caries than are the permanent teeth). The data presented in Figure 5.1 are crude prevalence rates that have not been standardised for differences between samples in adult age structure and in the proportions of anterior versus posterior teeth (standardisation is advisable when comparing rates between samples because

Table 5.1. *Changes in tooth size, dental wear and pathology during the transition from foraging to agriculture.*

	Foragers	Agriculturalists	Interpretation of Change	Reference
Tooth Size	Large	Small	Food preparation technology, and selection due to dental caries	Brace *et al.* 1987
Molar Wear Plane	Flat	Oblique	Chewing cycle predominates over puncture-crushing cycle	Smith 1984
Interproximal Wear Facets	Large	Small	Reduction in masticatory load	Hinton 1982
Microwear on Enamel	Deep pits; long, broad grooves	Shallow pits, short narrow grooves	Reduction in abrasion (pattern reversed if food is stone-ground)	Teaford 1991
Dental Caries	Low prevalence	High prevalence	Increase in cariogenic foods, and/or reduction in occlusal wear	Powell 1985
Dental Calculus	Low prevalence	High prevalence	Oral alkalinity from high-starch and/or high-protein diet; reduced abrasiveness of food particles	Hillson 1979; Turner 1979

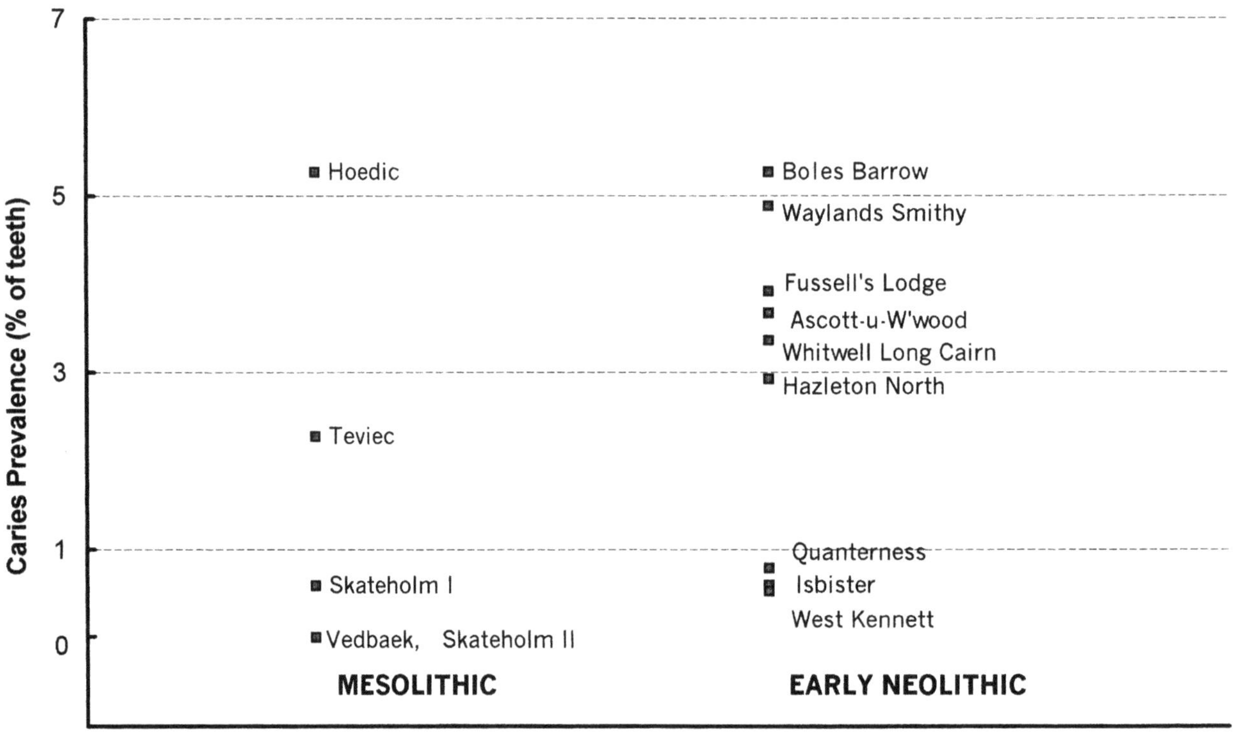

Figure 5.1. *Prevalence of dental caries in human skeletal samples from Mesolithic sites in western France and southern Scandinavia and Neolithic sites in Britain. Prevalence rates for caries in Mesolithic samples are taken from Meiklejohn and Zvelebil (1991) and caries rates for Neolithic sites are from published and unpublished excavation reports.*

caries is an age-progressive condition, and the posterior teeth are more liable to develop caries). Although the average caries prevalence is higher in the Neolithic than in the Mesolithic samples (2.88% versus 1.63%, unweighted averages of sample prevalence rates) the spread of values within each time period is substantial and the increase in average caries prevalence in the Neolithic is not statistically significant ($t = 1.14$, $df = 12$, $p = 0.14$, single tailed t-test). It must be concluded that this preliminary survey of dental caries rates does not reveal a distinction between the Mesolithic and the Neolithic samples.

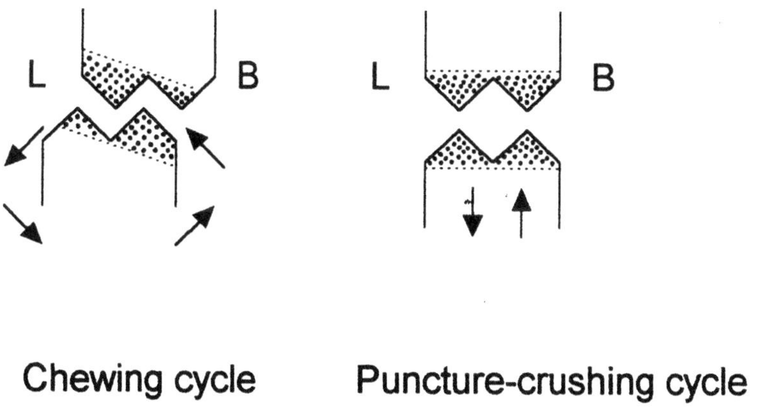

Figure 5.2. *Schematic diagram of the formation of molar occlusal wear resulting from the rotatory movements of the chewing cycle (left) and the vertical movements of the puncture-crushing cycle (right). Wear planes developing from the chewing cycle are bucco-lingually oblique, whereas those caused by puncture-crushing are horizontal.*

Dental wear

Dental wear is caused by a combination of attrition, abrasion and erosion. Attrition is wear caused by direct tooth-to-tooth contact; abrasion is from contact between teeth and abrasive particles in the diet, or wear caused by non-masticatory materials introduced into the mouth; and erosion is the loss of dental tissues through chemical dissolution (Larsen 1997: 247). The severity of wear on the occlusal surfaces of the chewing teeth is determined by the forces exerted during mastication and by the physical properties of the food, which may in turn be modified during extra-oral food preparation. The anterior teeth are more likely to be exposed to abrasion from the non-masticatory use of teeth as tools, while other cultural practices such as oral hygiene activities may also contribute to dental abrasion.

The transition from foraging to farming is thought to have produced a reduction in dental wear as a consequence of the change from consumption of fibrous plant materials to the use of domesticated cereal grains as well as associated developments in food preparation technology (Larsen 1997: 250). Though both Mesolithic and Early Neolithic populations may express a similar severity of wear, the detailed pattern of wear on the teeth is often different. The softer food consistency of agricultural diets produces a more oblique or cupped wear pattern than the flatter occlusal wear typical of hunter-gatherer populations (Smith 1984: 40). This difference in the wear pattern on the chewing teeth is due to the mechanics of mastication (Figure 5.2). The initial phase of mastication is that of puncture-crushing, a predominantly vertical movement in which the teeth in the opposing jaws do not contact each other while they repeatedly chop the interposed items of food. Puncture-crushing generates abrasion that is spread evenly across the occlusal surfaces of the teeth. In the second phase of mastication, the chewing cycle, the teeth move in a rotatory pattern and the opposing tooth cusps shear and grind across each other, generating oblique wear facets on their occlusal surfaces. The manner in which the cusps of the opposing teeth occlude ensures that the chewing cycle generates greater attrition on the lingual cusps of the upper teeth and on the buccal cusps of the lower teeth (Hillson 1996: 237). Consequently, diets with a high proportion of tough fibrous food require a longer phase of puncture-crushing than diets consisting of softer food, the former producing a flatter occlusal wear plane while the latter generate a more oblique plane.

An additional effect of high occlusal loads is the development of interproximal attritional facets between the crowns of the teeth that are adjacent to each other in the jaw. Interproximal wear is manifest in the formation of distinctive small oval highly polished facets that flatten the rounded mesial and distal margins of the premolar and molar tooth crowns (Figure 5.3). The polishing is produced by small lateral movements of the tooth crowns and the amount of interproximal wear depends on the magnitude of the mesial force vector exerted by loads on the teeth distal to the facet. Hunter-gatherers typically practice vigorous mastication which generates relatively greater amounts of interproximal attrition at a given level of occlusal wear (Hinton 1982).

Recording of the angle of occlusal wear can be undertaken using a method devised by Smith (1984). A modified protractor bearing a moveable arm is held parallel to the horizontal plane of the lower jaw, and the angle of orientation of the occlusal surface of a molar tooth is then measured in the lingual to buccal direction by positioning the moveable arm of the measuring device. The angle of wear tends to steepen with increasing occlusal wear, so the amount of occlusal tooth wear is also measured on an ordinal scale from one (no visible wear) to eight (complete attritional loss of the tooth crown). Figure 5.4 shows the results of these measurements obtained from six Early Neolithic individuals radiocarbon dated to about 3900 BC from the Whitwell Quarry long cairn (Witkin and Chamberlain forthcoming). The angles of occlusal wear measured on the lower first molars of these specimens are intermediate between the Mesolithic and Neolithic values recorded by Smith (1984) on European skeletons, and a linear regression through the Whitwell data most closely resembles the pattern observed in Smith's Late Mesolithic samples. The Whitwell sample exhibits a relatively flat angle of wear even when the occlusal wear has reached a maximum value of stage eight. This suggests that the diet consumed by the Whitwell Quarry individuals included a high proportion of tough fibrous food.

We also measured the interproximal attritional facet widths at the contact between the second premolar and

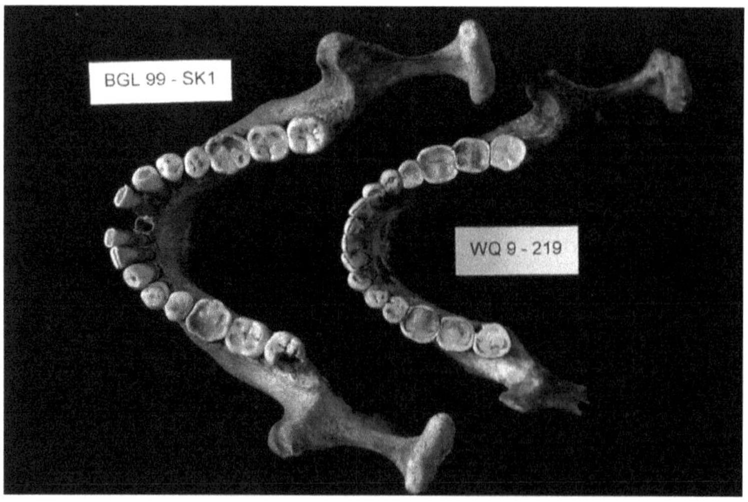

Figure 5.3. *Occlusal and interproximal dental wear on the lower teeth of a late medieval skeleton (left) and an Early Neolithic specimen from Whitwell Quarry long cairn. The occlusal wear is concentrated on the buccal cusps of the molars in the late medieval specimen but is evenly distributed across the cusps in the Early Neolithic specimen. Note also the severe interproximal wear between the molars in the Neolithic jaw.*

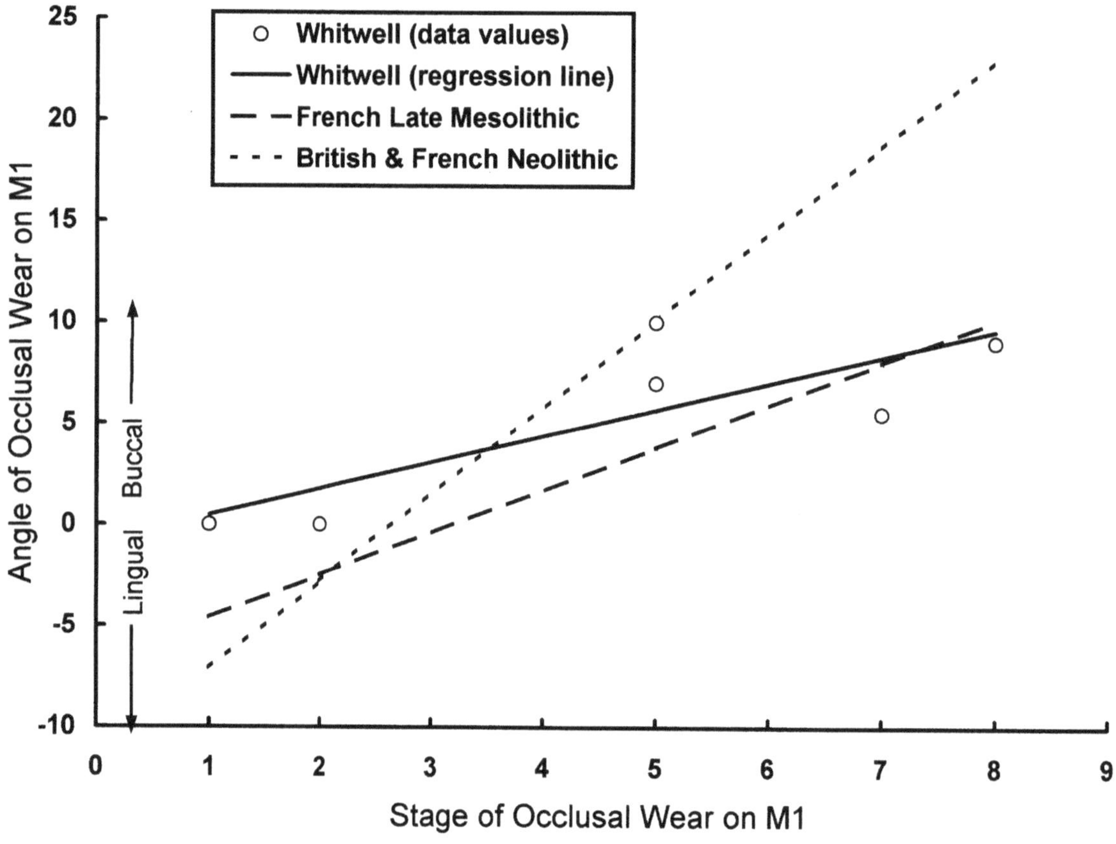

Figure 5.4. *The angle of occlusal wear measured on first molars plotted against the stage of occlusal wear for six individuals from the Early Neolithic Whitwell Quarry long cairn. The linear regression through the Whitwell data most closely resembles the pattern observed by Smith (1984) in Late Mesolithic samples.*

first molar on ten individuals from Whitwell Quarry and at the first molar/second molar contact in seven individuals, following the protocol described in Hinton (1982). The facet widths were plotted against molar occlusal wear stage, separately for the first and second molars (Figures 5.5 and 5.6). These figures show that the Whitwell Quarry sample exhibits a degree of interproximal wear that is intermediate between values established for North American hunter-gatherers (Archaic and Woodland periods) and agriculturalists (Mississippian period). Unfortunately there are no comparable data for interproximal wear in European Mesolithic samples, but the results shown here confirm that the Whitwell Quarry individuals exerted relatively high occlusal loads on their chewing teeth.

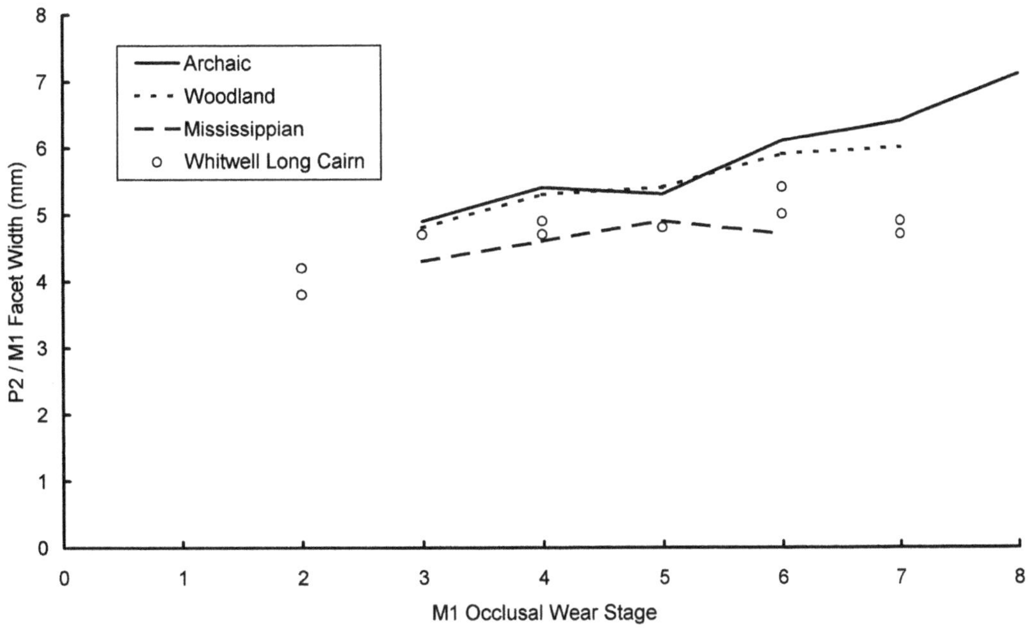

Figure 5.5. *Mesial interproximal facet width on the first molars of Early Neolithic specimens from Whitwell Quarry long cairn, compared to North American hunter-gatherers (Archaic and Woodland periods) and agriculturalists (Mississippian period).*

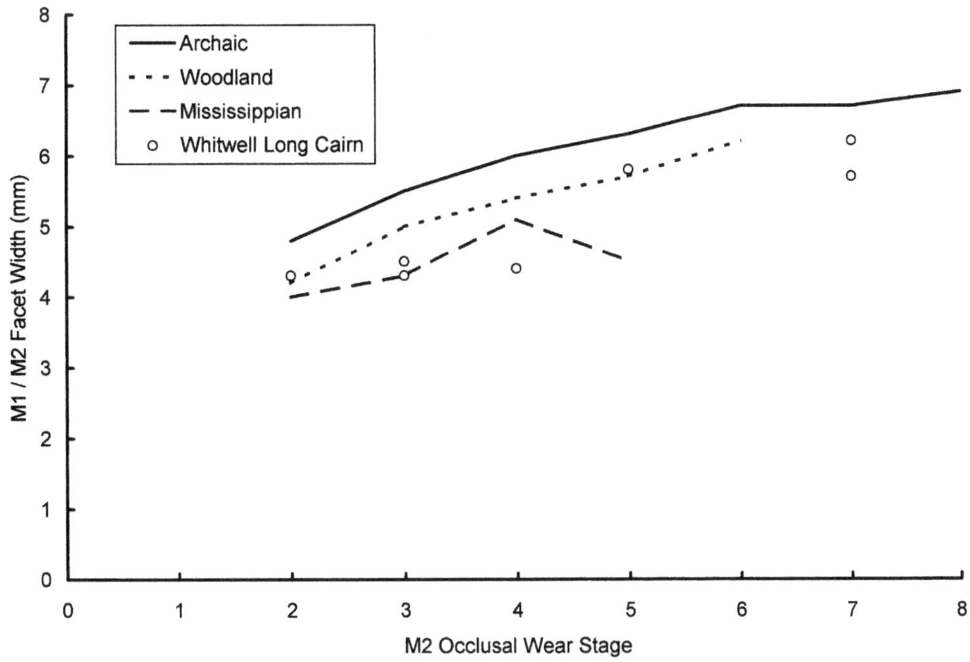

Figure 5.6. *Mesial interproximal facet width on the second molars of Early Neolithic specimens from Whitwell Quarry long cairn, compared to North American hunter-gatherers (Archaic and Woodland periods) and agriculturalists (Mississippian period).*

Discussion

The preliminary findings reported here do not support the prediction that the onset of agriculture in Britain was accompanied by marked changes in patterns of dental wear and pathology. No clear distinction is seen in dental caries prevalence rates between northwest European Mesolithic and British Neolithic populations. The combination of flat occlusal wear planes and well-developed interproximal wear facets seen in the Early Neolithic Whitwell Quarry sample indicates that in this population at least the masticatory loads were comparable with those seen in populations which depend on foraging. Other researchers working with archaeobotanical evidence have noted that the remains of wild plant foods are often encountered on Neolithic sites in Britain, and at some of these sites the collected plant resources appear to outnumber the cultivated ones (Moffett *et al.* 1989; Brown 1997). The wild plant resources identified at Neolithic sites include a broad range of tree and shrub fruits together with roots and tubers and, if consumed in significant quantities, these resources might be expected to result in dental wear patterns similar to those observed in the Whitwell Quarry sample.

Further investigation of patterns of dental wear and pathology on a wider range of Neolithic human remains, and in particular a comparison between Early and later Neolithic samples, are warranted in order to clarify our picture of Early Neolithic dietary preferences. Together with quantitative analysis of dental microwear, macroscopic studies of dentitions can usefully complement the evidence from stable isotopic studies in reconstructing past human diets.

Acknowledgements

We are grateful to Ian Wall of the Creswell Heritage Trust for the opportunity to study the human remains from Whitwell Quarry long cairn. Sharon Findlay assisted in the collation of data on prehistoric dental disease.

References

Brace, C.L., Rosenberg, K.R. and Hunt, K.D. 1987. Gradual change in human tooth size in the late Pleistocene and post-Pleistocene. *Evolution* 41: 191-204.

Bronk Ramsey, C., Pettitt, P.B., Hedges, R.E.M., Hodgins, G.W.L. and Owen, D.C. 2000. Radiocarbon dates from the Oxford AMS system: Archaeometry datelist 30. *Archaeometry* 42: 459-79.

Brown, N. 1997. A landscape of two halves: the Neolithic of Chelmer Valley / Blackwater Estuary, Essex. In P. Topping (ed.) *Neolithic Landscapes*. Oxford: Oxbow. 87-98.

Coles, B.J. 1998. Doggerland: a speculative survey. *Proceedings of the Prehistoric Society* 64: 45-81.

Davies, G., Buckland, P., Chamberlain, A., Richards, M., Tweddle, J. and Tyers, I. 2001. A palaeoenvironmental study and watching brief on borrow pits at Staythorpe power station, Staythorpe, Nottinghamshire. ARCUS Project Report 438f. Sheffield: ARCUS, University of Sheffield.

Hillson, S.W. 1979. Diet and dental disease. *World Archaeology* 11: 147-62.

Hillson, S. 1996. *Dental Anthropology*. Cambridge: Cambridge University Press.

Hillson, S. 2000. Dental pathology. In M.A. Katzenberg and S.R. Saunders (eds) *Biological Anthropology of the Human Skeleton*. New York: Wiley-Liss. 249-86.

Hinton, R.J. 1982. Differences in interproximal and occlusal tooth wear among prehistoric Tennessee Indians: implications for masticatory function. *American Journal of Physical Anthropology* 57: 103-15.

Larsen, C.S. 1997. *Bioarchaeology*. Cambridge: Cambridge University Press.

Lieverse, A.R. 1999. Diet and the aetiology of dental calculus. *International Journal of Osteoarchaeology* 9: 219-32.

Lukacs, J.R. 1989. Dental paleopathology: methods for reconstructing dietary patterns. In M.Y. Iscan and K.A.R. Kennedy (eds) *Reconstruction of Life from the Skeleton*. New York: Liss. 261-86.

Meiklejohn, C. and Zvelebil, M. 1991. Health status of European populations at the agricultural transition and the implications for the adoption of farming. In H. Bush and M. Zvelebil (eds). *Health in Past Societies*. Oxford: British Archaeological Reports (International Series) 567. 129-45.

Moffett, L., Robinson, M.A. and Straker, V. 1989. Cereals, fruit and nuts: charred plant remains from Neolithic sites in England and Wales and the Neolithic economy. In A. Milles, D. Williams and N. Gardner (eds) *The Beginnings of Agriculture*. Oxford: British Archaeological Reports (International Series) 496. 243-61.

Molleson, T. and Jones, K. 1991. Dental evidence for dietary change at Abu Hureyra. *Journal of Archaeological Science* 18: 525-39.

Powell, M.L. 1985. The analysis of dental wear and caries for dietary reconstruction. In R.I. Gilbert and J.H. Mielke (eds) *Analysis of Prehistoric Diets*. New York: Academic Press. 307-38.

Richards, M.P. and Sheridan, J.A. 2000. New AMS dates on human bone from Mesolithic Oronsay. *Antiquity* 74: 313-15.

Schulting, R.J. and Richards, M.P. 2001. Dating women and becoming farmers: new paleodietary and AMS dating evidence from the Breton Mesolithic cemeteries of Téviec and Hoëdic. *Journal of Anthropological Archaeology* 20: 314-44.

Smith, B.H. 1984. Patterns of molar wear in hunter-gatherers and agriculturalists. *American Journal of Physical Anthropology* 63: 39-56.

Teaford, M.F. 1991. Dental microwear: what can it tell us about diet and dental function. In M.A. Kelley and C.S. Larsen (eds) *Advances in Dental Anthropology*. New York: Wiley-Liss. 341-56.

Turner, C.G. 1979. Dental anthropological indications of agriculture among the Jomon people of central Japan. *American Journal of Physical Anthropology* 51: 619-36.

Witkin, A.V. and Chamberlain, A.T. forthcoming. Human remains from Whitwell Quarry long cairn. In B. Vyner (ed.) *Whitwell Quarry Long Cairn*. London: English Heritage.

The use of dental microwear to infer diet and subsistence patterns in past human populations: a preliminary study

Pia Nystrom[1] and Sue Cox[2]

[1] and [2] Department of Archaeology, University of Sheffield

Abstract

Dental microwear studies have demonstrated that microscopic wear patterns can be correlated with dietary differences. However, there has been little work on human material in comparison to the abundance of studies focused upon nonhuman primates and other animals. This study examines whether dental microwear analysis can successfully discriminate between human populations from different archaeological time periods, with particular interest in the dietary distinction between pre-agricultural and post-agricultural populations. The samples are from four British sites: an Early Neolithic site (~4455 - 3730 BC), a Bronze/Iron Age site (2700-2250 BC; 770-400 BC), a Medieval site (eighth to eleventh centuries AD) and an early modern site (AD 1753-1845).

For each individual, facet 9 of the occlusal surface on the first molar was examined at 500x magnification under a scanning electron microscope. Quantitative analytical techniques were applied to examine the data using both univariate and multivariate statistical analyses. In general, the prehistoric groups showed consistently larger pits and wider striation features than the historic populations. It is suggested that these differences indicate that the diet of the prehistoric populations comprised harder and tougher foods than the more processed and softer foods of the modern populations. The higher frequency of larger dental microwear features in the Bronze/Iron age group may be due to an increased inclusion of exogenous grit during food preparation.

Introduction

Reconstruction of diet and food preparation techniques of past human populations is of interest to palaeoanthropologists and archaeologists alike. A special focus has been on detecting when societies go through transitions such as the introduction of domestic cultivars or the use of vessels for cooking. The chronology of the introduction and adoption/integration of farming by hunter-gatherers in Europe has been discussed extensively (*e.g.* Harris and Hillman 1989; Schadla-Hall 1987; Thorpe 1996; Zvelebil 1994; 1996). Few researchers now subscribe to the idea put forth by Vere Gordon Childe of a Neolithic Revolution (Childe 1951:61), and the dichotomous division between a hunting-gathering and agricultural food procurement economy is falling at the wayside (Harris 1996; Smith 2001; Zvelebil 1996). Rather, there might have been many intermediate and transitory stages, which most likely were population and habitat specific. It is more probable that there was a gradual shift; as farming activities increased there was a concomitant decrease in hunting and gathering (in its most traditional sense) which continued throughout the Neolithic and Bronze Age and possibly into the Iron Age. In many European countries it is not until the Bronze Age and Iron Age that there is strong evidence of fully agricultural populations from evidence of domesticated animal bones and cultivated cereal crops (Harris 1996; Zvelebil 1994; 1996), although there appear to have been enclaves of immigrant peoples who brought along a fully fledged Neolithic culture at earlier times (*e.g.* Bakels 2000).

Many approaches have been used to investigate diet adaptation in prehistoric and historic human populations. The most direct evidence derives from botanical and animal remains at settlement sites (Jones 2000; Meadow 1989; O'Connor 1997; Robinson 2000), and isotopic signatures in bone, hair or teeth (*e.g.* O'Connell and Hedges 1999; Schutkowski *et al.* 1999). Furthermore, dental calculus may contain food remains, and coprolites have yielded interesting results (Dobney and Brothwell 1986; Poinar *et al.* 2001). Alternative approaches used to infer diet include macrowear pattern on teeth (Kaifu 1999; Smith 1984; see also Chamberlain and Witkin this volume), and evidence of wear or residues on tools (Clemente and Gibaja 1998; Kealhofer *et al.* 1999). More recently, microscopic wear on dental enamel has been used as evidence of diet (*e.g.* Gambarotta 1995; Hojo 1989; Molleson *et al.* 1993; Pastor 1992; Pérez-Pérez *et al.* 1994; Schmidt 2001; Teaford *et al.* 2001; Ungar and Spencer 1999). Dental microwear analysis is based on the principle that during mastication scars of pits and striations are left on the tooth surface. There are three factors most typically cited as responsible for microwear formation: 1) tooth-tooth contact, 2) tooth-food contact, and 3) plant phytoliths and other abrasives in or on the foods eaten (Teaford and Lytle 1996; Ungar *et al.* 1995).

In the present study, we examined the dental microwear pattern of four human populations from the northeast of England (from Early Neolithic, Bronze/Iron Age, Early Medieval and modern periods). Our aim was to record the dental microwear signature of the four populations, and to establish when the transition from a hunter-gatherer economy to a more agriculturally based economy might have taken place. It was expected that the microwear pattern in hunter-gatherer populations would reflect a hard-object diet with many abrasive inclusions. This would be typical of a population consuming mainly uncooked wild grains, fruits, roots, tubers, and roasted or dried meats. Such a diet would result in a high frequency of pits, especially large pits. In addition, with a highly abrasive diet, it is expected that the overall frequency of features would be increased (Hojo 1989; Molleson, Jones and Jones 1993; Puech, Serratrice and Leek 1983; Teaford and Lytle 1996). In contrast, the diet of agricultural societies would be much softer due to the introduction of pottery vessels for cooking. Hence, the microwear pattern would include fewer and smaller pits. However, not only food items contribute to tooth wear: differences in food processing techniques can also have a significant impact. Teaford and Lytle (1996) found significant differences in rates of tooth wear between populations consuming maize ground on igneous rocks (less wear) and maize ground on sandstone rocks (increased wear). Thus, depending on food preparation methods, the study populations might have included more abrasives in their diet, which may make their dental microwear signatures less distinguishable for those of pre-agricultural populations (Molleson *et al.* 1993; Teaford and Lytle 1996). Since the transition from a foraging to a farming way of life was a lengthy process, the foods consumed as well as the food preparation techniques might have been mixed, which would result in a more 'intermediate' stage of microwear pattern reflecting a more varied diet containing both hard and soft particles. We do, however, assume that the diets of human populations gradually became softer over time, and that it will be possible to discriminate between the four human populations in this study, presuming they had dissimilar diets and/or methods of food preparation.

Background

Inferential models for dental microwear are based on studies on extant mammals, especially primates (*e.g.* King *et al.* 1999; Nystrom *et al.* in press; Teaford and Glander 1991; Teaford and Robinson 1989; Teaford and Runestad 1992; Ungar 1990; 1994). Amongst living primates, the incidence of pit features and pit widths have been correlated with broad dietary habits (*e.g.* Teaford 1994). Pit density can be directly related to the abrasiveness of foods, whereas pit size is a reflection of the amount of crushing needed to comminute foods. The harder and more abrasive the diet, such as inclusion of nuts or hard seeds and seed coats, the greater the incidence of pits and the greater the pit widths tends to be (Teaford and Oyen 1989; Teaford and Runestad 1992). Primates that feed mainly on leaves have the lowest number of pits, and the pits tend to be small (Teaford 1988). Their teeth tend to be riddled with long, narrow striations, and it is assumed that plant phytoliths are mainly responsible for these features.

Differences in microwear patterns have been found for native human populations in the southeastern United States when comparing populations from the pre-contact period (before the arrival of Europeans) with contact and post-contact populations (Teaford 1991; Teaford *et al.* 2001). The pre-contact populations had significantly higher frequencies of enamel pitting and wider striations, and it has been suggested that they were hunter-gatherer populations who ate more hard-object foods. The microwear patterns reflect a progressive dietary change involving a decreased dependence on hunting, gathering and marine resources to an increasing dependence on maize agriculture. It is also possible to distinguish between populations that include a large portion of meat in their diet to those that eat a mainly vegetarian diet by examining the striation pattern on the buccal side of teeth (Lalueza *et al.* 1996).

It is widely accepted that the Neolithic people in Britain gained sustenance from both wild collected foods and cultivated cereals (Zvelebil 1994; Austin 2000; Robinson 2000). Peoples living in north Yorkshire in the eighth millennium BC had developed a complex hunter-gatherer society. Evidence includes stone-lined hearths and some faunal remains of mainly wild horse, pig, aurochs and deer (Schadla-Hall 1987). Owing to the poor preservation rate of cereal remains, it has been difficult to establish from archaeobotanical remains whether one food source was more important than another (Jones 2000). Even though the Derbyshire area in Neolithic times is not considered ideally suited to agriculture, there is evidence from Lismore Fields, for example, to suggest that substantial cereal cultivation occurred on a permanent basis and that the produce was stored (Jones 2000). It is assumed that foods were also collected but there has been no evidence to suggest the storage of wild plant foods (Jones 2000). By the time of the Bronze and Iron Ages, people were, on the whole, sedentary and relied mainly on cultivated crops and domesticated animals, although their diet might have been (at least seasonally) supplemented with wild plants and animals.

Material and methods

Samples from four British populations from different archaeological periods were used in this study: 15 individuals from an Early Neolithic population, four individuals from a Bronze/Iron Age population, 11 individuals from an Early Medieval population, and five individuals from a modern population (Table 6.1). The Neolithic sample is from a long cairn at Whitwell quarry, Derbyshire, and dates to ~4455-3730 BC. The Bronze Age sample is from Carsington Pasture cave, southeast of Brassington, Derbyshire. Dates from this site suggest two periods of deposition of human remains, one at 2700-2250 BC and one at 770-400 BC (Chamberlain 2001). The Early Medieval sample is from the Saxon/Norman cemetery of Blackgate, Newcastle upon Tyne (eighth-eleventh centuries AD; Boulter and Rega 1993). The modern sample is from the Newcastle Infirmary, Newcastle upon Tyne (AD 1753-1845; Nolan 1998).

The Neolithic site at Whitwell quarry is still under investigation. It is not known whether the site was linked to a permanent settlement area or used as a seasonal base for nomadic hunter-gatherers, or whether it was of ritual importance as is often the case with Neolithic burial chambers (A.Chamberlain pers.comm.). The Whitwell quarry material has an unusual pattern of dental wear that is thought to resemble hunter-gatherers rather than agriculturalists. The molar wear pattern for this Neolithic population is more closely similar to that of European

Time Period	Location	Age			
		Sub-Adult*	Young Adult	Mature Adult	Total
Neolithic	Whitwell Quarry, Derbyshire	9	3	3	15
Bronze Age or Early Iron Age	Carsington Pasture Cave, Derbyshire	1	1	2	4
Saxon-Norman	Blackgate Cemetery, Newcastle upon Tyne	6	3	2	11
Modern 18th C.	Newcastle Infirmary, Newcastle upon Tyne	1	3	1	5

Table 6.1. *Dental samples used in study. * Third molars unerupted.*

Mesolithic populations. Thus their diet might have included a high proportion of tough fibrous foods (Chamberlain and Witkin this volume).

From the existing evidence from Neolithic sites in Britain, and especially in Derbyshire, a likely scenario for the Neolithic community of the Whitwell area would be a population reliant on cereal cultivation, with an additional food source still maintained by hunting and the gathering of wild plant foods. Small finds from the site include animal bones, pottery sherds and a range of flint types including some very finely worked leaf and lozenge arrow heads. The presence of pottery sherds suggests that the population had the technology to produce ceramics, although they might have traded from an outside source. The purpose and functional use of the vessels is unknown. They could have been used for food storage or used for food preparation or cooking. If foods were cooked, the dental microwear pattern should reflect a softer diet owing to the softening effect that cooking would have upon abrasives in foods. The five fragments of burnt animal bone recovered from the site were possibly remains from cooking meats.

Carsington Pasture cave is located near Brassington, Derbyshire, approximately 25 miles southwest of Whitwell quarry. Two other cave sites, Rains cave and Harborough cave, both situated within 3 km of Carsington Pasture cave, have produced Neolithic and Bronze Age artefacts as well as human remains (Chamberlain 2001). The faunal finds from within the cave are dominated by domestic cattle, pig, dog, horse, sheep and goat. There are also a small number of wild species (red deer, roe deer, wild cattle, badger and red fox). Two faunal elements show evidence of butchery while there is no evidence to suggest food preparation or cooking (A. Chamberlain pers. comm.).

The origin of the people buried in the Blackgate cemetery is also shrouded in mystery, as it is unknown what settlement or settlements the cemetery served. It could have been the focal point for a number of agricultural communities or there could have been an urban centre nearby (Nolan 1998). By Early Medieval times, food preparation would have reached a more modern form, with all foods being well softened prior to consumption, although the volume of abrasives included in the food remains unknown. Even though the bulk of the food derived from domestic sources, it is very likely that most households supplemented their diet, at least seasonally, with wild plants and game. By the eighteenth century, the period of inception of the Newcastle Infirmary material, people relied mostly on agricultural produce, with some possible supplementation with wild plants and game. Most, if not all, foods would have been prepared and processed in a fairly modern fashion.

Sample preparation

Replicas of the occlusal surface of maxillary and mandibular first molars were prepared for each individual. Facet 9, a crushing/grinding facet, was chosen as it is considered to be most representative of microwear patterns that are directly associated with diet (Teaford and Runestad 1992). Negative impressions were made using a hydrophobic polyvinylsiloxane silicone based impression material (Coltène; President Jet, light body®), from which positive, high resolution Araldite replicas (Araldite MY 753, hardener HY 956, Ciba-Geigy) were made. The positive replicas were placed on aluminium SEM stubs, and sputter coated with a 20μm layer of gold palladium and examined with a high-resolution digital scanning electron microscope (Philips XL30) fitted with a TETRA backscatter electron detector. Accelerating voltage was set at 5kV, and the beam spot was set at 3.0. All images were recorded at 500x magnification. A single digitised image measuring approximately 0.031mm^2 of surface was collected from each tooth. All images were taken close to the base of facet 9.

For each digitised image, all microwear defects were counted and measured using a semi-automated image analysis system (Microware Version 3.0 $_{Beta;}$ Ungar, 1997). The microwear features were categorised into different variables based on size and shape parameters (see Cox 1999 for a detailed description; Solounias and Hayek 1993). The data were analysed using both univariate and multivariate statistics. Statistical significance was tested initially using the Mann-Whitney U test, followed by a One-Way Analysis of Variance (ANOVA) test. To reject the hypothesis that all population means were equal, alpha was set to ≤0.05 (2-tailed). To minimise Type I errors, a Scheffé correction for multiple comparisons was performed when comparing the four samples. A multivariate approach (Discriminant Function Analysis) was used to identify influential relationships between variables. For the discriminant function to be optimal certain assumptions must be met: the variables must be from multivariate normal distributions and the covariance matrices must all be equal. As these assumptions were not met by the original data, some variables had to be excluded, and the ones used for the Discriminant Function Analysis were log transformed (Tabachnick and Fidell 1996). All statistical analyses utilised the SPSS 6.0 statistical package (Norušis 1990).

Results

Feature density and frequency of pit features, two variables usually used to determine diet adaptation, did not show significant differences between the groups in the present study. All four groups had a moderate to high feature density, and all four groups had a fairly high pit frequency (Table 6.2). Several types of microwear features related to size and shape, however, appeared important for discriminating between the diets of the four groups. Three variables were related to pit dimensions (the average pit width, frequency of wide pits and the frequency of large pits based on length) and five variables related to size dimensions of striations (mean striation width, width and length of small striations, width and length of large striations) (Table 6.2). In general, the modern group had more small pits, and narrower and shorter striations, while the prehistoric groups had more large features (based on both width and length measurements). If there was a difference between the groups, it was more often than not the Bronze/Iron Age group that showed the greatest deviation. For most variables, the Medieval group tended to be more similar to the modern group than the prehistoric groups.

The two prehistoric populations (Neolithic and Bronze/Iron Age) had a higher frequency of large pits than the two historic groups. With the Scheffé correction it became apparent that it was the Bronze/Iron Age group that had significantly more larger pits than in the other three groups ($F= 3.7591$, $P= 0.021$). A similar trend was seen for the frequency of large pits based on length measurements ($F= 4.190$, $P= 0.014$). However, after correcting for multiple comparisons this difference became non-significant, as was the average pit width, which approached statistical significance ($F= 2.354$, $P= 0.092$; Table 6.2).

Several of the striation-related variables revealed group differences. The Bronze/Iron Age group had more large striations than the other three groups. Both the width and length of large striations approached statistical significance (width- $F= 2.965$, $P= 0.078$; length- $F= 2.383$, $P= 0.089$). This preponderance of big striations is also reflected in the average width of all striations ($F= 2.3538$, $P= 0.092$). The same distinction could be seen for the width and length measurements of small striations, but here the Scheffé correction proved statistically significant only in the case of the striation length with the Bronze/Iron Age group being significantly different from the two historical period groups ($F= 4.415$, $P= 0.011$; Table 6.2).

Discriminant functions analysis

Univariate tests provide basic information about the distribution of variables in the sample populations, and help identify some important differences among the groups. However, by using multivariate analysis, and including simultaneously as many independent variables as possible, important information about the relationships between variables is incorporated. This enables more diffuse differences between groups to be evaluated and detected, which is especially pertinent in this case where many variables approached statistical significance.

In the discriminant functions analysis, the independent variables included were the 27 microwear features listed in Table 6.2, and the dependent variable was the archaeological period of individual specimens (*i.e.* Neolithic, Bronze/Iron Age, Medieval and modern). As shown in Figure 6.1, by incorporating all 27 variables it was possible to separate out the four groups remarkably well, at a classification level of 100%. Function 1, which separates the prehistoric from the historic groups, is highly correlated with the frequency of pits, the width and length of small striations, and the average size of large pits (based on width and length). The variables with most influence on Function 2 were the frequency of round pits, small striations and long striations, and pit size and average width of striations. This function classified the Medieval group together with the Neolithic group, and the modern group with the Bronze/Iron Age group.

Discussion

Dental microwear has been used to infer dietary adaptations in past human populations (*e.g.* Gambarotta 1995; Molleson *et al.* 1993; Pastor 1992; Pérez- Pérez *et al.* 1994; Schmidt 2001; Teaford *et al.* 2001; Ungar and Spencer 1999). Such inferences are based on a correlation between the microwear pattern seen on extant animals (mainly primates) with known diets. Feature density and pit frequency are used to indicate the level of abrasives present in diet, while pit size indicates the hardness and toughness of the food (Teaford 1991). In the present study, no significant differences were found in feature density, and the proportion of pit features was similar for all four groups. These results suggest that on some level there was no difference over time in the abrasiveness of the foods consumed. Thus it was not possible to use feature frequency to discriminate diet adaptation between the different groups. Rather, it was amongst size-related variables that significant differences between the different human populations were to be found. Several pit and striation size variables showed important discrimination powers.

In general, the prehistoric groups had more larger microwear features (based both on width and length measurements) than the two historic groups. The Neolithic and Bronze/Iron Age groups tended to show a similar pattern, while the microwear pattern of the

Table 6.2. *Descriptive statistics (mean and standard deviation) and results of statistical comparisons between the four different human populations.*

Variable	Neolithic	Bronze or Early Iron Age	Saxon-Norman	Modern 18th c.	Mann Whitney U Z	Mann Whitney U P	ANOVA F	ANOVA P	ANOVA Scheffé
Feature Density	122.4 ± 58.8	119.5 ± 79.9	118.3 ± 43.9	156.0 ± 69.5					
Percent Pits	38.5 ± 11.7	37.8 ± 6.3	35.1 ± 11.6	32.8 ± 6.7					
Pits – mean width	3.0 ± 0.8	3.1 ± 0.7	2.6 ± 0.7	2.2 ± 0.4	-1.9596	0.050	2.3538	0.092	NS
Pits – mean length	7.9 ± 2.2	8.0 ± 1.8	7.1 ± 1.9	6.1 ± 1.1					
Striations- mean width	2.3 ± 0.4	2.7 ± 0.6	2.1 ± 0.4	1.9 ± 0.1	-1.7146	0.086	2.3538	0.092	NS
Striations- mean length	20.5 ± 3.3	26.4 ± 9.4	20.1 ± 4.9	18.2 ± 3.9					
Percent round pits	10.4 ± 9.5	8.2 ± 6.0	8.6 ± 5.3	4.0 ± 1.4					
Percent oval pits	28.1 ±6.2	29.6 ± 3.1	27.7 ± 8.3	28.8 ± 5.7					
Percent small striations	44.9 ± 10.2	45.8 ± 3.7	45.4 ± 8.8	45.5 ± 8.5					
Percent large striations	16.6 ± 7.8	16.4 ± 3.6	19.5 ± 7.3	21.7 ± 5.1					
Round pits – mean width	3.4 ± 1.6	3.5 ± 1.3	3.6 ± 1.8	3.1 ± 1.8					
Round pits – mean length	5.6 ± 3.0	5.1 ± 1.9	5.9 ± 3.6	4.5 ± 2.1					
Oval pits – mean width	2.9 ± 0.7	2.9 ± 0.6	2.5 ± 0.7	2.1 ± 0.3	-1.9596	0.050			
Oval pits – mean length	8.7 ± 2.4	8.6 ± 2.1	7.6 ± 2.1	6.4 ± 0.9	-1.9442	0.052			
Small striations – mean width	2.4 ± 0.5	2.8 ± 0.6	2.1 ± 0.4	2.0± 0.2			3.0806	0.042	NS
Small striations – mean length	15.1 ± 2.7	17.7 ± 4.0	13.0 ± 2.1	12.6 ± 0.6			4.4154	0.011	2 from 3,4
Large striations – mean width	2.0 ± 0.4	2.6 ± 0.8	2.3 ± 0.7	1.7 ± 0.2	-2.2045	0.028	2.9652	0.078	2 from 1,3,4
Large striations – mean length	34.6 ± 5.2	49.8 ± 23.4	36.4 ± 11.1	28.8 ± 7.0			2.3832	0.089	2 from 1,3,4
Percent pits – width < 3 µm	26.7 ± 10.9	23.6 ± 9.7	28.2 ± 9.6	28.3 ± 7.7					
Percent pits – width >3 µm	11.0 ± 7.4	13.5 ± 5.9	6.3 ± 4.2	4.4 ± 2.5	-2.2045	0.028	3.7591	0.021	2 from 1,3,4
Percent pits – length <10 µm	31.3 ± 12.3	29.9 ± 8.1	30.8 ± 10.3	29.9 ± 6.8					
Percent pits – length >10 µm	6.7 ± 3.3	7.4 ± 3.6	3.6 ± 3.2	2.9 ± 2.3	-1.9596	0.050	4.1904	0.014	NS
Percent striation – width < 3µm	48.9 ± 11.1	47.4 ± 5.0	53.5 ± 11.9	60.4 ± 5.6	-2.4495	0.014			
Percent striations – width >3 µm	12.5 ± 6.6	13.3 ± 12.2	11.3 ± 8.7	6.8 ± 2.1					
Percent striations – length <10 µm	15.2 ± 7.5	16.3 ± 6.2	19.4 ± 6.7	20.1 ± 7.2					
Percent striations – length >10 <30 µm	36.4 ± 8.5	31.3 ± 7.4	34.4 ± 8.6	38.9 ± 4.0					
Percent striations – length >30 µm	9.8 ± 5.0	14.6 ± 7.7	11.1 ± 6.3	8.2 ± 4.3					

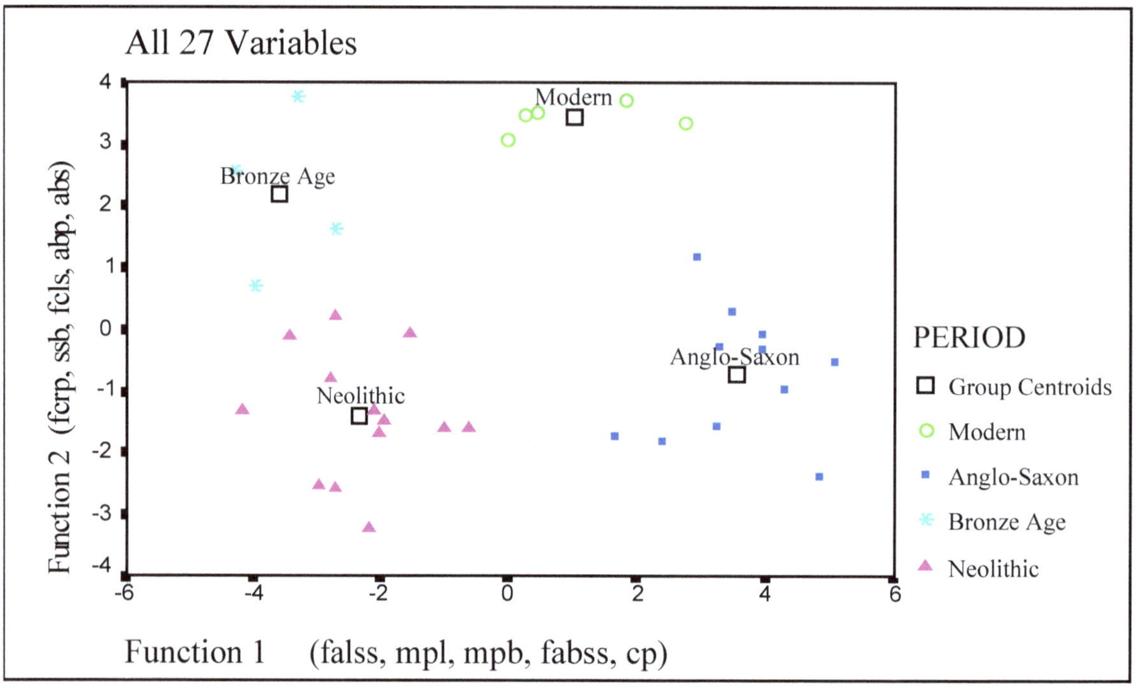

Figure 6.1. *This represents the discriminant analysis of all 27 variables (see Table 6.2 for a list). The four groups are defined by archaeological period. The group centroid is placed centrally to the distribution for that group. The variables most important for each function are listed along the appropriate axis. Function 1- average length of short striations (falss), percent of pits with long lengths (mpl), percent of pits with large breadths (mpb), average breadth of short striations (fabss) and percentage of pits (cp). Function 2 - percent of round pits (fcrp), percent of striations with short breadths (ssb), percent of long striations (fcls), average breadth of pits (abp) and average breadth of striations (abs).*

Medieval group was transitional. For some variables, the Medieval group was more similar to the modern group while for other variables it fell in between the prehistoric and modern groups. More often than not, the microwear patterns on the teeth of the Bronze/Iron Age group and the modern group showedthe greatest differences. Several of the pit size related variables showed that the prehistoric groups had larger pits than the historic groups. However, it was only the frequency of pits with a width >3µm which showed a significant difference, although the Bronze/Iron Age group had 3 times as many large-sized pits compared with the modern sample. Thus, it has to be concluded that the diet of both the prehistoric groups, but especially the Bronze/Iron Age group, contained a lot more hard and tough food items compared with the historic groups.

One of the few studies examining human molar microwear shows similar results as those found in the present study. Molleson *et al.* (1993) compared microwear features of a Late Neolithic population from northern Syria with those associated with more modern populations. The authors' aim was to investigate whether or not the microwear pattern associated with foods that had been cooked could be distinguished from the microwear pattern associated with uncooked foods. The two groups differed most greatly in pit diameter and pit density, with a greater proportion of smaller sized pits in modern populations compared to the Neolithic. It was suggested that the foods (in particular cereals) eaten by the modern populations had been rendered soft by cooking or by effective milling or sieving (Molleson *et al.* 1993: 466). Therefore a softer diet was associated with the occurrence of small pits.

In the present study, striation-related features were also important in discriminating between the groups. The prehistoric groups showed typically wider striations than the historical groups. However, it is not clear what differences in striation measurements signify. In many primate studies, striation width has shown few significant differences between genera (Teaford 1988; Teaford and Runestad 1992), and it has been suggested that abrasive particles of different sizes may produce striations of similar width (Ungar 1994:354). Where striation width has proven important is in intraspecific comparisons, for example between hunter-gatherers and agriculturalists in human populations (Teaford 1991). Teaford suggested that larger striation width may be associated with larger abrasives (such as sand particles) which might have been ingested with the foods of the hunter-gatherers, whereas the agriculturalists might have developed more advanced food preparation technologies to remove the abrasive particles from the foods, and thus reduce the hardness of their diet. Similar results were found in a comparison between a pre-agricultural population from western Japan and an early agriculturalist group in the same region (Hojo 1989). The teeth of the pre-agricultural population had larger and more frequent striations in comparison to the early agriculturalist group. The higher number of features was associated with a harder diet, leading to the conclusion that the amount of abrasives in the human diet has decreased over time (Hojo 1989).

Most researchers agree that the adoption of agriculture occurred over a lengthy transitional phase, possibly beginning in the Late Mesolithic and continuing throughout the Neolithic and into the Bronze Age (Zvelebil and Rowley-Conwy 1986). Even though it is accepted that hunter-gatherer activities probably continued to some extent into the Bronze Age, the Late Neolithic and Bronze Age populations in Britain are largely associated with more agricultural activities such as crop husbandry and domestication of animals and, in association with the new technology of ceramics, we see continually advancing methods of food preparation and cooking (Zvelebil and Rowley-Conwy 1986). Therefore, we would expect that the Early Neolithic population should exhibit more extreme microwear features resulting from a harder and tougher diet than the Bronze Age population, especially as food procurement for the Early Neolithic population was likely to be still largely dependent on hunter-gatherer activities. Surprisingly, the Bronze/Iron Age population consistently exhibited the most extreme dental microwear (more larger pits and wider striations) in comparison to the other three groups. Previous studies have most frequently related 'hard' diets involving the ingestion of large abrasives with a hunter-gatherer type subsistence, possibly associated with the presence of large amounts of exogenous grit within the diet on a regular basis (Teaford 1991; Hojo 1989; Molleson, Jones and Jones 1993). The results from the present study suggest that the Bronze/Iron Age population from Carsington Pasture cave were ingesting a very hard and abrasive diet. Why did the Bronze/Iron Age group have the largest features? At present there are no conclusive theories established which explain the precise cause of the observed differences in microwear. It is assumed that there were differences in the diet between pre-agricultural versus post-agricultural societies. Furthermore, the methods of food preparation used in association with an increased dependence on cooked foods (especially cereals) might also have had a significant influence on the dental microwear pattern. With increased food processing, the diet of more 'modern' humans groups has become progressively 'softer', and fewer abrasives particles were included into the diet compared with prehistoric populations. However, the quality of the querns used when grinding the food may play a part as an increased amount of exogenous grit might have been incorporated into the food while grinding cereals (*e.g.* Teaford and Lytle 1996).

Conclusion

Even though no significant difference was detected in the dental microwear frequency over time, there was a reduction in the size of the dental microwear features; the prehistoric groups had larger microwear features, including more large pit structures and wider striations, than the historic groups. This pattern mirrors the expected changes in diet and food processing procedures over time recorded from the archaeological record.

Compared with the Neolithic group, the Bronze/Iron Age group appears to have had a much more abrasive and rough diet. At first glance, this may appear to contradict the fact that Bronze Age peoples were mostly reliant on cultivated crops and domestic animals. A very likely explanation for the dental microwear pattern in this group may be food processing techniques using querns made of friable stone.

References

Austin, P. 2000. The emperor's new garden: woodland, trees, and people in the Neolithic of southern Britain. In A.S. Fairbairn (ed.) *Plants in Neolithic Britain and Beyond.* Neolithic Studies Group Seminar Papers 5. Oxford: Oxbow Books. 63-78.

Bakels, C. 2000. The Neolithization of the Netherlands: two ways, one result. In A.S. Fairbairn (ed.) *Plants in Neolithic Britain and Beyond.* Neolithic Studies Group Seminar Papers 5. Oxford: Oxbow Books. 101-6.

Boulter, S. and Rega, E. 1993. Report on human remains from Blackgate, Newcastle-upon-Tyne. ARCUS Project Report. Sheffield: ARCUS, University of Sheffield.

Chamberlain, A.T. 2001. Radiocarbon dates from Carsington Pasture cave, Brassington, Derbyshire. *Capra* 3.

Childe, V.G. 1951. *Man Makes Himself.* New York: The New American Library.

Clemente, I. and Gibaja, J.F. 1998. Working processes on cereals: an approach through microwear analysis. *Journal of Archaeological Science* 25: 457-64.

Cox, S.E. 1999. Dental microwear and diet in prehistoric and modern humans. Unpublished MSc dissertation, University of Sheffield.

Dobney, K. and Brothwell, D. 1986. Dental calculus: its relevance to ancient diet and oral ecology. In E. Cruwys and R.A. Foley (eds) *Teeth and Anthropology.* Oxford: British Archaeological Reports (International Series) 291. 55-82.

Gambarotta, J.P. 1995. Nombre, densité et dimensions des micro-traces d'usure dentaire chez des populations anciennes et actuelles du nord du Maroc. *Bulletin et Mémoires de la Société d'Anthropologie de Paris* 3-4: 159-76.

Harris, D. 1996. Introduction: themes and concepts in the study of early agriculture. In D. Harris (ed.) *The Origins and Spread of Agriculture and Pastoralism in Eurasia.* Washington DC: Smithsonian Institution Press. 1-11.

Harris, D. and Hillman, G.C. (eds) 1989. *Foraging and Farming: the evolution of plant exploitation.* London: Unwin Hyman.

Hojo, T. 1989. Dietary differences and microwear on the teeth of late stone age and early modern people from western Japan. *Scanning Microscopy* 3: 623-8.

Jones, G. 2000. Evaluating the importance of cultivation and collecting in Neolithic Britain. In A.S. Fairbairn (ed.) *Plants in Neolithic Britain and Beyond.* Neolithic Studies Group Seminar Papers 5. Oxford: Oxbow Books. 79-84.

Kaifu, Y. 1999. Changes of the pattern of tooth wear from prehistoric to recent periods in Japan. *American Journal of Physical Anthropology* 109: 485-99.

Kealhofer, L., Torrence, R. and Fullagar, R. 1999. Integrating phytoliths within use-wear/residue studies of stone tools. *Journal of Archaeological Science* 26: 527-46.

King, T., Aiello, L.C. and Andrews, P. 1999. Dental microwear of *Gripopithecus alpani. Journal of Human Evolution* 36: 3-31.

Lalueza, C., Pérez-Pérez, A. and Turbón, D. 1996. Dietary inferences through buccal microwear analysis of middle and upper Pleistocene human fossils. *American Journal of Physical Anthropology* 100: 367-87.

Meadow, R.H. 1989. Osteological evidence for the process of animal domestication. In J. Clutton-Brock (ed.) *The Walking Larder: the pattern of domestication, pastoralism and predation.* London: Unwin Hyman. 80-90.

Molleson, T., Jones, K. and Jones, S. 1993. Dietary changes and the effects of food preparation on microwear patterns in the Late Neolithic of Abu Hureyra, northern Syria. *Journal of Human Evolution* 24: 455-68.

Nolan, J. 1998. The Newcastle Infirmary at the Forth, Newcastle-upon-Tyne. Volume 1: The archaeology and history. Unpublished report. The International Centre for Life: Newcastle-upon-Tyne.

Norušis, M.J. 1990. *SPSS/PC+ Statistics™ 4.0 for the IBM PC/XT/AT and PS/2.* Chicago: SPSS Inc.

Nystrom, P., Phillips-Conroy, J.E. and Jolly, C.J. In press. The influence of age and sex on dental microwear patterns in baboons (*Papio* sp.) living in the Awash National Park, Ethiopia. *American Journal of Physical Anthropology.*

O'Connell, T.C. and Hedges, R.E.M. 1999. Isotopic comparison of hair and bone: archaeological analyses. *Journal of Archaeological Science* 26: 661-5.

O'Connor, T.P. 1997. Working at relationships: another look at animal domestication. *Antiquity* 71: 149-56.

Pastor, R.F. 1992. Dietary adaptations and dental microwear in Mesolithic and Chalcolithic South Asia. *Journal of Human Ecology* 2: 215-28.

Pérez-Pérez, A., Lalueza, C. and Turbón, D. 1994. Intraindividual and intragroup variability of buccal tooth striation pattern. *American Journal of Physical Anthropology* 94: 175-88.

Poinar, H.N., Kuch, M., Sobolik, K.D., Barnes, I., Stankiewicz, A.B, Kuder, T., Spaulding, W.G., Bryant, V.M., Cooper, A. and Pääbo, S. 2001. A molecular analysis of dietary diversity for three archaic Native Americans. *Proceedings of National Academy of Science (USA)* 98: 4317-22.

Puech, P.F., Serratrice, C. and Leek, F.F. 1983. Tooth wear as observed in ancient Egyptian skulls. *Journal of Human Evolution* 12: 617-29.

Robinson, M.A. 2000. Further considerations of Neolithic charred cereals, fruit and nuts. In A.S. Fairbairn (ed.) *Plants in Neolithic Britain and Beyond.* Neolithic Studies Group Seminar Papers 5. Oxford: Oxbow Books. 85-90.

Schadla-Hall, R.T. 1987. Recent investigations of the Early Mesolithic landscape and settlement in the Vale of Pickering, North Yorkshire. In P. Rowley-Conwy, M. Zvelebil and H.P. Blankholm (eds) *Mesolithic Northwest Europe: recent trends.* 46-54.

Schmidt, C.W. 2001. Dental microwear evidence for a dietary shift between two nonmaize-reliant prehistoric human populations from Indiana. *American Journal of Physical Anthropology* 114: 139-45.

Schutkowski, H., Herrmann, B., Wiedemann, F., Bocherens, H. and Grupe, G. 1999. Diet, status and decomposition at Weingarten: trace element and isotope analyses on Early Mediaeval skeletal material. *Journal of Archaeological Science* 26: 675-85.

Smith, B.H. 1984. Patterns of molar wear in hunter-gathers and agriculturalists. *American Journal of Physical Anthropology* 63: 39-56.

Smith, B.D. 2001. Low-level food production. *Journal of Archaeological Research* 9: 1-43.

Solounias, N. and Hayek, L.C. 1993. New methods of tooth microwear analysis and application to dietary determination of two extinct antelopes. *Journal of Zoology* (London.) 229: 421-45.

Tabachnick, B.G. and Fidell, L.S. 1996. *Using Multivariate Statistics.* Third edition. New York: HarperCollins.

Teaford, M.F. 1988. A review of dental microwear and diet in modern mammals. *Scanning Microscopy* 2:1149-66.

Teaford, M.F. 1991. Dental microwear: what can it tell us about diet and dental function? In M.A. Kelley and C.S. Larsen (eds) *Advances in Dental Anthropology.* New York: Wiley-Liss. 341-56.

Teaford, M.F. 1994. Dental microwear and dental function. *Evolutionary Anthropology* 3: 17-30.

Teaford, M.F. and Glander, K.E. 1991. Dental microwear in live, wild-trapped *Alouatta palliata* from Costa Rica. *American Journal of Physical Anthropology* 85: 313-19.

Teaford, M.F., Larsen, C.S., Pastor, R.F. and Nobel, V.E. 2001. Pits and striations: microscopic evidence of tooth use and masticatory behavior in La Florida. In C.S. Larsen (ed.) *Bioarchaeology of La Florida: Human biology in the northern frontier, New Spain.* Gainesville: University Press of Florida. 82-112.

Teaford, M.F. and Lytle, J.D. 1996. Diet-induced changes in rates of human tooth microwear: a case study involving stone-ground maize. *American Journal of Physical Anthropology* 100: 143-7.

Teaford, M.F. and Oyen, O.J. 1989. *In vivo* and *in vitro* turnover in dental microwear. *American Journal of Physical Anthropology* 80: 447-60.

Teaford, M.F. and Robinson, J.G. 1989. Seasonal or

ecological differences in diet and molar microwear in *Cebus nigrivittatus. American Journal of Physical Anthropology* 80: 391-401.

Teaford, M.F. and Runestad, J.A. 1992. Dental microwear and diet in Venezuelan primates. *American Journal of Physical Anthropology* 88: 347-64.

Thorpe, I.J. 1996. *The Origins of Agriculture in Europe.* London: Routledge.

Ungar, P.S. 1990. Incisor microwear and feeding behavior in *Alouatta seniculus* and *Cebus olivaceus. American Journal of Physical Anthropology* 20: 43-50.

Ungar, P.S. 1994. Incisor wear of Sumatran anthropoid primates. *American Journal of Physical Anthropology* 94: 339-63.

Ungar, P.S. 1997. Microware 3.0, a semiautomatic image analysis software package for the quantification of dental microwear, available for download on the Internet at http://Comp.uark.edu/~pungar.

Ungar, P.S. and Spencer, M.A. 1999. Incisor microwear, diet, and tooth use in three Amerindian populations. *American Journal of Physical Anthropology* 109: 387-96.

Ungar, P.S., Teaford, M.F., Glander, K.E. and Pastor, R.F. 1995. Dust accumulation in the canopy: a potential cause of dental microwear in primates. *American Journal of Physical Anthropology* 87: 93-9.

Zvelebil, M. 1994. Plant use in the Mesolithic and its role in the transition to farming. *Proceedings from the Prehistoric Society* 60: 35-74.

Zvelebil, M. 1996. The agricultural frontier and the transition to farming in the circum-Baltic region. In D. Harris (ed.) *The Origins and Spread of Agriculture and Pastoralism in Eurasia.* Washington DC: Smithsonian Institution Press. 323-45.

Zvelebil, M. and Rowley-Conwy, P. 1986. Foragers and farmers in Atlantic Europe. In M. Zvelebil (ed.) *Hunters in Transition: Mesolithic societies of temperate Eurasia and their transition to farming.* Cambridge: Cambridge University Press. 67-93.

You are where you ate: isotopic analysis in the reconstruction of prehistoric residency

Paul Budd[1], *Carolyn Chenery*[2], *Janet Montgomery*[3] *and Jane Evans*[2]

[1] Department of Archaeology, University of Durham
[2] NERC Isotope Geosciences Laboratory, British Geological Survey, Nottingham
[3] Department of Archaeological Sciences, University of Bradford

Introduction

The reconstruction of residency and mobility among prehistoric populations by scientific analysis of their skeletal remains is made possible by natural systematic variations in stable and radiogenic isotopes according to locality (Faure 1986). These become incorporated in teeth from the diet during tissue formation. Isotopes of strontium (Sr) have been used in this way since the mid 1980s (Ericson 1985). In recent years, however, the geographical resolution and reliability of the isotope approach has been greatly increased by the addition of highly complementary oxygen (O) isotope data (Barreiro *et al.* 1997; Budd *et al.* in press b). For pre-metallurgical populations, information relating to place of origin can be even further refined by additionally considering the lead (Pb) isotope composition of the tissue (Budd *et al.* in press b; Budd *et al.* 1999; Montgomery *et al.* 2000). For post-metallurgical - and especially post-industrial - populations, however, simple relationships between the isotopic compositions of Pb in foodstuffs and in teeth are complicated by exposure to manufactured products (Budd *et al.* 2000b).

Strontium (Sr) has four stable isotopes, one of which (^{87}Sr) is produced by the radioactive decay of rubidium, an element occurring naturally in many rocks and minerals. The abundance of ^{87}Sr (normally measured as a ratio to its stable sister, ^{86}Sr) is therefore dependent on both the rubidium content and age of the rock or mineral in which it is found. ^{87}Sr/^{86}Sr ratios can vary from as little as ~0.70 for basic volcanic rocks of recent age to ~0.74 in continental granites for instance (Åberg 1995). These variations are very large compared to routine measurement precision of *c.* ±0.00003 (2σ) (Budd *et al.* in press b). Sr is taken up by organisms, but the relative proportions of its isotopes are unaltered by such processes, so that soil, plant and animal strontium isotope ratios have all been shown to be related to those of the underlying geology and local hydrology (Blum *et al.* 2000). The characteristic Sr-isotope ratios of particular geographical regions and their persistence in local foods and in the tissues of feeding animals (including humans) provide the basis for the reconstruction of residency patterns.

Some Sr from the diet substitutes for calcium in the inorganic (biological apatite) mineral lattice of bones and teeth (Underwood 1977: 445). As the isotopic composition of dietary Sr incorporated in this way is unaltered by the process of transport and bioaccumulation, the isotope ratios of Sr found in hard tissues after death reflect a time-averaged signal representative of diet over some period of life, depending on the tissue under consideration.

Pb-isotope measurement of archaeological biomaterials to enhance and extend the Sr-based study of residency and mobility is now beginning to prove successful (Budd *et al.* 1997; Budd *et al.* in press b; Budd *et al.* 1999; Montgomery in prep.). In pre-metallurgical societies it is highly likely that lead incorporated into skeletal material was derived mainly, or solely, from the diet. As with Sr, the isotopic ratios of such dietary Pb primarily reflect those of the soil and therefore of the underlying geology. Like Sr, Pb-isotope ratios vary in a systematic manner in the geosphere as a result of radiogenic isotope evolution and therefore as a function of the age, parent isotope abundance, and geological history of crustal material. Human skeletal tissue accumulates Pb from the blood supply, although the mechanism is not properly understood. Pb turnover in bone is variable depending on tissue remodelling rates, density and function.

O-isotopes also vary naturally in a systematic way, but as a result of differences in climate and geography rather than underlying geology. O-isotope values are reported using δ notation as per mill (‰) deviations relative to the SMOW (standard mean ocean water) for ^{18}O/^{16}O ratios (R) where: $\delta = [(R_x/R_{sdt}) - 1] \times 10^3$. Within the body, the oxygen isotope composition of skeletal tissue is directly related to that of the oxygen consumed and this has been shown to be controlled primarily by drinking water (Longinelli 1984). Unlike Pb and Sr, biological processes alter O-isotopes quite readily. Metabolic processes elevate the level of ^{18}O in skeletal tissues relative to those of drinking water, increasing the value of δ^{18}O. However, among mammals skeletal tissues form at a relatively constant body temperature so that the oxygen isotope fractionation that does take place is very similar both within and between species (Longinelli 1984; Luz *et al.* 1990; Luz and Kolodny 1985; Luz *et al.* 1984; D'Angela and Longinelli 1990; Bryant and Froelich in press). Although there is some inter-species variation due to body mass, diet and metabolism, a number of researchers have developed calibrations to relate skeletal δ^{18}O to that of drinking water; that devised for humans by Levinson *et al.* (1987) is used here.

Calibrations of the fractionation between skeletal tissue δ^{18}O and drinking water δ^{18}O have been used for a number of years in studies of the climatological and hydrological environment of various species including humans (Fricke *et al.* 1998; Fricke and O'Neil 1996; Fricke *et al.* 1995; Bryant *et al.* 1996; Bryant *et al.* 1994; Iacumin *et al.* 1996a; Lecuyer *et al.* 1996; Luz *et al.* 1990). For much of the archaeological past, during which climatological parameters have been similar to those prevailing today, the isotopic measurement of biogenic oxygen also offers a potential insight into place of origin (White *et al.* 1998; Barreiro *et al.* 1997).

Diagenesis and tissue selection

The reconstruction of residential mobility by isotopic measurement rests on the presumption that the elements of interest preserved within archaeological tissue are those incorporated in vivo, and that these are, at least substantially, unaffected by subsequent contamination. Sr diagenesis has been addressed by various workers (Sillen 1986; Sillen and LeGeros 1991; Sealy *et al.* 1991; Budd *et al.* 2000a) who have found that the most highly mineralised tissues, especially dental enamel, are the most resistant to diagenesis. Methods for decontaminating less resilient material such as bone have been suggested, but remain in some doubt (Trickett *et al.* in prep.). Enamel, by contrast, proves to be a reliable reservoir of *in vivo* Sr in the great majority of cases (Budd *et al.* 2000a).

Caution is also needed in the consideration of Pb uptake from the burial environment. Previous studies comparing modern and pre-industrial bone Pb burdens (Grandjean 1988; Hisanaga *et al.* 1988; Patterson *et al.* 1991) appeared to show that modern bones have considerably elevated lead compared with their archaeological correlates, arguing for a lack of serious contamination of the latter. Studies of Roman individuals, on the other hand, appeared to indicate very high lead burdens (Mackie *et al.* 1975; Waldron *et al.* 1976). It is now understood that measured levels in bone were considerably elevated as a consequence of diagenetic accumulation (Waldron 1981). As with Sr, it appears that dental enamel is the most resistant tissue with respect to *post mortem* Pb diagenesis and this receives confirmation from recent pilot studies comparing the Pb content of archaeological and modern human teeth (Barreiro *et al.* 1997; Budd *et al.* 1997; Budd *et al.* 1998; Montgomery *et al.* 1999). Preliminary results suggest broad comparability between ancient and modern tooth enamel, although the same is not necessarily true of dentine. Pb distribution patterns across tooth components appear to be highly reproducible between ancient and modern individuals (Budd *et al.* 1998; Montgomery *et al.* 1999).

Dental enamel has always been the tissue of choice for O-isotope analysis because of the stability of the tissue and, in particular, of the tightly bound oxygen within its inorganic phase. The biological apatite within enamel has the approximate formula $Ca_{10}(PO_4)_6(OH)_2$, but in humans about 2%-3% of the mineral is carbonate of which about 90% substitutes for the phosphate group and the remainder for the hydroxyl group (Bryant *et al.* 1996). There are therefore three 'compartments' in which oxygen can be found, phosphate, carbonate and hydroxyl. These also have varied oxygen contents, containing 35, 3.3 and <1.6 wt. % oxygen respectively.

Although phosphate is clearly the dominant reservoir of enamel oxygen, the strong bonding which makes it so resistant to diagenesis also makes it difficult to extract for analysis. Conventional methods involve procedures to precipitate the sample as either $AgPO_4$ or $BiPO_4$, which is then reacted with a fluorinating reagent to liberate O_2 for conversion to CO_2 for analysis. However, the procedure is time consuming and can demand relatively large initial sample sizes, typically 15-30mg. By contrast, carbonate oxygen extraction is a simple process involving a reaction with orthophosphoric acid which liberates CO_2 and can be used directly for both oxygen and carbon isotope analysis. This method requires an even larger sample size, typically ~100 mg. Although both carbonate and phosphate phases of biological apatites have been shown to preserve biogenic oxygen, the carbonate phase is generally considered more susceptible to diagenesis (Iacumin *et al.* 1996b). The current study addresses the sampling constraints associated with phosphate analysis using the laser fluorination technique. In this method a high-power laser is focused onto the solid sample, heating it directly in the presence of ClF_3. Oxygen is thus extracted from phosphate, carbonate and hydroxyl groups. About 90% of the oxygen liberated is derived from the phosphate phase.

Teeth are particularly advantageous for study not just because of the excellent preservation of biogenic elements in enamel, but also because different teeth preserve elements ingested at particular stages of life. Deciduous tooth formation is instigated in the developing foetus within 14-19 weeks of fertilisation and enamel mineralisation is complete within a year of birth. Once formed, the enamel is not remodelled so its composition is a reliable indicator of *in utero* or neonatal exposure. Although development of the permanent first molar is initiated *in utero*, most of the permanent dentition is formed in childhood from three to four months after birth until about 12 or more years of age. As with deciduous teeth, permanent enamel is thought to preserve lead ingested only during the period of formation. Differences between the formation periods of different teeth make the dentition a potential archive of both the recent and more remote exposure of a single individual. This is especially so in the case of juveniles retaining both deciduous and permanent teeth.

The Monkton-up-Wimbourne case study

This paper presents a case study in which the combined Pb-, Sr- and O-isotope method has been used to attempt to reconstruct the pattern of residential mobility of four Neolithic individuals, an adult female and three juveniles, recently recovered from the excavation of a henge-type monument in southern England. The burials date from the mid to late fourth millennium BC, towards the end of the earlier Neolithic in England. Although small-scale, the study is made topical by current interest in the nature of settlement and residential mobility associated with early agriculture and domestication. After a long tradition of archaeological thought in which the arrival of a Neolithic cultural assemblage was considered synonymous with the establishment of permanent settlement and adoption of a sedentary life style, recent years have seen a significant revision of opinion. An absence of permanent settlement remains and evidence for broad-based subsistence strategies has promoted a consensus in which the earlier Neolithic is seen to have strong continuity with the preceding Mesolithic in terms of residential mobility (Whittle 1999: 59). Faunal and lithic evidence also suggests a high degree of mobility in the Neolithic of lowland southern England (Thomas 1999: 29) although Thomas (1999: 223) also stresses the difficulty of assessing changes in subsistence and residency within the Neolithic. The direct scientific approach outlined here offers the possibility to do just that and the results also suggest what could be achieved with more widespread application.

The four burials examined were excavated by Martin Green in 1997 on farmland near the village of Monkton-up-Wimbourne, Dorset, England. Dug into the chalk of Cranborne Chase, the site resembles a Neolithic henge monument (M. Green pers. comm.). Neolithic Peterborough Ware pottery was recovered from the excavated area and radiocarbon determinations on the skeletons gave a calibrated date range of 3500-3100 BC (M. Green pers. comm.). The four individuals were recovered from a single, oval grave pit cut into the chalk bedrock and backfilled with chalk blocks and rubble. The form of the single grave pit and the articulated and undisturbed nature of the bones were interpreted as one burial event at the time of initial construction of the site. Macroscopic skeletal preservation was excellent, although there was no osteological evidence to indicate cause of death (J. McKinley pers. comm.). Permanent and, where present, deciduous tooth samples were taken from all four individuals: a young-middle aged adult female (C) and three juveniles aged 5, 8-9 and 9-10 years (A, B and D respectively).

Methods

Each tooth was cleaned ultrasonically for five minutes in high-purity water and then in acetone. Each was then divided longitudinally using a flexible diamond-edged rotary dental saw to produce half-tooth samples for mass spectrometry. Enamel and dentine samples were separated using a tungsten carbide dental bur. Enamel was cleaned of all adhering dentine and abraded from the surface to a depth of >100μm using the bur. The surface material was discarded to produce a sample of core enamel tissue. Samples of primary, crown dentine were obtained by removing all secondary dentine from the pulp cavity. Sub-samples of enamel from each tooth were isolated for O-isotope analysis and the remaining tissues transferred to a clean (class 100, laminar flow) working area at NIGL for further preparation for Sr- and Pb-isotope analysis.

Samples of both the underlying chalk and soil excavated from the shaft on site were collected for Pb- and Sr-isotope analysis to assess the isotopic composition of metals available to aqueous phases in the burial environment. Aqueous leaches failed to produce sufficient Pb for analysis. Approximately 2g of chalk - from the centre of a large block taken from the site - was isolated in the low-lead laboratory, ground in a tungsten carbide mortar and pestle and placed in an acid-leached Teflon beaker. A finely divided soil sample of similar size was placed in a second beaker. Both samples were then leached overnight at room temperature in 5ml of 10% acetic acid and the leachates removed for analysis.

At NIGL, samples for Pb- and Sr-isotope analysis were first cleaned ultrasonically in high purity water to remove dust, rinsed twice, dried in high-purity acetone and then weighed into pre-cleaned Teflon beakers. Solid samples were dissolved in Teflon-distilled 16M HNO_3 and an aliquot (20% by volume) was transferred to a second pre-cleaned Teflon beaker and spiked with ^{208}Pb tracer solution for Pb concentration determination using the isotope dilution method. Pb was collected from all samples using conventional anion exchange methods and the washes from the larger aliquot of each sample were spiked with ^{84}Sr tracer solution. Sr was collected from this fraction using conventional ion exchange methods.

The Pb-isotope compositions were determined by Plasma Ionisation Multi-collector Mass Spectrometry (PIMMS) using a VG Elemental P54 mass spectrometer. Data were corrected for mass discrimination using the Tl spiking technique (Ketterer *et al.* 1991; Walder *et al.* 1993). Errors on the Pb isotope ratios were determined from replicate analysis of the NBS 981 standard as 0.014% for $^{207}Pb/^{206}Pb$ and 0.024% for the $^{208}Pb/^{206}Pb$ (n=27, 2σ. The Pb and Sr concentrations and the Sr isotope composition were determined by Thermal Ionisation Mass Spectrometry (TIMS) using a Finnegan Mat 262 multi-collector mass spectrometer. The international standard for strontium gave a value of $^{87}Sr/^{86}Sr = 0.710200 \pm 32$ (2σ, n=10) during the period of analysis. All Sr ratios have been corrected to the accepted value for NBS 987 of 0.710235. The mean Sr blank was 300pg (range 33-400pg).

Isotope dilution elemental concentration accuracy can exceed 1% (2σ) for homogeneous samples in favourable circumstances (Dickin 1995) and this value is widely quoted for Sr concentration determinations on rock powders (Evans 1996). However, dental tissues have not been well characterised with respect to compositional heterogeneity and preliminary LA-ICP-MS studies suggest a degree of Pb and Sr variation within tooth enamel (Montgomery *et al.* 1999). Estimates of reproducibility for the current study are based on a preliminary study of three replicate tooth enamel measurements undertaken for both a permanent and a deciduous tooth. These suggest errors of 14% for Pb and 10% for Sr (2σ, n=3).

Tissue samples selected for O-isotope analysis were cleaned in dichloromethane, soaked overnight in concentrated (100 volume) H_2O_2, and rinsed in double deionised water (<0.5μS cm^{-1}). Samples were allowed to dry and then heated in a vacuum for one hour at 600°C. Oxygen extraction was undertaken by laser fluorination to maximise efficiency and minimise sample size.

Approximately 2mg of dental enamel was weighed and placed into one of 16 wells of a nickel sample holder. The holder was placed into a stainless steel sample chamber fitted with a barium fluoride window. The chamber was evacuated overnight and pre-fluorinated with ClF_3 for one hour at room temperature. Oxygen released during pre-fluorination was below detection limits. The samples were reacted with 1.5×10^{-4} moles of ClF_3 using a 10 watt CO_2 laser beam (10.6μm wavelength) until a clear melt-bead was formed and ceased to react. Cryogenic and KBr traps were used to purify the sample oxygen prior to conversion to CO_2 over hot platinised carbon. The yield was then measured and the CO_2 transferred cryogenically to sample collection vessels to be analysed off line. A V.G. Isotech Optima dual inlet isotope ratio mass spectrometer, operating Micromass DI2.47 software, was used to determine the enamel oxygen isotope composition $\delta^{18}O$.

Apatite reference material certified with respect to O-isotope composition is not available. However, a number of researchers have made use of two NBS phosphate mineral standards, 120b and 120c, as a laboratory standard. Although the laser fluorination and purification line is a relatively new configuration, our initial assessment of external reproducibility based on NBS 120b (21.54 ± 0.21‰, 2 σ, n=6) and NBS 120c (20.70 ± 0.23‰, 2 σ, n=2) are comparable with previously reported data (Budd *et al.* in press a). Initial assessments of reproducibility based on replicate analyses of single tooth enamel specimens gave $\delta^{18}O$ values of 15.75 ± 0.14‰ (2σ, n=3) and 18.69 ± 0.15‰ (2σ, n=3) for a modern and an archaeological tooth respectively.

Results

Sr isotope ratios for all of the Neolithic tooth samples are shown in Figure 7.1. O-isotope data for the enamel from each tooth have been converted to drinking water values after Levinson *et al.* (1987). Pb-isotope ratios are reported as the non-standard $^{207}Pb/^{206}Pb$ and $^{208}Pb/^{206}Pb$ ratios that are widely used in archaeological literature. Although it is desirable to facilitate the wider comparison of data by use of the standard ^{204}Pb ratios, the very low Pb concentrations of the teeth in this case resulted in ^{204}Pb ratios of relatively poor precision. Since the main purpose of the study involves the inter-comparison of the teeth and local soil, ^{206}Pb ratios were adopted. Pb-isotope ratios of the soil leachates are shown in Figure 7.2.

Sr-isotope ratios for the samples are plotted in Figure 7.1. In all cases the dentine Sr-isotope values are considerably closer to those of the soil values than those of the enamel. Taken together with the elevated Sr content of the dentine in comparison with the enamel, the results suggest diagenetic addition of soil Sr in these cases. The evidence for diagenesis of the dentine is discussed in detail elsewhere (Budd *et al.* 2000a). Enamel Sr concentrations however are similar to those of modern people suggesting that the enamel is more representative of the biogenic signal (Budd *et al.* 2000a).

It is immediately apparent that there are very significant differences between enamel values for deciduous and permanent dentition for each of the juveniles and that they are very different from soil values. The enamel of the adult female (C), formed in her early childhood, has a Sr-isotope composition very different to that of the Monkton geology. Her childhood strontium, with an $^{87}Sr/^{86}Sr$ ratio over 0.71, is far more radiogenic than that associated with the Cretaceous chalk of the burial environment (typically ~0.707 consistent with the values shown in Figure 7.2). Although diagenesis of tissue samples can never be wholly dismissed, it is evident that any soil contamination of her teeth could only cause an underestimate of the radiogenic signal within the enamel. In fact, in this case the excellent state of preservation and the low values of both Sr and Pb in the enamel strongly argue for its integrity. This woman's 'non-local origin' pattern is confirmed by the Pb-isotope data (Figure 7.2), which show her enamel to have considerably elevated $^{207}Pb/^{206}Pb$ and $^{208}Pb/^{206}Pb$ ratios with respect to both her dentine and the soil leachate. The enamel Pb-isotope ratios are not sufficiently low to represent the local (chalk) geology, whereas the dentine values are intermediate between those of the enamel and those typical of the Cretaceous chalk.

The opportunity to analyse both deciduous and permanent tooth enamel from the three juveniles - and the accuracy with which it is possible to estimate their ages at death - make it possible to extract particularly detailed information with respect to changes in their dietary intake of Pb and Sr in life. Figure 7.1 shows that all three juveniles experienced very significant changes in the isotopic composition of the Sr they ingested between birth and early childhood. In the case of the eldest two children, B and D, the change appears to have been from less (both have similar deciduous values) to more radiogenic Sr. For the youngest child, A, the change is in the opposite direction from a more radiogenic birth signal to a more 'chalk-like' early childhood value. The Pb data (Figure 7.2) are relatively ambiguous in this case. No change is indicated between the deciduous and permanent enamel for B, although both A and D would appear to have deciduous enamel Pb-isotope ratios plotting at the edge of the chalk field and in the direction of C's childhood (enamel) values.

Given the very high probability that all four individuals died at the same time, and the known periods of formation of the enamel of the teeth concerned, it is possible to estimate the relative time at which each tissue was formed. The Sr data may thus be plotted so as to relate each sample on a single time scale (Figure 7.3). This shows the variation between each juvenile's diet at birth, represented by the deciduous enamel, and in early childhood, represented by the permanent enamel. Error bars associated with the permanent teeth represent the maximum period of enamel mineralisation (Hillson 1996), as the exact period during which metal entrapment take places in forming enamel is not yet fully understood. The adult female (C) is represented by dashed lines indicating the values of her enamel and dentine. All of the samples are very significantly different from one another and the pattern of variation is systematic over

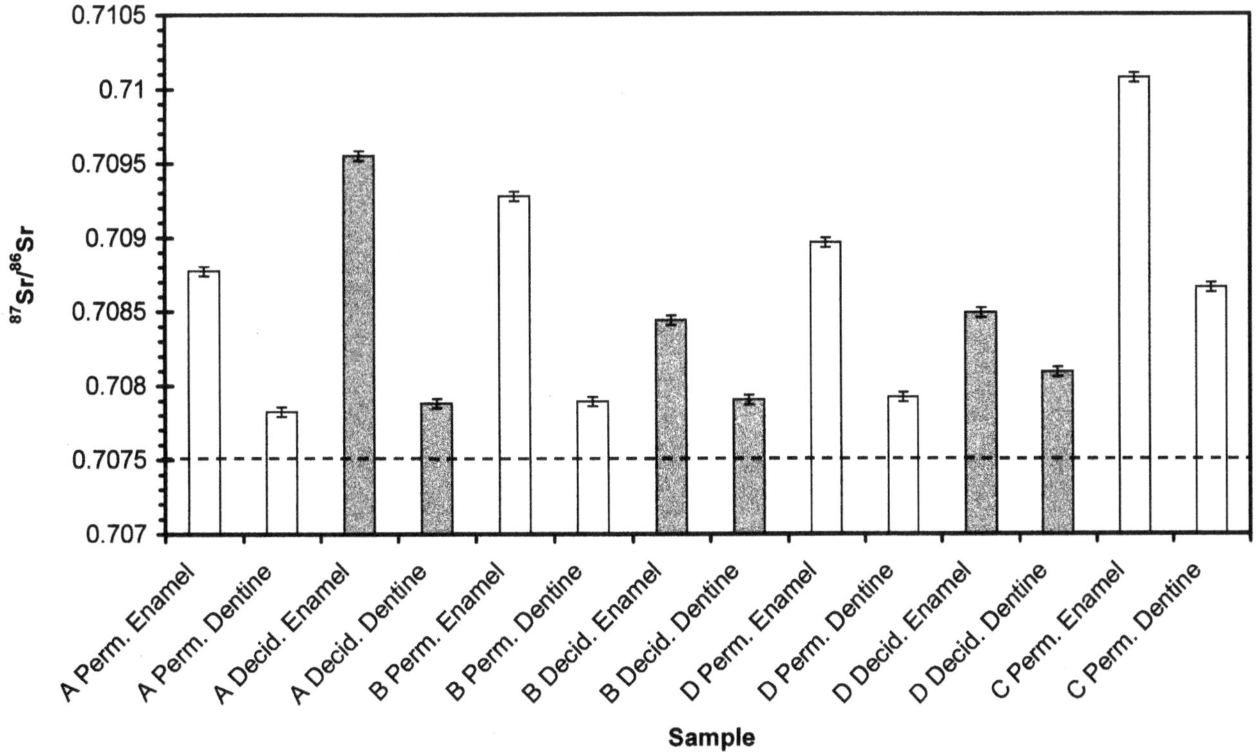

Figure 7.1. *Sr isotope ratios for the deciduous (shaded) and permanent (clear) tooth enamel and dentine of the three juveniles (A, B and D) and permanent tooth enamel and dentine for the adult female (C). Error bars represent 2δ errors calculated from 10 replicates of the NBS 987 standard. Dashed line indicates soil values.*

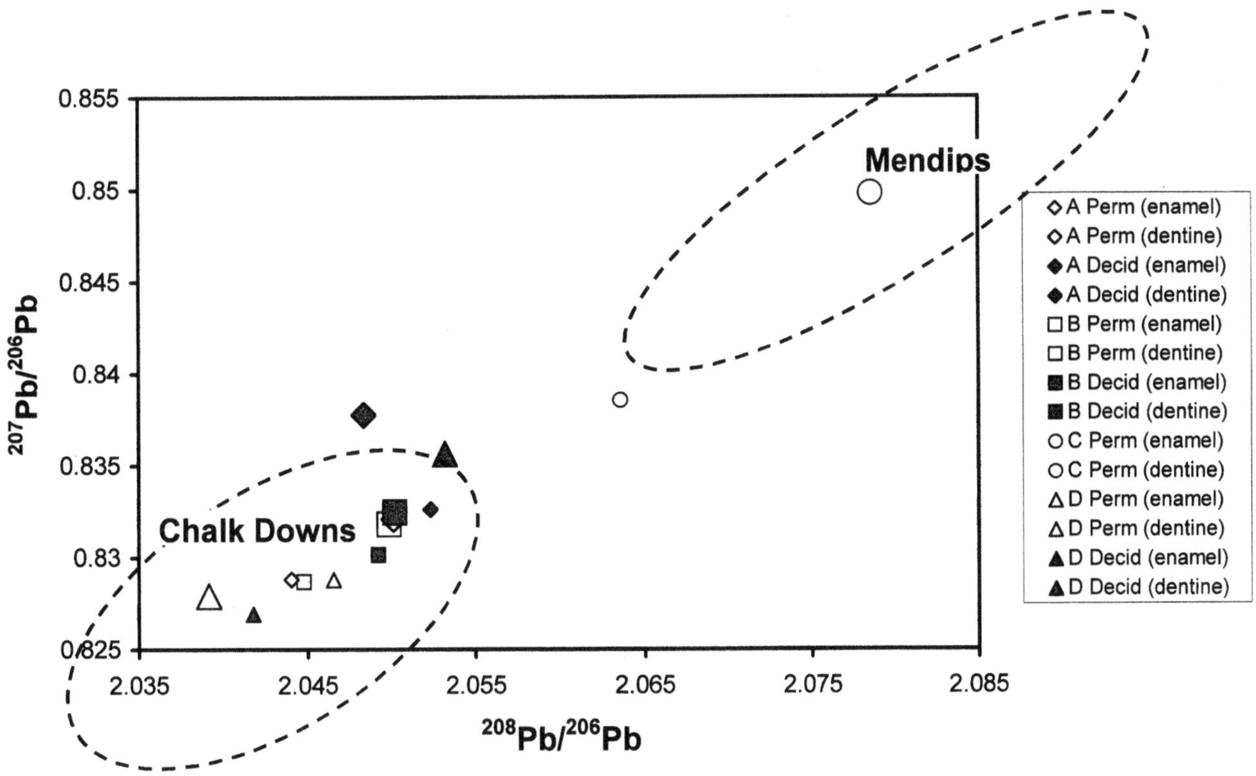

Figure 7.2. *Pb isotope ratios for the deciduous and permanent tooth enamel (large symbols) and dentine (small symbols) of the three juveniles (A, B and D) and permanent tooth enamel and dentine for the adult female (C). Errors on the Pb isotope ratios were determined from replicate analysis of the NBS 981 standard as 0.014% for $^{207}Pb/^{206}Pb$ and 0.024% for the $^{208}Pb/^{206}Pb$ (n=27, 2σ). These are smaller than the symbols at this scale.*

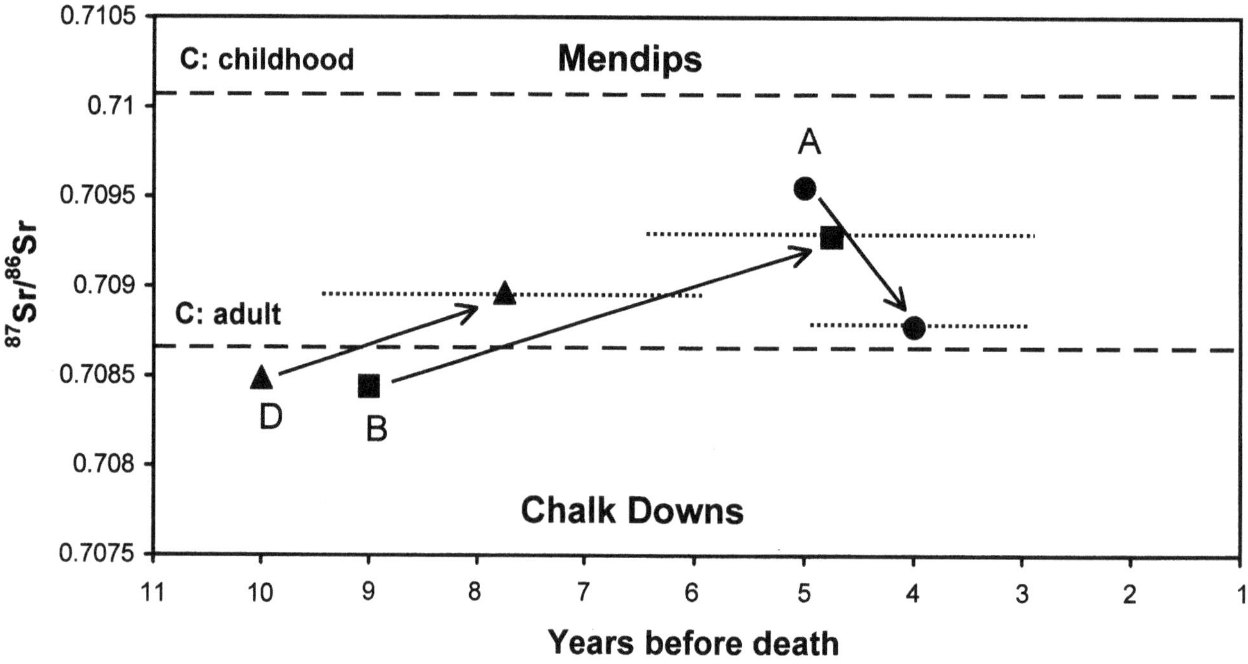

Figure 7.3. *Sr isotope ratios for the juveniles' tooth enamel relative to its time of formation in years before death. Error bars associated with the permanent teeth represent the maximum period of enamel mineralisation. The adult female, C, is represented by dashed lines indicating the values of her enamel and dentine.*

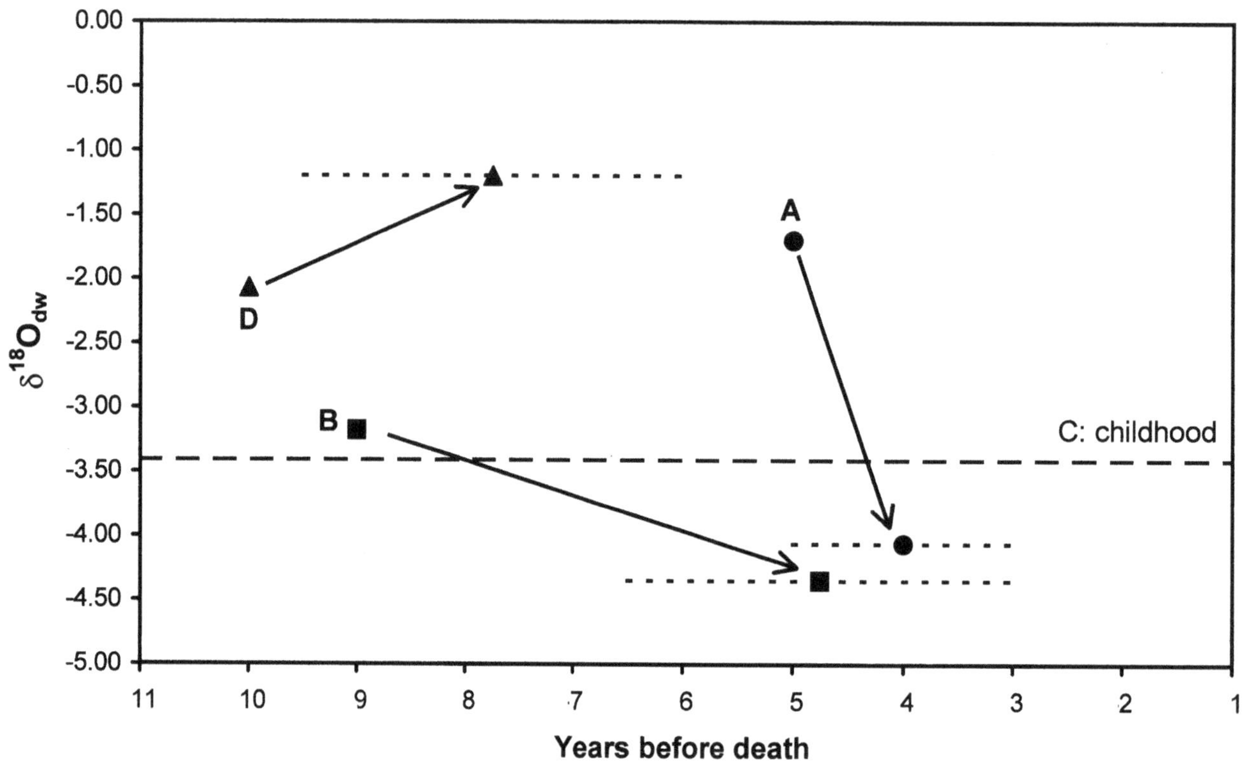

Figure 7.4. *O-isotope ratios for drinking water calculated from enamel measurements. Values for the juveniles are plotted relative to the time of enamel formation in years before death. The adult female, C, is represented by a dashed line representing her childhood drinking water at the time of permanent tooth enamel formation.*

time. O-isotope measurements of tooth enamel from all four individuals have been used to calculate drinking water values using the calibration of Levinson *et al.* (1987). Like the Sr-isotope data these may be plotted on a single time scale for comparison (Figure 7.4). Variations between the individuals, with respect to both deciduous and permanent teeth, are considerably greater than expected, varying between -4.35‰ and -1.20‰. This range of values is considerably more positive (enriched in ^{18}O) than that which we would expect from meteoric water falling in southern England today (typically *c.* -7‰ to -6‰), and is similar to rainwater in southern France or Spain. In part these relatively ^{18}O-enriched values may be explained by the warmer climatic conditions still prevailing in the region in the mid-fourth millennium BC at the close of the Holocene climatic optimum.

The comparatively large variation between individuals may reflect the highly localised (in time and space) nature of the palaeo-precipitation signal obtained from teeth. Unlike many climatic proxy records which reflect long-term averages, enamel oxygen is entrapped within a few years at most and probably reflects a predominantly surface water contribution to the diet. The potential of tooth analysis to provide a high resolution signal is demonstrated by the difference between the data obtained here and that recently published by Dennis *et al.* (2001) for speleotherm fluid inclusions from the Mendips. These suggest that the long-term average (~150 year intervals over 5000 years) groundwater signal for southern England is remarkably uniform after *c.* 3000 BC, with ^{18}O values around -6.5‰. Prior to this, preliminary estimates suggest ^{18}O is likely to have been more positive, but probably no more than about -5.25‰ (P. Dennis pers. comm.). The tooth data suggest that shorter-term and seasonal variation might have been considerably greater than the speleotherm data, taken in isolation, would suggest.

Aside from their large variance from modern meteoric water values, another striking feature of the O-isotope data is the large apparent differences between neo-natal and childhood drinking water values for the juveniles. In each case, the deciduous tooth examined was an upper canine. Cusp formation for this tooth is initiated about 15 to 18 weeks after fertilisation and the crown is generally 30% complete at birth; crown mineralisation being complete by nine months of age (Lunt and Law 1974). Under these circumstances, it might be expected that a significant part of oxygen entrapment during crown mineralisation might have taken place during the period of infant breastfeeding. We have argued elsewhere (Budd *et al.* in press a) that breastfeeding causes significant fractionation resulting in shifts of approximately 2‰ in the calculated value of 'drinking water'. This trophic effect results in calculated values of $\delta^{18}O_{dw}$ for breastfed infants which are ~2‰ more positive (enriched in ^{18}O) than those of the mother's actual drinking water. The effect of breastfeeding may go someway to explaining the significant differences in apparent drinking water O-isotope compositions between deciduous and permanent teeth for individuals A and B, but could not explain the opposite trend observed with D. It is, of course, possible that this latter individual was weaned well before nine months of age.

Conclusions

We conclude that the Sr and Pb isotope signature of C's dental enamel cannot have arisen by diagenesis and is representative of her biogenic signal during early childhood. The radical differences between these ratios and those of the local chalk geology show that she cannot have spent her early childhood in that locality. Rather, her childhood diet between the ages of about two and six years (after which the crown of the permanent maxillary second premolar is fully mineralised) appears to have contained Sr with a $^{87}Sr/^{86}Sr$ ratio of ~0.71 and $^{207}Pb/^{206}Pb$ and $^{208}Pb/^{206}Pb$ ratios of ~0.85 and ~2.08 respectively. The combination of Pb- and Sr-isotopes gives us a considerable degree of resolution with which to identify her place of origin. Although there are a number of localities within the UK with this combination of Pb and Sr isotopes, the nearest to Monkton-up-Wimbourne is the area of the Mendip orefield about 80km to the northwest. The next nearest matching locality is in the north Pennines, some 350km to the north. Although such long-range migration cannot be excluded, it seems more reasonable to postulate that individual C spent at least a part of her early childhood in the Mendips area.

The O-isotope composition of tooth enamel from the permanent teeth analysed can be used to calculate a range of drinking water O-isotope compositions for individuals A, B and C ranging from -4.35‰ to -3.41‰. Such values are significantly more positive (enriched in ^{18}O) than those which we would expect for long-term averages of meteoric water falling during this period in southern England (*c.* -5.25‰). Individual D's permanent tooth enamel is even more divergent, giving an apparent drinking water value of -1.20‰. Although the reasons for this are not yet clear, especially with regard to D, it is possible that a significant proportion of ingested water was derived from surface waters which tend to show much greater seasonal variation than ground water. It is also possible that some dietary water might have been ingested following cooking or boiling, which tend to enrich water in ^{18}O.

Analysis of the enamel from both deciduous and permanent teeth for the three juveniles has been highly revealing. The eldest two, B and D, have deciduous enamel Pb and Sr isotope ratios that are similar to one another and similar to the chalk geology. Their permanent enamel on the other hand has far more radiogenic Sr isotope ratios. It seems likely that both children experienced a change of diet between birth and early childhood, certainly with respect to Sr. This may reflect movement away from the typical chalk geology to an area with more radiogenic Sr, but perhaps not radically different Pb. The youngest child, A, shows the opposite trend with a deciduous tooth that contains more radiogenic Sr than that found in the permanent tooth. This suggests a move towards the chalk from elsewhere between birth and early childhood. Differences in the apparent O-isotope ratios of drinking water calculated from deciduous and permanent teeth suggest that A and B might have been breast-fed up until at least nine months of age, although there is no evidence to suggest that this was the case for D.

Taken as a whole, the data suggest a considerable level of mobility for all of the individuals, even over the short lives of the juveniles. The adult female must have moved at least 80km between her early childhood and her death. Considering the relative times of tissue formation between the various individuals, the Sr data at least are consistent with a similar pattern of movement for all three juveniles; from the chalk, to a more radiogenic area and back again. It is tempting to invoke the idea of all four individuals travelling together, perhaps as part of a larger group. Whilst this aspect of the interpretation remains purely speculative, the study has produced clear evidence for fairly long distance movement among a small group of Neolithic people. The results are consistent with the idea of the earlier Neolithic lifestyle in southern England as largely mobile involving, at most, only short-term sedentism.

Acknowledgements

The authors would like to thank Jackie McKinley and Charlotte Roberts for their help and support and Martin Green, Julian Richards and Ian Potts for their encouragement and for providing the site information and the samples for analysis. Thanks also to Barbara Barreiro and Geoff Nowell for analytical support at NIGL and Paul Dennis for his helpful comments on O-isotope palaeo-precipitation data. Two of the authors (PB and JM) were supported by the Natural Environment Research Council (NERC). NIGL Publication No. 453.

References

Åberg, G. 1995. The use of natural strontium isotopes as tracers in environmental studies. *Water, Air and Soil Pollution* 79: 309-22.

Barreiro, B., Budd, P., Chenery, C. and Montgomery, J. 1997. Combined Pb-, Sr- and O-isotope compositions of human dental tissues in life history reconstruction. *Abstracts. Applied Isotope Geochemistry Conference, Lake Louise, Canada, September 1997.*

Blum, J.D., Taliaferro, E.H., Weisse, M.T. and Holmes, R.T. 2000. Changes in Sr/Ca, Ba/Ca and 87Sr/86Sr ratios between two forest ecosystems in the northeastern U.S.A. *Biogeochemistry* 49: 87-101.

Bryant, J. D. and Froelich, P.N. In press. A model of oxygen isotope fractionation in body water of large mammals. *Geochimica et Cosmochimica Acta* 1995.

Bryant, J.D., Koch, P.L., Froelich, P.N., Showers, W.J. and Genna, B.J. 1996. Oxygen isotope partitioning between phosphate and carbonate in mammalian apatite. *Geochimica et Cosmochimica Acta* 60: 5145-8.

Bryant, J.D., Luz, B. and Froelich, P.N. 1994. Oxygen isotopic composition of fossil horse tooth phosphate as a record of continental paleoclimate. *Palaeogeography, Palaeoclimatology, Palaeoecology* 107: 303-16.

Budd, P., Chenery, C., Morphet, K. and Montgomery, J. In press a. Laser fluorination oxygen isotope analysis of human dental enamel as an indicator of infant weaning. *American Journal of Physical Anthropology.*

Budd, P., Christensen, J., Haggerty, R., Halliday, A.N., Montgomery, J. and Young, S.M.M. 1997. The measurement of biogenic lead within archaeological mammalian dental enamel for life history reconstruction and pollution exposure monitoring. *Abstracts. 213th ACS National Meeting, San Francisco April 13-17 1997*: 35-GEOC.

Budd, P., Montgomery, J., Barreiro, B. and Thomas, R.G. 2000a. Differential diagenesis of strontium in archaeological human dental tissues. *Applied Geochemistry* 15: 687-94.

Budd, P., Montgomery, J., Cox, A., Krause, P., Barreiro, B. and Thomas, R.G. 1998. The distribution of lead within ancient and modern human teeth: implications for long-term and historical exposure monitoring. *The Science of the Total Environment* 220: 121-36.

Budd, P., Montgomery, J., Evans, J. and Barreiro, B. 2000b. Human tooth enamel as a record of the comparative lead exposure of prehistoric and modern people. *The Science of the Total Environment* 263:1-10.

Budd, P., Montgomery, J., Evans, J. and Chenery, C. In press b. Combined Pb-, Sr- and O-isotope analysis of human dental tissue for the reconstruction of archaeological residential mobility. In J.G. Holland and S.D. Tanner (eds) *Plasma Source Mass Spectrometry: the new millennium.* Cambridge: Royal Society of Chemistry.

Budd, P., Montgomery, J., Rainbird, P., Thomas, R. and Young, S. 1999. Pb- and Sr-isotope composition of human dental enamel: an indicator of Pacific islander population dynamics. In J.-C. Galipaud and I. Lilley (eds) *The Pacific from 5000 to 2000 BP: colonisation and transformations.* Paris: Institut de Recherche pour le Développement. 301-11.

D'Angela, D., and Longinelli, A. 1990. Oxygen isotopes in living mammal's bone phosphate: further results. *Chemical Geology (Isotope Geosciences Section)* 86: 75-82.

Dennis, P.F., Rowe, P.J. and Atkinson, T.C. 2001. The recovery and isotopic measurement of water from fluid inclusions in speleotherms. *Geochimica et Cosmoschimica Acta* 65: 871-84.

Dickin, A.P. 1995. *Radiogenic Isotopes.* Cambridge: Cambridge University Press.

Ericson, J.E. 1985. Strontium isotope characterization in the study of prehistoric human ecology. *Journal of Human Evolution* 14: 503-14.

Evans, J.A. 1996. Dating the transition of smectite to illite in Palaeozoic mudrocks using the Rb-Sr whole-rock technique. *Journal of the Geological Society of London* 153: 101-8.

Faure, G. 1986. *Principles of Isotope Geology*. Second edition. New York: John Wiley & Sons Inc.

Fricke, H.C., Clyde, W.C and O'Neil, J.R. 1998. Intra-tooth variations in $\delta^{18}O$ (PO_4) of mammalian tooth enamel as a record of seasonal variations in continental climate variables. *Geochimica et Cosmochimica Acta* 62:1839-50.

Fricke, H.C. and O'Neil, J.R. 1996. Inter- and intra-tooth variation in the oxygen isotope composition of mammalian tooth enamel phosphate: implications for palaeoclimatological and palaeobiological research. *Palaeogeography, Palaeoclimatology, Palaeoecology* 126: 91-9.

Fricke, H.C., O'Neil, J.R. and Lynnerup, N. 1995. Oxygen isotope composition of human tooth enamel from medieval Greenland: linking climate and society. *Geology* 23: 869-72.

Grandjean, P. 1988. Ancient skeletons as silent witnesses of lead exposures in the past. *CRC Critical Reviews in Toxicology* 19: 11-21.

Hillson, S. 1996. *Dental Anthropology*. First edition. Cambridge: Cambridge University Press.

Hisanaga, A., Eguchi, Y., Hirata, M. and Ishinishi, N. 1988. Lead levels in ancient and contemporary Japanese bones. *Biological Trace Element Research* 16: 77-85.

Iacumin, P., Bocherens, H., Mariotti, A. and Longinelli, A. 1996a. An isotopic palaeoenvironmental study of human skeletal remains from the Nile Valley. *Palaeogeography, Palaeoclimatology, Palaeoecology* 126: 15-30.

Iacumin, P., Bocherens, H., Mariotti, A. and Longinelli, A. 1996b. Oxygen isotope analyses of co-existing carbonate and phosphate in biogenic apatite: a way to monitor diagenetic alteration of bone phosphate. *Earth and Planetary Science Letters* 142: 1-6.

Ketterer, M.E., Peters, M.J. and Tisdale, P.J. 1991. Verification of a correction procedure for measurement of lead isotope ratios by inductively coupled plasma mass spectrometry. *Journal of Analytical Atomic Spectrometry* 6: 439-43.

Lecuyer, C., Grandjean, P., Paris, F., Robardet, M. and Robineau, D. 1996. Deciphering 'temperature' and 'salinity' from biogenic phosphates: the $\delta^{18}O$ of coexisting fishes and mammals of the Middle Miocene Sea of western France. *Palaeogeography, Palaeoclimatology, Palaeoecology* 126: 61-74.

Levinson, A.A., Luz, B. and Kolodny, Y. 1987. Variations in oxygen isotope compositions of human teeth and urinary stones. *Applied Geochemistry* 2: 367-71.

Longinelli, A. 1984. Oxygen isotopes in mammal bone phosphate: a new tool for paleohydrological and paleoclimatological research? *Geochimica et Cosmochimica Acta* 48: 385-90.

Lunt, R.C., and Law, D.B. 1974. A review of the chronology of calcification of deciduous teeth. *Journal of the American Dental Association* 89: 599-606.

Luz, B., Cormie, A.B. and Schwarcz, H.P. 1990. Oxygen isotope variations in phosphate of deer bones. *Geochimica et Cosmochimica Acta* 54: 1723-8.

Luz, B. and Kolodny, Y. 1985. Oxygen isotope variations in phosphate of biogenic apatites. IV: Mammal bones and teeth. *Earth and Planetary Science Letters* 75: 29-36.

Luz, B., Kolodny, Y. and Horowitz, M. 1984. Fractionation of oxygen isotopes between mammalian bone-phosphate and environmental drinking water. *Geochimica et Cosmochimica Acta* 48: 1689-93.

Mackie, A, Townshend, A. and Waldron, H.A. 1975. Lead concentrations in bones from Roman York. *Journal of Archaeological Science* 2: 235-7.

Montgomery, J. In preparation. Lead and strontium isotope compositions of human dental tissues as an indicator of ancient exposure and population dynamics. University of Bradford.

Montgomery, J., Budd, P., Cox, A., Krause, P. and Thomas R.G. 1999. LA-ICP-MS evidence for the distribution of lead and strontium in Romano-British, medieval and modern human teeth: implications for life history and exposure reconstruction. In S.M.M. Young, A.M. Pollard, P. Budd and R.A. Ixer (eds) Metals in Antiquity. Oxford: BAR Publishing/British Archaeological Reports (International Series) 792. 290-6.

Montgomery, J., Budd, P. and Evans, J. 2000. Reconstructing lifetime movements of ancient people: a Neolithic case study from southern England. *European Journal of Archaeology* 3: 407-22.

Patterson, C.C., Ericson, J.E., Manea-Krichten, M. and Shirahata H. 1991. Natural skeletal levels of lead in *Homo sapiens sapiens* uncontaminated by technological lead. *The Science of the Total Environment* 107: 205-36.

Sealy, J.C., van der Merwe, N.J., Sillen, A., Kruger, F.J. and Krueger, H.W. 1991. $^{87}Sr/^{86}Sr$ as a dietary indicator in modern and archaeological bone. *Journal of Archaeological Science* 18: 399-416.

Sillen, A. 1986. Biogenetic and diagenetic Sr/Ca in Plio-Pleistocene fossils of the Omo Shungura formation. *Palaeobiology* 12: 311-23.

Sillen, A. and LeGeros, R. 1991. Solubility profiles of synthetic apatites and of modern and fossil bones. *Journal of Archaeological Science* 18: 385-97.

Thomas, J. 1999. *Understanding the Neolithic*. London: Routledge.

Trickett, M., Budd, P., Montgomery, J. and Evans, J. In preparation. An assessment of the effectiveness of solubility profiling in the removal of diagenetic Sr from archaeological dental tissues. *Applied Geochemistry*.

Underwood, E.J. 1977. *Trace Elements in Human and Animal Nutrition*. London: Academic Press.

Walder, A.J., Platzner, I. and Freedman, P.A. 1993. Isotope ratio measurement of lead, neodymium and Neodymium-samarium mixtures, hafnium and hafnium-lutetium mixtures with a double focusing multiple collector inductively coupled plasma mass spectrometer. *Journal of Analytical Atomic*

Spectrometry 8: 19-23.

Waldron, H.A. 1981. Postmortem absorption of lead by the skeleton. *American Journal of Physical Anthropology* 55: 395-8.

Waldron, H.A., Mackie, A. and Townshend, A. 1976. The lead content of some Romano-British bones. *Archaeometry* 18: 221-7.

White, C.D., Spence, M.W., Stuart-Williams, H.L.Q. and Schwarcz, H.P. 1998. Oxygen isotopes and the identification of geographical origins: the valley of Oaxaca versus the valley of Mexico. *Journal of Archaeological Science* 25: 643-55.

Whittle, A. 1999. The Neolithic period, c. 4000 - 2500/2000 BC: changing the world. In J. Hunter and I. Ralston (eds) *The Archaeology of Britain*. London: Routledge. 58-76.

Diet and culture in southern Britain: the evidence from Yarnton

Gill Hey[1], *Jacqui Mulville*[2] *and Mark Robinson*[3]

[1] Oxford Archaeology, Oxford
[2] School of History and Archaeology, Cardiff University
[3] Natural History Museum, Oxford

Introduction

In 1989 Lisa Moffet, Mark Robinson and Vanessa Straker wrote a paper in which they questioned the reliance of the Neolithic population on cereal cultivation and, instead, pointed to the importance of gathered foods in the Neolithic diet, and the reliance on domestic rather than wild animals. More than 10 years on, a considerable amount of fieldwork has been undertaken, especially in advance of development in south-east England, and the collection of environmental evidence has become a routine and important part of archaeological investigation. It is timely to review how their hypothesis has stood the test of time, especially in the light of evidence emerging from Yarnton, one of the major landscape projects undertaken in southern Britain during the 1990s. Concurrently, there has been debate about the character of Neolithic settlement, its possible mobility and the extent to which Neolithic populations relied on pastoralism, at least in southern England (*e.g.* Whittle 1997; Barclay and Hey 1999). Can the Yarnton evidence contribute to this discussion?

Yarnton is a rare Neolithic and Early Bronze Age landscape, in that it is an area that was repeatedly inhabited over this period of time and yet it does not appear to have had a major ceremonial focus. This is not to say that it would have been unusual. Indeed, it can be argued that it is more typical of the everyday, forming the type of landscape that has seldom been investigated on any scale. The site lies on the floodplain of the River Thames where, largely sealed beneath alluvium, remains of earlier prehistoric buildings, posthole clusters, pits and burial sites have been revealed. These represent activity from the Early Neolithic (*c.* 3800 BC) to the end of the Bronze Age around 3000 years later, as demonstrated by the presence of all major pottery styles and a number of radiocarbon determinations. During the course of the excavations, evidence has gradually been accumulating about the environment, farming strategies and social practices of the population and how these changed through time. From this information some useful remarks can be made on earlier prehistoric diet.

Background

The study area is situated in the upper Thames valley eight kilometres north of Oxford, between the villages of Yarnton and Cassington, where the River Thames loops around the Corallian Ridge (Figure 8.1). It lies approximately mid way between the ceremonial sites of the Stanton Harcourt area focused on the Devil's Quoits, and the Abingdon/Barrow Hills complex at Radley. Previous Neolithic and Early Bronze Age finds in the immediate vicinity of Yarnton include Peterborough Ware and Grooved Ware pits examined in advance of gravel extraction in the parish of Cassington, and the Beaker flat grave cemetery excavated by E.T. Leeds in the 1930s (Leeds 1934; 1940; Case and Whittle 1982).

Investigations by Oxford Archaeology at Yarnton and Cassington were prompted by extraction of gravel by Hanson Aggregates, and work in the 200-hectare study area has been funded by English Heritage. A small percentage (around 10%) of the quarry lies on the higher, second gravel terrace, where the medieval and modern villages of Yarnton, Worton and Cassington are located, but most extraction is taking place on the floodplain of the Thames. Archaeological excavation was initially undertaken on the second gravel terrace, and around 6 hectares have now been examined in detail on this higher ground, but the focus of study moved to the floodplain when the potential of this area became apparent in 1992. A further 10.5 hectares have been excavated on the floodplain gravel islands, within the palaeochannels and on adjacent low-lying areas (Figure 8.2). In addition, *c.* 10 hectares have been observed during topsoil stripping, 393 evaluation trenches have been dug (providing a 2% sample), 51 test-pits have been sieved and 182 hectares of farmland have been fieldwalked.

The excavation areas were chosen on the basis of evaluation results to examine a range of Neolithic and Bronze Age activities and, although the shortcomings of standard evaluation techniques for recovering these kinds of sites is readily acknowledged (*cf* Hey and Lacey 2001), we believe that we have a representative and extensive sample of the archaeological evidence. The results presented here derive largely from post-excavation assessment, as full-scale analysis is in progress (Hey in prep.).

Evidence of Neolithic environment, land-use strategies and diet comes from sites and features with diverse characters, including a rectangular building and associated features, pits found individually and in clusters, and burnt stone spreads and middens. A long enclosure on the floodplain and a U-shaped enclosure on the higher terrace formed the focus of burial until the Bronze Age; unmarked graves away from funerary monuments have also been found. These features have provided a variety of contexts for deposition (Hey 1997). The differing topographical locations on which people were active, and the varied localised environments within the features they dug, have led to the survival of a range of material which provides evidence of the food eaten by the communities who visited Yarnton, and the way that this was acquired.

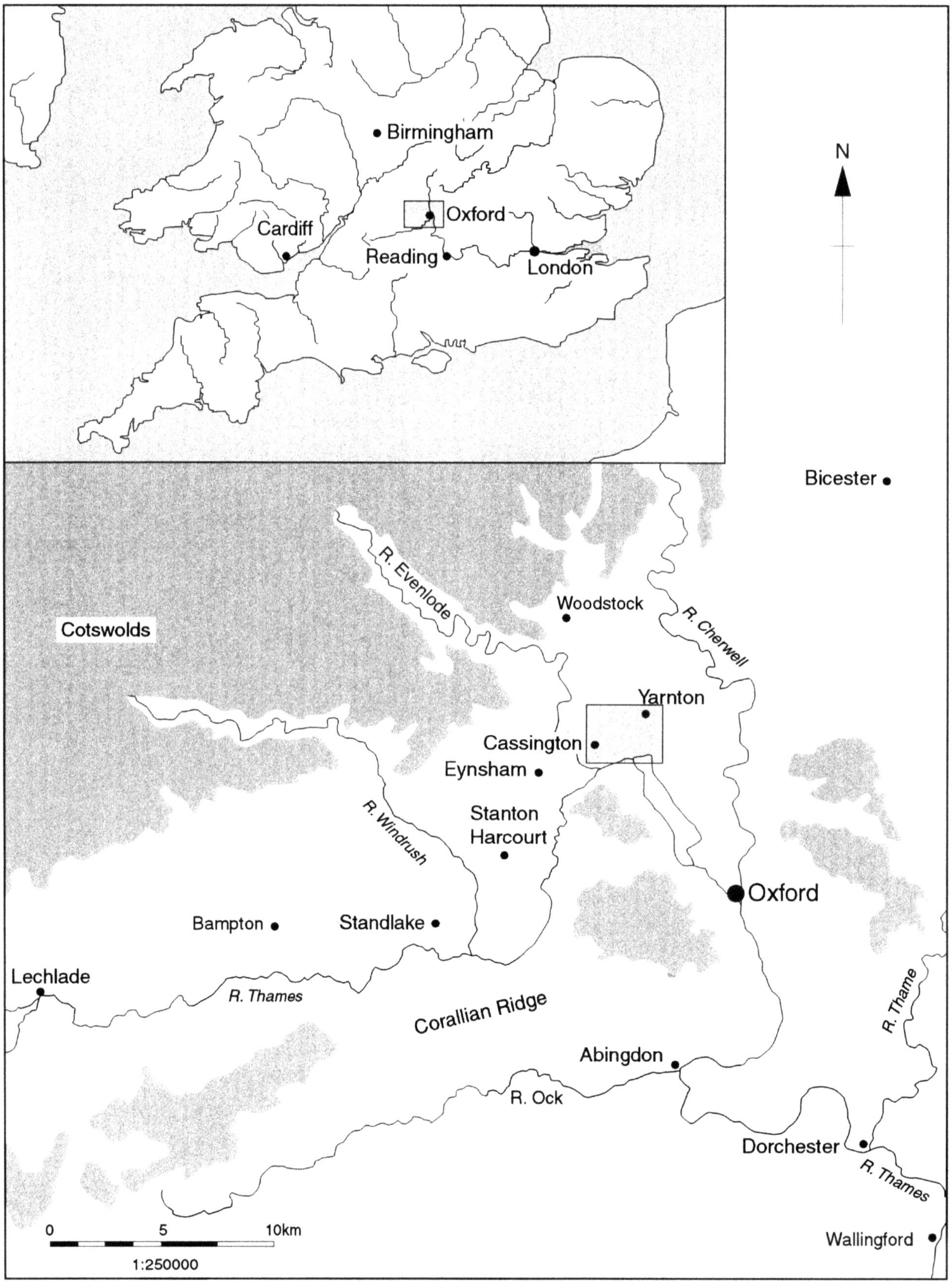

Figure 8.1. *Location of the Yarnton study area*

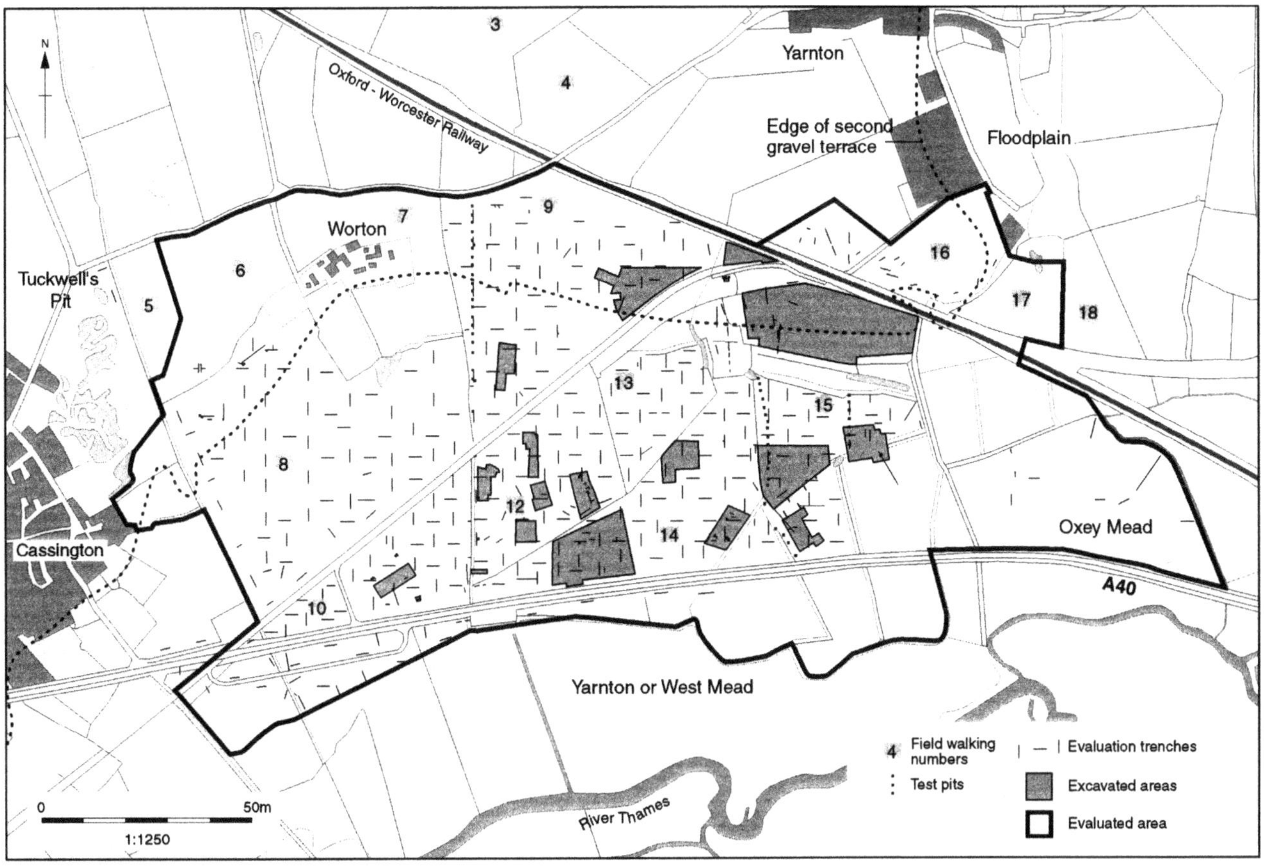

Figure 8.2. *Excavation and evaluation areas in the Yarnton study area*

Animal bones, burnt and unburnt, were recovered during hand excavation of features and also in residues from soil flotation. In general they are poorly preserved, especially on the highest parts of the floodplain where a regime of alternating waterlogging and drying, and increasing soil acidity, has resulted in accelerated decay. Charred plant remains were recovered from samples taken from a full range of features, but principally pits and postholes of structures. In total 20,714 litres of soil were floated, 1,668 litres coming from the total sampling of the postholes belonging to the Neolithic rectangular building. Larger features, especially ditches, yielded useful mollusc sequences and some deeper features contained waterlogged plants and pollen. The absence of waterlogged remains in the Neolithic period is explained below.

Domestic contexts

The discovery of Bronze Age buildings is relatively common on this topography and more than 15 have been excavated at Yarnton, the earliest being of Early Bronze Age date associated with Biconical Urn pottery. These buildings are small, circular structures of posthole construction and demonstrate that even insubstantial house remains can survive in this landscape. Sometimes they are found singly, sometimes in small groups, and they are usually associated with pits, waterholes, fencelines and individual postholes perhaps representing tethering posts and the like.

The direct evidence for Neolithic settlement is much more ephemeral but, in 1996, a Neolithic building was revealed in the excavation area known as Site 7. This was a substantial rectangular structure, some 20m x 10m in size (Figure 8.3). Large posts might have supported a roof, with the outer lines of smaller postholes representing the wall line or a lean-to roof. The whole suggests a major investment of effort. Mixed human and animal bone was recovered from three associated contexts, including the post packing of one large post pit which contained both burnt and unburnt bone and flint flakes, including a heavily calcined flake which might have come from a funeral pyre. A charred tuber of onion couch grass, particularly characteristic of pyre tinder, was also present. The animal bone included a partial pig burial with articulating hind limbs, whilst other identifiable material came from prime meat-bearing bones of domestic animals. Overall, the assemblage suggests the remains of meals, possibly feasting. Pomoideae charcoal produced a radiocarbon date of 3950-3640 BC (at 95% confidence [OxA-6772; 4970 ±60]).

The main line of posts was interrupted at a later stage by the digging of a hearth pit from which Peterborough Ware was recovered, and a pit with Grooved Ware and a shale bead was excavated in the eastern portion of the building. Some postholes may be of this date, explaining the radiocarbon result of 2900-2580 BC (OxA-6773; 4180±55) on *Prunus* charcoal from one feature. Whatever its original purpose, this structure seems to have retained special significance, in much the same way as the long enclosure did for burial.

Evidence for contemporary activity should be revealed

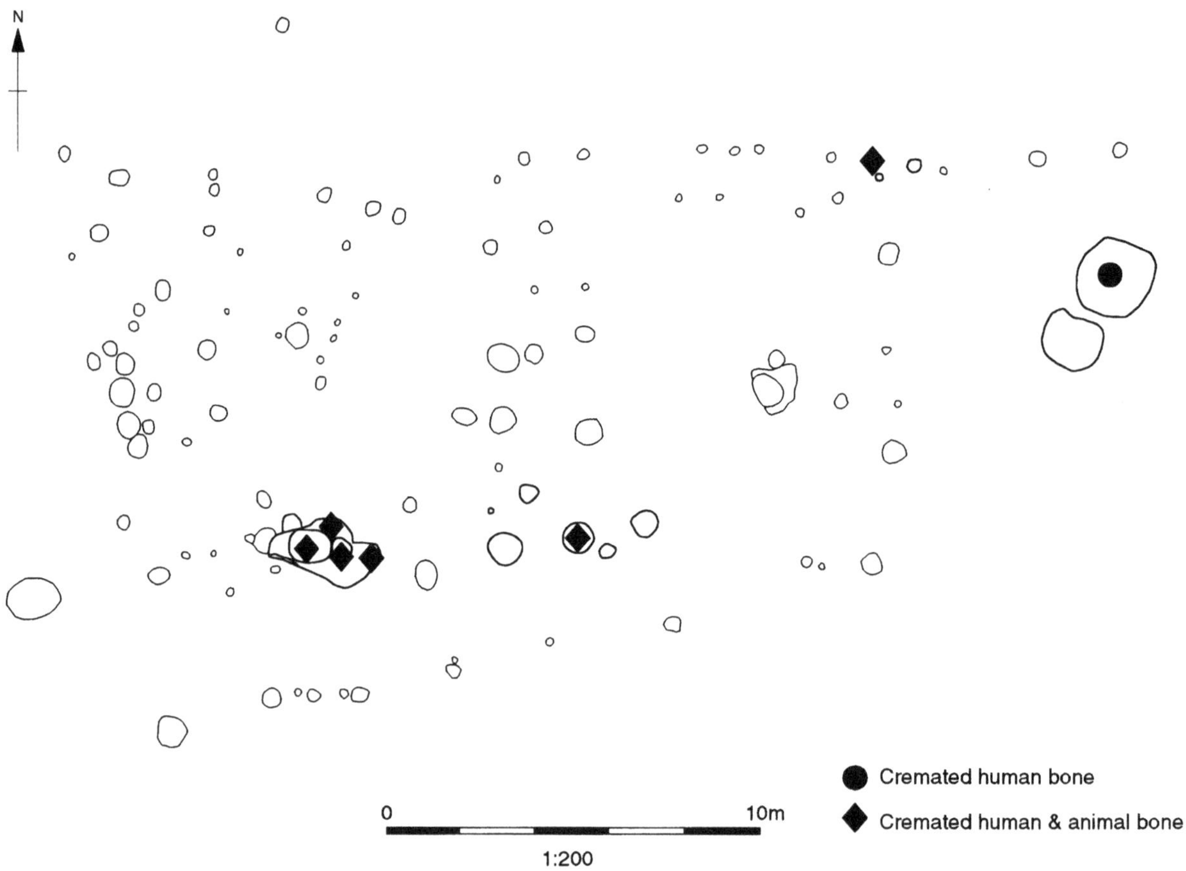

Figure 8.3. *Plan of the Neolithic building on Site 7, Yarnton*

by material falling into postholes as wooden posts began to decay, if this activity was of a type to produce durable material (Reynolds 1995). For example charred cereals, derived from the burning of these foods during their preparation, are commonly found in the postholes of Iron Age buildings in the region, such as at Gravelly Guy, Oxfordshire (Lambrick and Allen in press). No charred cereals have been recovered from Neolithic or Bronze Age house contexts at Yarnton, despite near 100% sampling for flotation, indicating that activities resulting in charring of grain were not taking place. We consider cereal preparation and storage was not a major activity within their walls.

Middle to Late Neolithic and Early Bronze Age buildings are rare in this landscape. Settlement of these periods at Yarnton is represented by clusters of pits and postholes, and such groups have been exposed over large areas of the Yarnton floodplain, with some distinct concentrations of activity (Figure 8.4). Although postholes are often found in small clusters, no houses have been identified and the small scale of these feature groups and the absence of visible structures (which cannot be attributed to poor survival, given the presence of Early Neolithic and Bronze Age buildings) strongly suggest that they represent short-term living sites, perhaps small encampments. Pits are usually associated with these clusters, but were also found in groups or alone. It is from a number of these pits that the majority of food remains from Yarnton have been recovered. Isolated discoveries of later Neolithic pits are now a common occurrence in the Thames valley and on adjacent slopes; these are probably the more damaged equivalents of the Yarnton groups.

Other contexts

Other contexts have proved less useful for the recovery of plant and animal remains. Concentrations of pottery (Neolithic Bowl, Peterborough Ware and Grooved Ware) with contemporary flintwork were exposed on a buried ground surface and these appear to be the remains of small middens. Although animal bone was retrieved from them, including cattle, pig, sheep, horse, dog and red deer, the insecure context of this material (which was only sealed by alluvium in the Iron Age) makes its attribution problematic. Charred plant remains were absent from the samples taken from these deposits, probably owing to the long period of weathering prior to burial, but their context would, in any case, have been insecure. The same difficulties pertain to the burnt stone spreads.

The ditches of a long enclosure, which seems to have been first cut in the Early Neolithic, had filled gradually through time, but contained very little material until the upper fills were deposited. Nevertheless, an important sequence of molluscs was recovered. Pits associated with this enclosure, mostly lying just beyond its entrance, did yield comparatively substantial quantities of charred remains and bone.

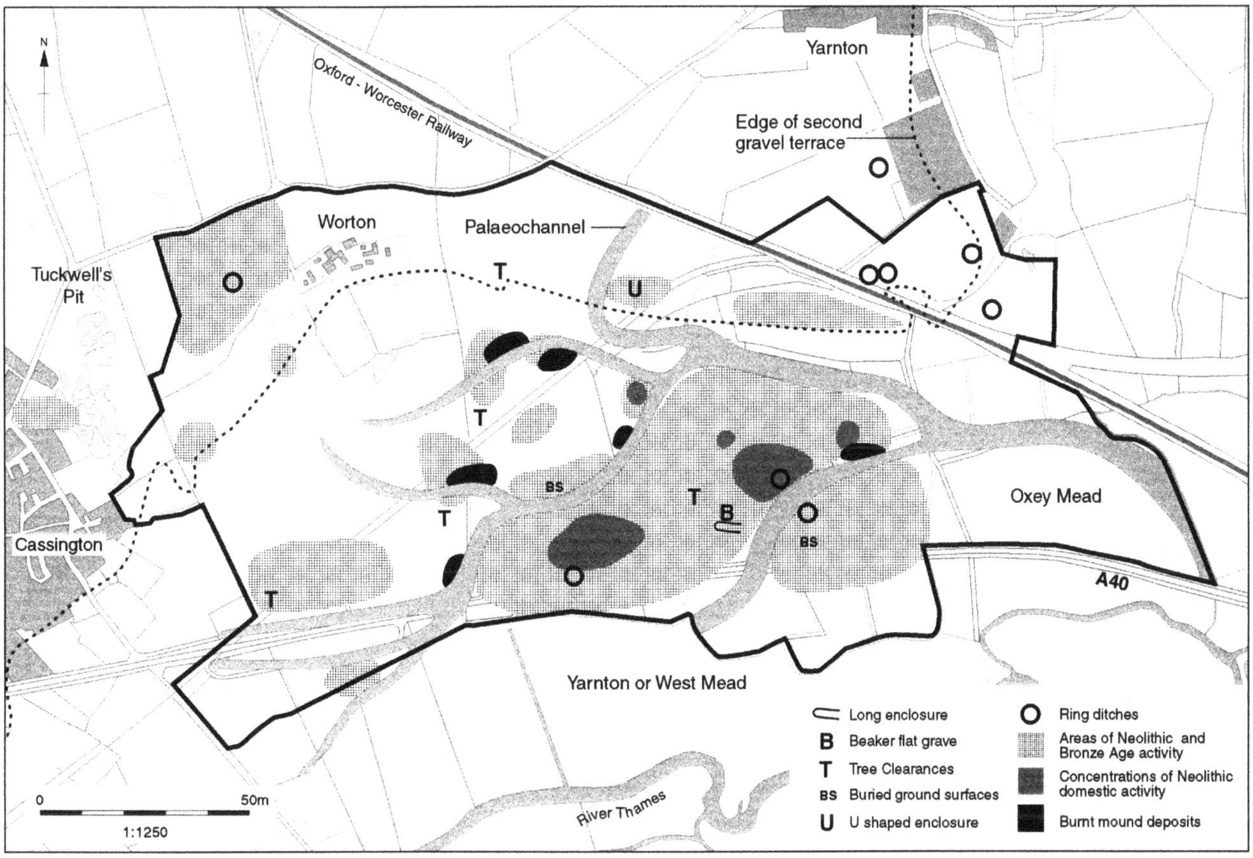

Figure 8.4. *The distribution of activity areas and monuments in the Yarnton study area*

The Neolithic topography, environment and land use

In the fourth and third millennia BC the floodplain topography, now lying beneath alluvium and Roman ploughsoils, was much more varied than it is today. The floodplain was crossed by many small watercourses, part of the braided river system of the Thames, which had created islands in the floodplain gravel in the Late Devensian. Where these streams migrated, they eroded the gravel, leaving lower ground surfaces exposed. An interesting factor to emerge from comparing the depth and date of excavated waterlogged deposits was the relatively low level of the water table before the Middle Bronze Age, when ditches on the lowest-lying parts of the floodplain first became permanently waterlogged. Indeed, the occurrence of tree-throw holes in the beds of palaeochannels suggests that they were dry throughout most of the year, only to become permanently active in the later Bronze Age.

The absence of waterlogged macrofossils and pollen limit our evidence for the general environment in the immediate area. However, pollen cores taken by Petra Day at Sidlings Copse, eight kilometres away, indicate that adjacent areas were heavily wooded with lime, oak, hazel and alder (Day 1991). At Yarnton, snails of woodland species found their way into some of the postholes of the Neolithic building, although the assemblage is too small to provide a good picture of the surrounding area.

There was, in fact, a good deal of variability in the degree of woodland clearance across the upper Thames valley at this time; around the Abingdon causewayed enclosure woodland had been substantially cleared at the start of the Neolithic (Parker 1995), and the Devil's Quoits henge was constructed in the mid third millennium BC in grazed grassland (Barclay *et al.* 1995). Yet parts of the floodplain in the Stanton Harcourt remained wooded until at least the Late Bronze Age (Robinson 1993: 14-15). The site of the Drayton cursus was cleared for the construction of the monument in the mid fourth millennium BC, and experienced another episode of clearance towards the end of the Neolithic (Barclay *et al.* in press).

The long enclosure at Yarnton was similarly constructed within a clearing, for snails of grazed grassland dominate the species, although a background of woodland snails is present. By the time the ditch was partially silted, probably after the deposition of Peterborough Ware in nearby pits, woodland had encroached onto this site. By the Early to Middle Bronze Age (when a pit and a burial were cut into the top silts of the ditch) the site was once again part of an open landscape. It is suggested that this represents shifting woodland clearance within a forested environment and is a pattern that may be supported by the general disposition of living sites across the Yarnton floodplain.

Although more than 300 pits and 500 postholes have been exposed, they lay within discrete groups, with no intercutting of features. There are a few cases where pits seem to have been deliberately sited near to, or within, special structures such as the long enclosure and the Neolithic house. In one instance, a house with mid second millennium BC Biconical Urn pottery lay immediately adjacent to a pit with late third millennium Grooved Ware. However, pits containing more than a single pottery style were extremely rare. This strongly indicates that the people who used this landscape seldom revisited the same habitation sites and the clearings so laboriously carved out of the forest by their predecessors. Indeed, such clearings might have become overgrown, as in the demonstrable case of the long enclosure. The physical evidence of tree-throw holes themselves, which are ubiquitous on the floodplain, suggests episodes of clearance. They are found in clusters, some groups demonstrating evidence of the burning of stumps and containing contemporary finds.

Thus, Yarnton probably represents an area in the upper Thames valley to which populations returned on a frequent basis, but perhaps not annually over this period of time. There is no indication that there were areas of permanently grazed grassland, as found around cursus monuments in the area (Barclay and Hey 1999) or henge monuments (Barclay *et al.* 1996).

Although the absence of permanent woodland clearings at Yarnton precludes the possibility of long-term arable farming, it may indicate that people were regularly grazing their animals in the area. An increase in dung beetles in Neolithic woodland settings in southern Britain has been demonstrated (Robinson 2000a) and may reflect the utility of these environments to Neolithic populations. If the communities visiting Yarnton were of small size, their browsing animals might have only very gradually turned wooded environments into open clearings by the consumption of new growth, unlike ceremonial landscapes whicht would potentially have been visited by larger groups with substantial herds.

The importance of floodplain locations primarily for grazing is underlined by the absence of cultivation soils and plough marks at Yarnton, a telling point in an area so intensively investigated. In addition, the fills of features are routinely stone-free, indicating no mixing of the gravel substrate with topsoils during ploughing. The physical evidence, therefore, supports the hypothesis that arable agriculture was not an important activity in this landscape.

Foods

Charred plant remains recovered in flotation at Yarnton show that cereals were consumed from the earlier Neolithic period onwards. One pit on the floodplain yielded charred bread, made of partly crushed grains of barley (*Hordeum* sp.) in a vesicular matrix. It was dated to 3640-3350 cal BC (at 95% confidence [OxA-6412; 4675±70 and NZA-8679; 4672±57]). Some emmer wheat (*Triticum dicoccum*) grains were also found in the pit along with hazel nutshells and part of an apple core. Cereal grains, including bread or rivet wheat (free-threshing *Triticum* sp.) and six-row hulled barley (*Hordeum vulgare*), have been found as well as emmer wheat, but cereals represent only 7% of charred food remains at Yarnton. In total, the site has yielded 201 Neolithic cereal grains, more than half of which came from the pit containing the bread. Floating of all soil from the postholes of the Neolithic house yielded only three hazel nutshells and no cereals. A similar building recently excavated at Whitehorse Stone in Kent on the route of the Channel tunnel rail link (Oxford Archaeological Unit 2000) was also fully sampled and yielded very little evidence of plant foods. The low level of cereal use suggested by the evidence from soil flotation is backed up by the paucity of grinding stones (Rowe forthcoming). Thus it appears that the longevity of the Neolithic and Early Bronze Age settlement record and the presence of buildings in some phases does not necessarily demonstrate a fully settled and arable economy.

Plant remains are dominated by gathered fruits and nuts. In all Neolithic and Early Bronze Age contexts at Yarnton these represent 93% of the food remains recovered (2728 fragments), comprising hazel nutshells, apples and sloe, hazel nuts being by far the most abundant (Robinson 2000b). This is a similar pattern to that found by Moffet *et al.* (1989), and is comparable to other sites excavated recently in the region, for example Eton Rowing Lake and Cople in Bedfordshire (Robinson 2000b).

The role of wild food plants in the Neolithic is still a source of debate (Jones 2000; Robinson 2000b). Most Neolithic sites have a high proportion of gathered wild foods alongside those of cereal, as shown by Moffet, Straker and Robinson in 1989, in contrast to sites from the Late Bronze Age onwards. Did these foods comprise a major part of the diet or is their predominance merely an artefact of taphonomic factors? Systematic sieving at Yarnton and these other recently excavated sites has helped to reduce the problems associated with recovery, but the relative importance of different plant foods remains enigmatic. A new line of enquiry is being opened by the analysis of organic residues within pottery fabrics: plant waxes have been found on three Peterborough Ware vessels during ongoing analysis of Yarnton pottery sherds at Bristol University (Copley *et al.* 2002).

In contrast to the plant remains, the animal bone assemblage at Yarnton is dominated by domestic animals with only a few wild species present. Cattle, sheep, pig, horse, dog and deer were recorded. Cattle is the most abundant species and the complete burial of a prime individual was found in a Late Neolithic/Early Bronze Age pit (Figure 8.5). The predominance of cattle is common on the majority of Neolithic sites in this area (Barclay et al. 1996; Serjeantson nd), but at Yarnton pig bone makes up a significant component of the identifiable assemblage, particularly in some Peterborough and Grooved Ware pits. This may indicate the importance of pigs in feasting, although the repeated occurrence of pig bone with human bone in the Early Neolithic building and some later Neolithic pits, and the presence of a partial pig cremation in a Peterborough Ware pit on Site 3, may suggest a more special significance. Sheep were present from the Early Neolithic and became more important over time, perhaps as open areas of grassland became more extensive. The preliminary results of work on organic residues suggest that most vessels with degraded animal fats contained a mixture of ruminant and porcine fats.

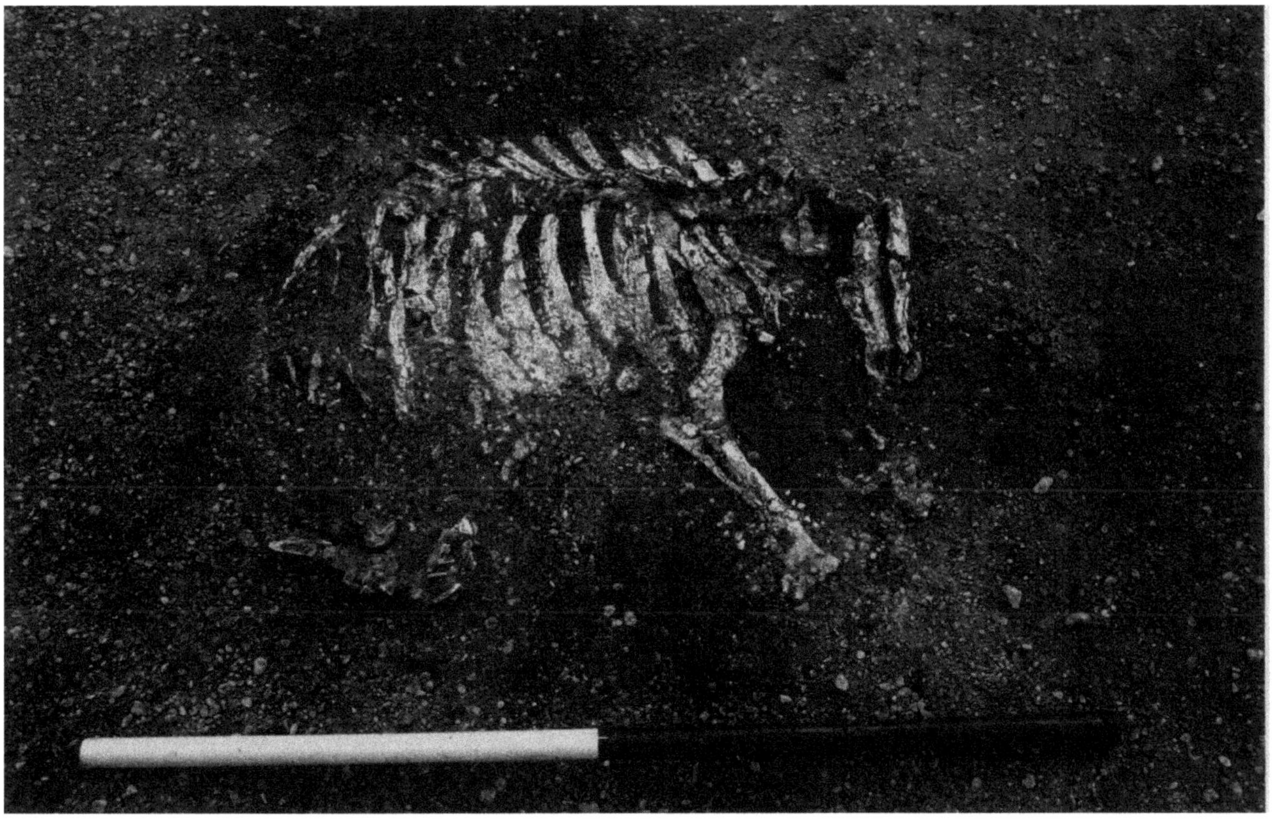

Figure 8.5. *The cow burial in a Late Neolithic/Early Bronze Age pit at Yarnton*

Evidence of dairy fats was present in several vessels (44% of those where residues were present), including in a number of Grooved Ware sherds (Copley *et al.* 2002).

As with charred food, most animal bone is derived from pits but an interesting assemblage came from a buried ground surface associated with probable midden deposits of Plain Bowl, Peterborough Ware and Grooved Ware pottery. This contained an unusual predominance of skulls and included articulating horse bones and a possible heads and hooves deposit similar to those recorded from Fussell's Lodge and Hemp Knoll (Grigson 1966; Robertson-Mackay 1980).

The paucity of wild animal remains is of note. Whilst a single fish bone was recovered from Bronze Age deposits, neither fish, nor fowl, nor any other wild food species was recovered from Neolithic contexts, with the exception of one tooth and bone fragment of deer. The floodplain should have provided ample fishing and fowling in the immediate area. The wooded environment would have provided cover for red and roe deer, but apart from antler fragments (most of which are worked and worn) and the occasional amphibian caught in a pit, wild animal bones are almost completely absent. This avoidance of non-domestic foods does not extend to plants, and it is possible that grazing herds made deer move away from settlements, but the absence of fish and wild fowl are more difficult to assign to environmental changes. Certainly by the Middle Bronze Age, the placing of wild animals and antler tools within waterholes seems to have had special significance, as it had within the Barrow Hills barrow cemetery at an earlier date (Barclay and Halpin 1999).

The predominance of domestic animals in conjunction with the presence of grazed grassland, the absence of ploughsoils and the lack of any evidence for widespread cultivation of cereals must lead to the conclusion that the people visiting the Yarnton floodplain in the Neolithic were pastoralists. The absence of structures or other evidence of intensive occupation indicates that the duration of single episodes of habitation was brief.

The context of food remains

The context within which food remains were deposited merits closer scrutiny as it has important implications for interpretation of the evidence. For example, the only food remains near to the long enclosure came from pits which appeared to contain material that was deliberately placed there. One pit just outside the entrance to the enclosure was lined with sherds of early third millennium BC Fengate Ware, placed with alternating interior and exterior faces. This group of pits also contained cremated human and animal bone, and some fine implements, including part of a polished flint macehead and another flake apparently taken from the same piece of flint.

However, the vast majority of pits at Yarnton with placed deposits, and hence food remains, were found on sites where domestic features predominated. Site 7 is a typical example, although feature density was comparatively high here (and included the Neolithic building, pit and posthole clusters and Bronze Age circular

Figure 8.6. *The distribution of pits containing substantial assemblages of Peterborough Ware and Grooved Ware on Site 7, Yarnton*

houses). An overwhelming number of finds came from just a few pits which contained substantial assemblages which appear to have been formally deposited. These are Middle to Late Neolithic in date; individual pits contained either Peterborough Ware or Grooved Ware pottery with little spatial overlap between the groups (Figure 8.6). If all the features containing datable sherds are plotted the distribution is less dramatic, but discrete feature clusters can still be seen.

What characterises these pits is their large assemblages of pottery which is often decorated. These sherds represent the major part of a few vessels but none are complete. The pits also contain a wider range of vessels than normally recovered, including items such as dishes and cups. In one pit, three large cobbles sat in its base and pottery sherds lay around and over them, suggesting that the pots had been deliberately smashed *in situ*. Varying amounts of flint were found in these contexts, although fine collections are not common on this site. The pottery and flint debris were not burnt but, nevertheless, sat within the matrix of charcoal-rich soil from which the charred plant remains and burnt animal bone were recovered. The deposits were, thus, selected from different sources.

Plant foods from both Peterborough Ware and Grooved Ware contexts on this site were almost exclusively forest foods: hundreds of hazel nutshells, a few sloes and apples and just six cereal grains, including three of wheat. The animal bone was also mostly burnt and, overall, pig predominated over cattle, with very few remains of sheep or deer. In some pits there appears to have been an attempt to represent a wide range of species. We have examined two of these groups in detail, one with Peterborough Ware and the other with Grooved Ware finds. In one pair of Peterborough Ware pits cattle, pig, deer and a burnt human incisor were recovered, in addition to 100 hazel nutshells and one wheat grain. There are interesting differences in the proportion of species in the Peterborough Ware and Grooved Ware pits. Pig bone is twice as common as cattle with Peterborough Ware, and cattle are twice as common as pigs with Grooved Ware. This is the reverse of the expected pattern, and stands in contrast to the provisional results from lipids from Yarnton (examined by Richard Evershed, Stephanie Dudd and Anna Mukherjee at Bristol University) which suggest mainly beef in Peterborough Ware pots and pork in the single Grooved Ware sample with sufficiently preserved residue.

We suggest that the food remains contained within these pits had been specially selected and may not be representative of an everyday diet. Their clustering within just a few pits, which also contained substantial assemblages of highly decorated pottery, some of which was placed, does not suggest mundane refuse disposal. The pottery and flint do not derive from a burnt context and do not indicate sweepings from the hearth or the kitchen floor. However, these features do lie within and adjacent to the pits and postholes of habitation sites and they are not associated with ceremonial monuments, even though such sites were accessible. Thus it is unlikely that they represent large-scale ritual events. They may much more probably be the result of small, household rituals commemorated by acts of deposition within the ground. As such, they are unlikely to contain the exotic, or to be the result of conspicuous consumption. They probably represent a midway point between ceremonial deposits and everyday debris.

Conclusions

We do not conclude that cereals were unimportant in the Neolithic in this area. Recent pollen work adjacent to the Abingdon causewayed enclosure shows that this monument was constructed in a cleared landscape in which cereals were cultivated, or at least processed, on a regular basis (Parker 1995). Abingdon was a special site, however. Other ceremonial sites in the upper Thames valley were located within extensive areas of grazed grassland, suggesting pastoral farming (Barclay and Hey 1999). Yarnton, with its pits and postholes and no major ceremonial focus, is much more typical of the sites with scattered pit groups that are increasingly coming to light in this area. Whatever the conclusions to be drawn from the Neolithic house at Yarnton, the evidence for the Middle Neolithic to the Early Bronze Age period shows the persistent use of this landscape by mobile pastoralists, exploiting the rich floodplain grasslands in addition to the higher terraces and slopes, perhaps on a seasonal basis. In other words, the Yarnton evidence represents one part of a broader subsistence strategy, similar to that suggested in the very different context of Windmill Hill by Fairbairn (1999: 155).

It could be suggested that the shifting clearings indicate a desire to create browse rather than grassland, since cattle and pigs are essentially woodland creatures. Ring-barking of trees encourages secondary growth which can easily be reached by grazing animals, and ultimately leads to the death of trees which can then be felled with a minimum of effort (Robinson 1992: 51-3). Burning of stumps would have cleared the area, leaving the characteristic burning often found on these floodplain sites, and enriching the soil, enabling small plots of cereals to be grown and areas of grassland to be extended. This seems to have occurred in the context of a gradual opening-out of the woodland, as communities or, perhaps more accurately, herds grew in size and sheep became a more important component of animal husbandry. By the Middle Bronze Age the Yarnton floodplain landscape was mainly grazed grassland.

As for the pits within which the majority of the finds were recovered, they have remained as a lasting symbol of individual events, the small-scale ceremonies that punctuated everyday life, the births, comings of age, 'marriages' and deaths of the Neolithic, the things with which these communities were most intimately concerned.

Acknowledgements

Funding for this project was provided by English Heritage and we are very grateful to them for their support. We would also like to thank the finds and environmental specialists who worked on this project, particularly Alistair Barclay and Pippa Bradley, for providing us with their preliminary results on the pottery and flint. Caroline Chissell provided the results of analysis for several of the floodplain sites. The staff from Oxford Archaeology who worked at Yarnton are too numerous to mention here but we should like to acknowledge the hard work that they put into this project. Particular thanks go to Chris Bell, however, who supervised and later managed several of the excavations. The illustrations are by Peter Lorimer.

References

Barclay, A., Bradley, R., Hey, G. and Lambrick, G. 1996. The earlier prehistory of the Oxford region in the light of recent research. *Oxoniensia* 61: 1-20.

Barclay, A., Gray, M. and Lambrick, G. 1995. *Excavations at the Devil's Quoits, Oxfordshire 1972-3 and 1988*. Thames Valley Landscapes: the Windrush Valley Monograph 3. Oxford: Oxford Archaeological Unit.

Barclay, A. and Halpin, C. 1999. *Excavations at Barrow Hills, Radley, Oxfordshire. Volume I: the Neolithic and Bronze Age monument complex.* Thames Valley Landscapes Monograph 11. Oxford: Oxford Archaeological Unit.

Barclay, A. and Hey, G. 1999. Cattle, cursus monuments and the river: the development of domestic and ritual landscapes in the Upper Thames Valley. In A. Barclay and J Harding (eds) *Pathways and Ceremonies: the cursus monuments of Britain and* Ireland. Neolithic Studies Group Seminar Papers 4. Oxford: Oxbow. 67-76.

Barclay, A., Lambrick, G., Moore, J. and Robinson, M. In press. *Lines across the Land: excavations at the Drayton cursus.* Thames Valley Landscapes Monograph 14. Oxford: Oxford Archaeological Unit.

Case, H.J. and Whittle, A.W.R. 1982. Cassington, 1950-2: Late Neolithic pits and the big enclosure. In H.J. Case and A.W.R. Whittle (eds) *Settlement Patterns in the Oxford Region: excavations at the Abingdon causewayed enclosure and other sites*. York and Oxford: CBA and Department of Antiquities, Ashmolean Museum. 118-51.

Copley, M., Mukherjee, A., Aillaud, S. and Docherty, G. 2002. Was dairying an important element of animal husbandry in prehistoric Britain?: organic residue analysis of Neolithic/Early Bronze Age pottery vessels from Yarnton floodplain. Unpublished report. Oxford: Oxford Archaeological Unit.

Day, S.P. 1991. Post-glacial vegetational history of the Oxford region. *New Phytologist* 119: 445-70.

Fairbairn, A. 1999. Charred plant remains. In A.W.R. Whittle, J. Pollard and C. Grigson, *The Harmony of Symbols: the Windmill Hill causewayed enclosure.* Oxford: Oxbow. 139-56.

Grigson, C. 1966. The animal remains from Fussell's Lodge long barrow. In P. Ashbee, The Fussell's Lodge long barrow excavations 1957. *Archaeologia* 100: 63-73.

Hey, G. 1997. Neolithic settlement at Yarnton, Oxfordshire. In P. Topping (ed.) *Neolithic Landscapes.* Oxford: Oxbow. 99-111.

Hey, G. In prep. *Yarnton: Neolithic and Bronze Age settlement and landscapes.* Thames Valley Landscape Monograph. Oxford: Oxford Archaeological Unit.

Hey, G. and Lacey, M. 2001. *Evaluation of Archaeological Decision-making Processes and Sampling Sstrategies.* Oxford: Oxford Archaeological Unit for Kent County Council.

Jones, G. 2000. Evaluating the importance of cultivation and collecting in Neolithic Britain. In A.S. Fairbairn (ed.) *Plants in Neolithic Britain and Beyond.* Neolithic Studies Group Seminar Papers 5. Oxford: Oxbow. 79-84.

Lambrick, G. and Allen, T. In press. *Gravelly Guy, Stanton Harcourt: the development of a prehistoric and Romano-British community.* Thames Valley Landscapes Monograph. Oxford: Oxford Archaeological Unit.

Leeds, E.T. 1934. Recent Bronze Age discoveries in Berkshire and Oxfordshire. *Antiquaries Journal* 14: 264-76.

Leeds, E.T. 1940. New discoveries of Neolithic pottery in Oxfordshire. *Oxoniensia* 5: 1-12.

Moffett, L., Robinson, M.A. and Straker, V. 1989. Cereals, fruit and nuts: charred plant remains from Neolithic sites in England and Wales and the Neolithic economy. In A. Milles, D. Williams and N. Gardener (eds) *The Beginnings of Agriculture.* Oxford: British Archaeological Reports (International Series) 496. 243-61.

Oxford Archaeological Unit. 2000. White Horse Stone: a Neolithic longhouse. *Current Archaeology* 168: 450-3.

Parker, A.G. 1995. Late Quaternary environmental change in the upper Thames basin, central-southern England. Unpublished DPhil thesis, University of Oxford.

Reynolds, P.J. 1995. The life and death of a posthole. *Interpreting Stratigraphy* 5: 21-5.

Robertson-Mackay, M.E. 1980. A 'head and hoofs' burial beneath a round barrow, with other Neolithic and Bronze Age sites, on Hemp Knoll, near Avebury, Wiltshire. *Proceedings of the Prehistoric Society* 46: 123-76.

Robinson, M. 1992. Environmental archaeology of the river gravels: past achievements and future directions. In M. Fulford and E. Nichols (eds) *Developing Landscapes of Lowland Britain: the archaeology of the British gravels, a review.* Occasional Papers 14. London: Society of Antiquaries. 47-62

Robinson, M.A. 1993. Pre-Iron Age environment and finds. In T.G. Allen and M.A. Robinson (eds) *The Prehistoric Landscape and Iron Age Enclosed Settlement at Mingies Ditch, Hardwick-with-Yelford, Oxon.* Thames Valley Landscapes: the Windrush Valley Monograph 2. Oxford: Oxford Archaeological Unit. 7-19.

Robinson, M.A. 2000a. Coleopteran evidence for the elm decline, Neolithic activity in woodland, clearance and the use of the landscape. In A.S. Fairbairn (ed.) *Plants in Neolithic Britain and Beyond.* Neolithic Studies Group Seminar Papers 5. Oxford: Oxbow. 27-36.

Robinson, M.A. 2000b. Further considerations of Neolithic charred cereals, fruit and nuts. In A.S. Fairbairn (ed.) *Plants in Neolithic Britain and Beyond.* Neolithic Studies Group Seminar Papers 5. Oxford: Oxbow. 85-90.

Rowe, F. Forthcoming. Corn grinding in southern England: what can the querns tell us? In G.J. Barclay and K. Brophy (eds) *Regional Diversity in the Neolithic of Britain and Ireland.* Neolithic Studies Group Seminar Papers. Oxford: Oxbow.

Serjeantson, D. nd. Review of environmental archaeology in southern Britain. Animal bones. Neolithic and Early Bronze Age (4000 BC - 1500 BC): the development of agriculture and animal husbandry. Unpublished report. Department of Archaeology, University of Southampton.

Whittle, A. 1997. Moving on and moving around: Neolithic settlement mobility. In P. Topping (ed.) *Neolithic Landscapes.* Oxford: Oxbow. 16-22.

Dairying, dairy products and milk residues: potential studies in European prehistory

Oliver E. Craig

Department of Fossil Fuels, University of Newcastle

Introduction

Identification of the original contents of archaeological pottery has fascinated archaeologists and scientists alike throughout the 20th century. In 1933 Johannes Grüss first identified an amorphous black residue in a ceramic vessel from the Hallstatt period as overcooked milk by conducting some basic chemical tests (Grüss 1933). Although these analyses may have lacked the necessary rigour which is now demanded from archaeological scientific investigations, this early work helped pioneer a new approach to artefact analysis. In the last decade, the development of biochemical techniques such as gas and liquid chromatography, mass spectrometry, immunology and stable isotopic analysis has provided much greater scope for the reliable identification of residues. Issues of contamination from the burial environment have been addressed by careful sampling methods (Heron *et al.* 1991), simulated degradation of replica pots has led to greater understanding of post-depositional processes (Dudd *et al.* 1998; Craig *et al.* 2000) and numerous procedural controls have been implemented to rule out post-excavation contamination. Finally, a greater appreciation of the many and varied uses of archaeological pottery, including multiple inputs, re-use and a range of non-culinary uses, has tempered interpretations based on residue analysis alone.

In keeping with Grüss's pioneering work of the 1930s, it is the identification of milk residues that has returned to become a focus for more recent studies and with good reason. Not only does milk contain a unique range of lipids and proteins, which are suspended in solution and free to adsorb to the ceramic surface, but also residue analysis may provide the only way to unequivocally identify the production, preparation and consumption in prehistory of this important and versatile product. Defining the origins of dairying in Europe and the Near East has provoked much debate over the last twenty years. Whether dairying was a principal component of a neat package of innovations that spread widely across Europe in the third millennium BC (Sherratt 1981; 1983; Harrison 1985) - the so-called secondary products revolution - or was present earlier in the Neolithic, perhaps on a smaller scale (*e.g.* Bogucki 1984a; 1984b; Chapman 1982), is a question that remains largely unresolved.

Until the appearance of pictorial tablets depicting milking scenes (Green 1980), evidence for dairying is restricted to interpretations of archaeozoological data (*e.g.* Bogucki 1984a; Milisauskas and Kruk 1989; Greenfield 1988; 1991; Harrison 1985; Legge 1981a; 1981b) or from distinctive vessel forms (Sherratt 1981; Bogucki 1984b). Residue analysis of pottery could clearly contribute to this debate. But how can a technique which provides direct information on pottery use, such as food preparation and consumption practices (*e.g.* Evershed *et al.* 1992; 1997), be used to study widespread changes in food production? Also, what is meant by 'dairying', and what is the economic, social and cultural significance of this practice? This paper will first review some of the theories proposed for the development of dairying, specifically in Europe, and then examine the potential of new methods for analysis of milk residues, before considering the role that residue analysis should play in future studies of dairying and dairy products in the past.

The development of dairying in Europe: general perspectives

The identification of dairying as an economic activity poses a number of challenges to archaeological researchers. As evidence of exploitation of domestic ruminants can be found in faunal assemblages from sites across Europe, often dating to the earliest phases of the Neolithic, it is reasonable to assume that milk would have been available for human consumption from these animals at this time. Dairying is not a specific technology nor is it necessarily limited to specific strains of livestock (Sherratt 1997: 206). Whether or not the inhabitants of Early Neolithic Europe had the necessary genetic adaptation to digest the lactose in fresh milk (Sherratt 1981) is ambiguous from modern phylogenetic data (Akoi 1986; Holden and Mace 1997; Bodmer and Cavalli-Sforza 1976: 279) but the question is largely irrelevant considering the wide range of low lactose, easily stored products that can be made from milk. Therefore the key issue in defining a dairying economy is not only to provide evidence for the utilisation of dairy products but also to assess the scale of production and the significance of this activity.

The intensification of dairying in the third millennium BC in many parts of Europe was first proposed over twenty years ago (Sherratt 1981) as a component of the secondary product complex, an economic package that had far-reaching social and cultural implications and represented a unilinear progression of farming practice. Although dairying was initially inferred, rather tentatively (see Chapman 1982), from the appearance of new ceramic vessels associated with manipulations of liquids, weight has been given to this hypothesis by analysis of post-Neolithic faunal assemblages, for example from Britain (Legge 1981a), Spain (Harrison 1985) and the Balkans (Greenfield 1988). In these cases, kill-off patterns, with cattle bone assemblages dominated by juveniles and/or adult females, compare favorably with those obtained from modern herds optimised for milk production (produced by Payne 1973). A cattle-based dairy economy, optimised for maximum production with the potential for exchange, is thus implied.

However, similar evidence for an optimised dairy economy can be found at earlier Neolithic sites; for example, Bogucki (1984a) has argued that the mortality profiles of Linear Pottery (LBK) faunal assemblages from temperate Europe are not inconsistent with Payne's models and evidence from the Early Neolithic layers of Arène Candide, in the west Mediterranean, shows that sheep kill-off patterns also fit well with the optimised milk production strategies (Rowley-Conwy 2000). Legge has also noted female bias in adult cattle at Neolithic 'ceremonial' sites in the UK and at other Neolithic sites in Switzerland (1981b).

The use of faunal kill-off patterns alone to define the scale and specifics of prehistoric animal husbandry practice has been criticised. Besides problems inherent with preservation and recovery of animal bones, ancient livestock may have had different productivity compared to their modern counterparts (Halstead 1998). In addition, Halstead demonstrates, with reference to modern pastoralism in northern Greece, that the decision to utilise livestock for meat or milk and when to cull is dependent on a range of economic and environmental factors which may fluctuate from season to season (Halstead 1998). Furthermore, high juvenile culling need not indicate a dairy economy but could imply also a mechanism to preserve fodder (McCormick 1998); indeed, the presence of calves may be a prerequisite of early cattle dairying in order to assist milk 'let down' (McCormick 1992). Equally, high frequencies of adult females in a cattle bone assemblage, rather than indicating an optimised dairy strategy, may reflect a sexual bias in maturity rates and thus indicate a strategy of meat productivity (McCormick 1992).

A number of ecological and economic arguments are often cited in relation to this debate. It is often inferred that dairying must have been practiced as soon as domesticated ruminants were available, as it offers by far the most energy-efficient use of uncultivable land (Holmes 1970; Ingold 1980:176; Legge 1981a) and results in products that are suitable for storage, thus possessing additional economic value (Bogucki 1987). Others point out that the large herds/flocks implied by a strategy optimised for dairy production (Dahl and Hjort 1976: 220) are labour-intensive and economically untenable in landscapes without easily accessible pasture (Halstead 1996), such as the relatively narrow, riverside environments inhabited by the Early Neolithic farmers of central Europe. Halstead argues that a mixed farming strategy, where small numbers of a variety of animals are kept for a mixture of products (*e.g.* meat, milk, wool), principally for domestic use, not only seems more economically plausible in such environments, but is also evident in the considerable heterogeneity that exists in Neolithic faunal assemblages (Halstead 1996). Even at sites which show other supposed indicators of dairying, such as high numbers of juveniles, other domesticates are always present (*e.g.* Rowley-Conwy 2000).

These arguments support the concept that a specialised dairy economy could only develop towards the end of the Neolithic, after substantial areas of primary forest were cleared which, in turn, fits within the framework of a 'secondary products revolution' (Sherratt 1981; 1983). However, this does not necessarily preclude the independent localised development of large-scale milk-based pastoralism at earlier periods in the Neolithic. For sedentary communities, provision of cultivated or collected fodders and stalling of animals could have compensated for limited pasture. Large-scale seasonal movement of animals to exploit fresh pastures offers an alternative strategy. For example, the lowland LBK communities of the north European plain seem to have focused their economic activities on cattle herding rather than on grain cultivation (Bogucki 1987). Settlement patterns here suggest a high degree of residential mobility consistent with a semi-mobile, milk-based pastoralist economy; the presence of perforated ceramic sherds (putatively interpreted as cheese-strainers) recovered from many of these 'camps' has been taken as further evidence of dairying (Bogucki 1984a; 1984b). The socio-economic relationship between these semi-mobile intensive dairy farmers and the sedentary grain cultivators of the loess is unclear but it is interesting to speculate upon the potential for exchange, which would further complicate economic inferences made from faunal analysis alone.

Whether economic factors favour dairying or meat production may be less significant than the cultural values that might pertain to pastoralism. In addition to 'allocative' resources such as milk and meat, animals may provide an 'authoritative' resource in terms of their socio-political significance (Hall 1986). The requirements of animals for ritual use cannot be overlooked (Keswani 1994), and caution has to be exercised when interpreting economic regimes from animal assemblages recovered from ceremonial centers (Entwhistle and Grant 1989). The association between cattle and power through conspicuous display, exchange transactions and redistribution is exemplified by the complex pastoralist societies of southern and eastern Africa (Reid 1996). In particular, mobile pastoralism is likely to enhance social intercourse between neighbouring households and villages through contact and exchange. Thus, the social, ideological and political demands for large herds or flocks may provide the impetus for a specialised economy, such as dairying, and may explain the resulting economic risks. Equally, the existence of taboos may prevent dairy consumption, irrespective of the economic benefits.

Defining the scale of dairying is complicated by these cultural, economic and ecological factors. The available evidence neither demonstrates nor disproves a unilinear progression from small-scale household dairying as part of a mixed economic strategy to the development of a large milk-based pastoral sector in the third millennium

BC, as originally proposed by Sherratt (1981; 1983). Whether dairying was practiced on a large scale, with large numbers of animals on expanses of open pasture, or as a daily activity by individual homesteads may be of little consequence for the study of consumption and the consumers of dairy produce. Nevertheless it does imply a difference in the way that animals were perceived and cared for. As Mike Parker Pearson has discussed in the introduction to this volume, food studies offer new scope for the integration of archaeology, social theory and science. Identification of milk residues from ceramics may contribute greatly to our understanding of many aspects of milk production, its preparation and consumption in the past. The key issue is how these new methods can be most appropriately integrated into the expanding discussion of food in archaeology and where the archaeological emphasis should lie.

Methodologies for the identification of milk residues on ceramics

Milk is unique to mammals and is only produced by the lactating mammary gland. It contains the necessary nutrients to support the infant during development and thus has a distinctive range of proteins, lipids and sugars, which are dissolved or suspended in solution. During the processing of raw milk into different dairy products, the relative amounts of these biomolecules will change and some molecules may become chemically altered. Residue analysis has focused on the detection of the lipid and protein components of milk because lactose, the principal sugar in milk, is thought to be lost rapidly to bio-degradation and leaching in the burial environment.

Identification of milk lipids

Lipids can be extracted from archaeological ceramics with organic solvents and analysed by high-temperature gas chromatography or gas chromatography/mass spectrometry (see Heron and Evershed 1993 for a review). Fresh milk contains a characteristic distribution of lipids, a range of fatty acids with carbon chain lengths of between four and 20 including branched chain and monounsaturated species and a complex range of triacyglycerol (TAG) species (Mottram and Evershed 2001). However, over time many of the shorter-chain fatty acids which are characteristic of milk are lost and the TAG distribution is altered to more closely resemble animal adipose fat rather than milk fat (Dudd and Evershed 1998). Therefore, although lipid yields from archaeological pottery are generally high, there are no single lipid compounds that are unique to milk.

Dudd and Evershed (1998) overcame this problem by measuring the stable carbon isotope ratio (^{12}C:^{13}C) of the most prominent unsaturated fatty acids (with carbon chain lengths of 16 [C16:0] and of 18 [C18:0]) using gas chromatography combustion isotope ratio mass spectrometry (GC-c-IRMS). The relative proportion of light (^{12}C) and heavy (^{13}C) carbon atoms in these molecules reflects the source of carbon from which the molecule was synthesised. The ratio between these two carbon molecules - the carbon isotope ratio ($\delta^{13}C$) expressed in parts per thousand relative to an international standard - in the C16:0 and C18:0 fatty acids is different in various animal products. Fats from non-ruminant animals have similar isotope values for both acids but in ruminants the C18:0 fatty acid contains less ^{13}C than the C16:0. It is therefore said to be isotopically 'lighter' or 'depleted'; significantly this difference is even more pronounced in ruminant milk fats. The absolute isotope ratios of these fatty acids in milk are a function of the animal's diet, but in all cases the C18:0 fatty acid in ruminant milks is between 3%-5‰ depleted in the heavier (^{13}C) stable carbon isotope compared to the C16:0.

This new approach has a number of methodological advantages. The C16:0 and C18:0 fatty acids are the most commonly encountered fats on archaeological pottery. They can be easily extracted and analysed and have been shown to be derived from pottery usage rather than the depositional matrix (Heron *et al.* 1991). Furthermore, as the absolute isotopic measurements made on these fats are direct indicators of animal diet, information concerning types of pasture and fodder provision may be obtained and directly related to husbandry strategy (*i.e.* milk/meat exploitation). However, there are problems of interpretation of the isotopic values derived from only two fatty acids. Both are present at variable concentrations in most animal and vegetable products; consequently, multiple inputs in pottery will result in a mixed isotope signal, which may be hard to interpret. This problem can be overcome by taking into account other diagnostic lipids present in the sample (Dudd and Evershed 1998).

Identification of milk proteins

A number of proteins are unique to milk; these include whey proteins and caseins (α-, β-, γ-casein). Of these, α-casein is the most attractive for study: not only is it the most abundant protein in milk (13.7 g l^{-1} in bovine milk; Jenness 1970) but it is also thermostable (Wells 1908). Its stability is demonstrated by the fact that bovine α-casein can be detected in baby vomit (Sato 1992) and in milk that has been heated at 75°C for 2 days; once complexed with ceramic it can survive an order of magnitude longer (Table 9.1). The immunological methods employed for detecting milk proteins on ceramic involve using indirectly labelled antibodies, which will bind tightly to specific regions of the α-casein structure. The regions that are targeted by antibodies are unique to that particular protein, and can even be used to distinguish α-caseins that are derived from different species of animal. This is in part due to the random structure of the casein molecule, which facilitates the production of antibodies against its primary amino acid sequence (Enomoto *et al.* 1990), and because of a number of amino acid substitutions that exist between α-caseins from different species.

However the application of immunological methods to archaeological materials poses a number of problems. Perhaps the greatest methodological challenge has been the removal of proteins, such as caseins, which have survived by strong association with the mineral (ceramic)

Sample	Treatment	DACIA response[1] (ng g^{-1} sample)
Simulated pot[5] (Bovine milk)	None	>500
	Buried 2 years in controlled plots[6]	100-500
	Heated 75°C, 336 days, anoxic, 95% humidity	100-500
Bovine milk[2]	No Treatment	>500
	Heated 75°C, 2 days, anoxic, 95% humidity	>500
Ethnographic milk pot[7]	None	>500
	5 day continuous leaching experiment[3]	>500
Blank ceramics	None	n.d.
	Buried 2 years in controlled plots	n.d.
Simulated pots[5] (Bovine adipose)	None	n.d.
	Buried 2 years in controlled plots	n.d.
Soil samples[4]		n.d.

Table 9.1. *Identification of milk proteins on control samples using Digestion and Capture Immunoassay.*

n.d. - not detected; [1]DACIA using anti-bovine α-casein monoclonal antibody. Two separate extractions all standard deviation <10%. Positive twice background values; [2]against $100\mu l$; [3]experiment performed by Carl Challinor, University of Bradford; [4]range of soil samples (clay, loam, sand); [5]simulated pot 'open' fired high porosity and 'kiln' fired low porosity (Craig 2000); [6]burial experiments repeated in two plots with measured hydrology, microbial biomass/activity and pH; [7]obtained from N.E. India, used for *c.* 10 years for heating bovine milk and discarded 1988.

surface (9.1). Extraction with conventional solvents yields insufficient quantities of protein for analysis. An improved method has been developed (digestion and capture immunoassay: DACIA) to increase yields. This involves digesting the mineral surface with hydrofluoric acid and simultaneously capturing any released molecules for subsequent immunological detection. The sensitivity and specificity of this technique has been demonstrated by testing a range of simulated and ethnographic ceramics with known input (Table 9.1; Craig and Collins 2000).

Another problem with immunological-based approaches is the use of inappropriate methods, which result in non-specific cross-reactions with contaminants or compounds derived from the burial environment generating a positive response in the assay even when the target protein is absent. Such issues have been widely discussed in connection with earlier immunological studies of residues, notably those which have attempted to identify species of blood on stone tools (Gurfinkel and Franklin 1988; Smith and Wilson 1992; Cattaneo *et al.* 1993; Manning 1994; Downs and Lowenstein 1995; Child and Pollard 1992; Leach and Mauldin 1996; Tuross *et al.* 1996; Leach 1998). To address these criticisms, experimental design has to include an appropriate range of negative controls to confirm the specificity of the test; these may include blank ceramic, blank ceramics that have been exposed to the burial environment, ceramics with non-milk input also exposed to the burial environment, and soil samples (Table 9.1).

Preservation of organic residues on ceramics

The potential for preservation for the different classes of organic residues encountered on archaeological pottery is highly variable and dependent on a number of factors. These include mode of use, physical properties of the ceramic, depositional environment and post-excavation treatment. Organic residues present within the porous matrix are likely to be protected from microbial degradation and are less susceptible to leaching. In addition, organic-mineral interactions at the ceramic surface stabilise organic structures and retard chemical and biological degradation mechanisms (Evershed 1993; Heron and Evershed 1993; Craig and Collins 2002). Charring of residues may also provide a mechanism for protection (Oudemans and Boon 1991) but is likely to result in substantial structural modification preventing routine extraction and identification, especially using immunological methods.

The length of time in the depositional environment may not be as important as the nature of the environment itself (*i.e.* pH, redox potential, hydrology, microbial activity). Burial experiments have shown that whilst the majority of the protein component of absorbed milk residues is lost within the first few months of burial probably by leaching, subsequent degradation of the remaining fraction is greatly reduced (Craig 2000). Lipids, which are hydrophobic and less susceptible to leaching, may stay at high concentrations for thousands of years (*e.g.* Charters *et al.* 1993) and are less susceptible to structural modification than proteins. Immunological methods rely on the preservation of large, hydrophilic protein sub-units (polypeptides), which undergo side chain modification and hydrolysis (chain splitting). Lipids are therefore likely to survive over longer timescales and in a wider range of environmental conditions than proteins (*e.g.* Regert *et al.* 1998).

Contamination of potsherds post-deposition and post-excavation is more of a problem for lipid residue analysis than for protein residue analysis. Although proteins are mobile in the burial environment and are easily transferred during handling, contaminating proteins will not usually be detected owing to the specificity of the immunological tests (of course, extra precautions would have to be taken if research is directed towards the detection of human proteins). Post-excavation contamination with lipids is particularly significant especially when no provision has been made for residue analysis prior to excavation and post-excavation analysis. Contamination with plasticizers, glues and skin lipids are commonly encountered and can interfere with analysis, especially if they are in high concentration. Washing and brushing of sherds is also likely to result in loss of information, although no systematic study of this has been carried out. It is recommended that newly excavated sherds are as handled as little as possible, air-dried together with any adhering soil and wrapped in aluminium foil or acid-free tissue and then bagged.

The cost of analysis is likely to be a decisive factor when assessing the impact of residue analysis to archaeology. Generally, lipid analysis is expensive, especially when combined with isotopic measurements needed to identify dairy products, but provides more information into pottery use per analysis (*e.g.* analysis of other plant and animal lipids and contaminants). Protein analysis is cheaper but only a single compound can be detected per analysis (*e.g.* bovine milk protein). These factors will obviously affect the sampling strategy and ultimately the type of questions that can be addressed. Furthermore the costs associated with residue analysis highlight the need for these techniques to be utilised efficiently, for example to augment less expensive forms of pottery use analysis such as determination of the form/function relationship and use-wear analysis. One final but important consideration is the discrimination between food and non-food residues. For precisely the same reason that products such as milk and blood are the most prepossessing products for residue analysis, they also provide excellent products to seal pots that lack the required porosity. For example, Ethiopian potters use milk to seal ceramic vessels immediately after firing (Rice 1987: 163). Discrimination between sealants and food residues is methodologically challenging but characteristics such as vessel permeability may be used to aid interpretations.

In conclusion, given the relative strengths and weaknesses of both approaches, there is clear advantage in combining the two different methods of residue analysis for the study of dairying. The two methods are entirely independent, in that they target different biomolecules using different analytical methods, and provide complementary information.

Development of dairying in Europe: the potential of residue analysis

There is no doubt that the successful identification of milk residues on ceramics may contribute to the study of the origins and significance of dairying and dairy products, but at present only limited studies have been undertaken. Milk residues have been detected at a number of UK sites: milk proteins in the Late Bronze Age-Early Iron Age site of Cladh Hallan in the Western Isles of Scotland (Craig *et al.* 2000), milk lipids in Iron Age and Early Medieval ceramics from Northamptonshire (Dudd and Evershed 1998) and milk lipids in Late Neolithic sherds from the Welsh borders (Dudd *et al.* 1999). These cases unequivocally demonstrate that milk was an element of the prehistoric economy. In the Western Isles, where faunal assemblages are exceptionally well preserved, the detection of milk residues supports the interpretation of a large optimised, dairy economy based on the high number of associated neonatal cattle bones, a feature of later prehistoric settlements in this area (Parker Pearson *et al.* 1996; 1999). However, without other palaeodietary indicators, differential preservation, contamination and multiple uses of ceramics (such as sealing) complicate a more quantitative approach to residue analysis, which aims to define the scale of animal husbandry practices. Furthermore, biases associated with small sample sizes (implicit considering the amount of ceramic generally recovered from sites and the costs of analysis) limits inter-site and temporal comparisons.

One way to overcome biases involved in sampling a random selection of domestic pottery is to target specific ceramic artefacts. Many debates involving an early or late origin for dairying have involved speculation on the techno-function of pottery shapes or specific ceramic artefacts. The appearances of new forms of Bronze Age and Copper Age vessels across Europe which have been associated with dairying (Sherratt 1981; 1983) provide obvious avenues for a targeted investigation. In central Europe, both cattle and sets of distinctive drinking and pouring vessels were included in the graves of the Copper Age Baden culture (Whittle 1996.123). The increased significance of cattle, together with the appearance of these new vessels, may imply new patterns of dairy consumption, perhaps reflecting the new economic importance of dairy products and dairy animals as a result of intensification or innovation. Similarly, a systematic study of the function of LBK ceramic sieves, interpreted as dairy-processing utensils (Bogucki 1984a &b) provides a clear hypothesis to test using residue analysis. The identification of milk on these artefacts would also indicate the existence of fermented dairy products, ideal commodities for storage and exchange and perhaps with added non-economic value derived by the consumers of these protein-rich, 'special' products.

At most sites residue analysis is confined to the analysis of pottery with unknown techno-function, often from domestic contexts. Identification of milk residues in these cases is of limited value to our wider understanding of the significance of dairying, unless this is integrated with other lines of enquiry. Assessment of settlement distribution and size, and availability of surrounding resources, combined with analysis of plant and animal remains and stable isotopes extracted from human bone can provide some scope for broadly determining the maximum scale of dairy production and consumption. Evidence for the scale and significance of dairying may also be obtained through a greater understanding of the material culture involved in preparing and consuming dairy products. Associating vessel contents with broad classifications of ceramic typologies (*e.g.* storage vessels, cooking pots, serving wares), use-wear characteristics, and depositional context provides one way of achieving this. Identification of vessels that have been dedicated for milk use, evident from the fatty acid isotopic signal, would also be useful in this respect. In addition, information concerning specific consumption practices, such as feasting, may be gained by relating food residues to distinctive pottery styles, methods of deposition and context.

Finding evidence for fodder provision is also crucial in defining specific animal husbandry practices. If dairying is suggested in marginal environments or at sites where surrounding pasture is limited, such as many Early Neolithic European settlements, then provision of fodders must be envisaged. Besides the economic implications, the deliberate provision of gathered and cultivated fodders implies a different relationship between humans and their livestock. Evidence of cattle byres (Nielsen *et al.* 2000) and twig foddering of sheep and goats (Rasmussen 1993) have been found at the earliest Neolithic settlement sites in Switzerland. Stable carbon and nitrogen isotopic analysis of animal bone collagen and of sectioned teeth can be used to reveal variations in cattle diets, including seasonal changes and weaning (Balasse *et al.* 1997; 1999). When combined with direct isotopic measurements of adipose and milk fats in pots, this approach may be particularly applicable to the identification of foddering regimes, such as the supplement of marine ($\delta^{13}C$) or $\delta^{4}C$ fodders like maize and millet, for specific husbandry practices.

Dairy products have also been implicated as important commodities for exchange (Bogucki 1987). The identification of milk-containing vessels not produced locally, at sites with no other evidence of indigenous dairy production, may be the only way of substantiating this interpretation. Similarly, targeting storage vessels, suitable for transport, may facilitate the identification of producer and consumer sites, especially if the products transported can be related to economic practices at specific sites. Finally, it is worth noting that identifying the original species of dairy products, using an immunological approach, is vital to the interpretation of dairying in the past. Although cattle, sheep and goats can all be exploited to obtain a product with similar nutritional properties, the implications of these activities are very different.

Conclusions

Dairying can only be placed in its proper economic, social and cultural context by knowing its scale and significance. Although the identification of milk residues provides unequivocal evidence of dairying, it provides little information relating to these aspects. Whilst residue analysis of specific ceramic artefacts may be used to give weight to theories such as the 'secondary products revolution', it is hard to see how these methods can be used to track the supposed intensification of dairying through the Neolithic when used in isolation. As Bogucki has pointed out, new ways of food production were not introduced uniformly throughout Europe (1987); hence it is vital that regional and site-specific studies are undertaken. In this respect, it is crucial that residue analysis is integrated with other forms of cultural, dietary, economic and environmental evidence to look at specific production, preparation and consumption practices. When used in this way, information additional to pottery use may be gained, such as the nature of exchange networks, seasonal mobility, ritual consumption and strategies for fodder provision.

Acknowledgements

The impetus for this paper came from a meeting in Ljubljana entitled the '8th Neolithic Seminars: perspectives from pottery' and I thank Michael Budja for organising the meeting and for his invitation. A similar version of this paper is also published in the proceedings of that conference. I also thank Dr Matthew Collins, Dr Nicky Miner and Ms Jennifer Hiller for help with all aspects of this manuscript. The research was supported with a grant from the UK National Environmental Research Council (GR3/12827).

References

Akoi, K. 1986. A stochastic-model of gene culture co-evolution suggested by the culture historical hypothesis for the evolution of adult lactose absorption in humans. *Proceedings of the National Academy of Sciences of the USA* 839: 2929-33.

Balasse, M., Bocherens, H., Tresset, A., Mariotti, A. and Vigne, J.D. 1997. Emergence of dairy production in the Neolithic? Contribution of isotopic analysis of cattle archaeological bones. *Comptes Rendus de l'Académie des Sciences Série 2 Fascicule A: Sciences de la Terre et des Planètes* 325: 1005-10.

Balasse, M., Bocherens, H. and Mariotti, A. 1999. Intra-bone variability of collagen and apatite isotopic composition used as evidence of a change of diet. *Journal of Archaeological Science* 26: 593-8.

Bodmer, W.F. and Cavalli-Sforza, L.L. 1976. *Genetics, Evolution and Man.* San Francisco: W.H. Freeman.

Bogucki, P. 1984a. The antiquity of dairying in temperate Europe. *Expedition* 28: 51-8.

Bogucki, P. 1984b. Ceramic sieves of the Linear Pottery culture and their economic implications. *Oxford Journal of Archaeology* 3: 15-30.

Bogucki, P. 1987. The establishment of agrarian communities on the north European plain. *Current Anthropology* 28: 1-24.

Cattaneo, C., Gelsthorpe, K., Phillips, P. and Sokol, R.J. 1993. Blood residues on stone tools – indoor and outdoor experiments. *World Archaeology* 25: 29-43.

Chapman, J.C. 1982. 'The secondary products revolution' and the limitations of the Neolithic. *University of London Institute of Archaeology Bulletin* 19: 107-22.

Charters, S., Evershed, R. P., Goad, L. J., Leyden, A.,

Blinkhorn, P.W. and Denham, V. 1993. Quantification and distribution of lipid in archaeological ceramics: implications for sampling potsherds for organic residue analysis and the classification of vessel use. *Archaeometry* 35: 211-23.

Child, A.M. and Pollard, A.M.. 1992. A review of the application of immunochemistry to archaeological bone. *Journal of Archaeological Science* 19: 39-47.

Craig, O.E., Mulville, J., Parker Pearson, M., Sokol, R.J., Gelsthorpe, K., Stacey, R. and Collins, M.J. 2000. Detecting milk proteins in ancient pots. *Nature* 408: 312.

Craig, O.E. 2000. The identification and detection of proteins in archaeological potsherds. Unpublished PhD thesis, University of Newcastle upon Tyne.

Craig, O.E. and Collins, M.J. 2000. An improved method for the immunological detection of mineral bound protein using hydrofluoric acid and direct capture. *Journal of Immunological Methods* 236: 89-97.

Craig, O.E. and Collins, M.J. 2002. The removal of protein from mineral surfaces: implications for residue analysis of archaeological materials. *Journal of Archaeological Science* 29: 1077-82.

Dahl, G. and Hjort, A. 1976. *Having Herds: pastoral herd growth and household economy.* Stockholm: University of Stockholm.

Downs, E.F. and Lowenstein, J.M. 1995. Identification of archaeological blood proteins - a cautionary note. *Journal of Archaeological Science* 22: 11-16.

Dudd, S.N. and Evershed, R.P. 1998. Direct demonstration of milk as an element of archaeological economies. *Science* 282: 1478-81.

Dudd, S.N., Regert, M. and Evershed, R.P. 1998. Assessing microbial lipid contributions during laboratory degradations of fats and oils and pure triacylglycerols absorbed in ceramic potsherds. *Organic Geochemistry* 295: 1345-54.

Dudd, S.N., Evershed, R.P. and Gibson, A.M. 1999. Evidence for varying patterns of exploitation of animal products in different prehistoric pottery traditions based on lipids preserved in surface and absorbed residues. *Journal of Archaeological Science* 26:1473-8.

Enomoto, A., Shon, D., Aoki, Y., Yamauchi, K. and Kamingawa, S. 1990. Antibodies raised against peptide fragments of bovine αs1-casein cross-react with the intact protein only when the peptides contain both B and T cell determinants. *Molecular Immunology* 27: 581-6.

Entwhistle, R. and Grant, A. 1989. The evidence for cereal cultivation and animal husbandry in the southern British Neolithic and Bronze Age. In A. Milles, D. Williams, and N. Gardner (eds) *The Beginnings of Agriculture.* Oxford: British Archaeological Reports (International Series) 496: 203-15.

Evershed, R.P. 1993. Biomolecular archaeology and lipids. *World Archaeology* 25: 74-93.

Evershed, R.P., Heron, C., Charters, C. and Goad, L.J. 1992. The survival of food residues: new methods of analysis, interpretation, and application. In A.M. Pollard (ed.) *New Developments in Archaeological Science.* Oxford: Oxford University Press for the British Academy. 187-208.

Evershed, R.P., Mottram, H.R., Dudd, S.N., Charters, S., Stott, A.W., Lawrence, G.J., Gibson, A.M., Conner, A., Blinkhorn, P.W. and Reeves, V. 1997. New criteria for the identification of animal fats preserved in archaeological pottery. *Naturwissenschaften* 84: 402-6.

Green, M.W. 1980. Animal husbandry at Uruk in the Archaic period. *Journal of Near Eastern Studies* 39: 1-35.

Greenfield, H.J. 1988. The origins of milk production in the Old World - a zooarchaeological perspective from the central Balkans. *Current Anthropology* 29: 573-93.

Greenfield, H.J. 1991. Fauna from the late Neolithic of the central Balkans - issues of subsistence and land-use. *Journal of Field Archaeology* 18: 161-86.

Grüss, J. 1933. Über Milchreste aus der Hallsattzeit und andere Funde. *Forschungen und Fortschrifte* 9: 105-6.

Gurfinkel, D.M. and Franklin, U.M. 1988. A study of the feasibility of detecting blood residue on artefacts. *Journal of Archaeological Science* 15: 83-97.

Hall, M. 1986. The role of cattle in southern African agropastoral societies. *South African Archaeological Society Goodwin Series* 5: 83-7.

Halstead, P. 1996. Pastoralism or household herding? Problems of scale and specialization in early Greek animal husbandry. *World Archaeology* 28: 20-42.

Halstead, P. 1998. Mortality models and milking: problems of uniformitarianism, optimality and equifinality reconsidered. *Anthropozoologica* 27: 3-20.

Harrison, R.J. 1985. The 'policultivo gandero' or secondary products revolution in Spanish agriculture, 5000-1000bc. *Proceedings of the Prehistoric Society* 51: 75-102.

Heron, C. and Evershed, R.P. 1993. The analysis of organic residues and the study of pottery use. *Archaeological Method and Theory* 5: 247-86.

Heron, C., Evershed, R.P. and Goad, L.J. 1991. Effects of migration of soil lipids on organic residues associated with buried potsherds. *Journal of Archaeological Science* 186: 641-59.

Holden, C. and Mace, R. 1997. Phylogenetic analysis of the evolution of lactose digestion in adults. *Human Biology* 69:605-28.

Holmes, W. 1970. Animals for food. *Proceedings of the Nutritional Society* 29: 273-81.

Ingold, T. 1980. *Hunters, Pastoralists and Ranchers.* Cambridge: Cambridge University Press.

Jenness, R. 1970. Protein composition of milk. In H.A. McKenzie (ed.) *Milk Proteins: chemistry and molecular biology. Volume I.* New York: Academic Press. 17-43.

Keswani, P.S. 1994. The social context of animal husbandry in early agricultural societies -

ethnographic insights and an archaeological example from Cyprus. *Journal of Anthropological Archaeology* 13: 255-77.

Leach, J.D. 1998. A brief comment on the immunological identification of plant residues on prehistoric stone tools and ceramics: results of a blind test. *Journal of Archaeological Science* 25: 171-5.

Leach, J.D. and Mauldin, R.P. 1996. Immunological residue analysis: results of recent archaeological and experimental studies. *Texas Journal of Science* 48: 25-34.

Legge, A.J. 1981a. The agricultural economy. In R. Mercer (ed.) *Grimes Graves, Norfolk: excavations 1971-72*. London: HMSO. 79-118.

Legge, A.J. 1981b. Aspects of animal husbandry. In R. Mercer (ed.) *Farming Practice in British Prehistory*. Edinburgh: Edinburgh University Press. 169-81.

Manning, A.P. 1994. A cautionary note on the use of hemastix and dot-blot assays for the detection and confirmation of archaeological blood residues. *Journal of Archaeological Science* 21: 159-62.

McCormick, F. 1992. Early faunal evidence for dairying *Oxford Journal of Archaeology* 11: 201-9.

McCormick, F. 1998. Calf slaughter as a response to marginality. In C.M. Mills and G. Coles (eds) *Life on the Edge: human settlement and marginality*. Oxford: Oxbow. 49-53.

Milisauskas, S. and Kruk, J. 1989. Neolithic economy in central Europe. *Journal of World Prehistory* 3: 403-46.

Mottram, H.R. and Evershed, R.P. 2001. Elucidation of the composition of bovine milk fat triacylglycerols using high-performance liquid chromatography-atmospheric pressure chemical ionisation mass spectrometry. *Journal of Chromatography A* 926: 239-53.

Nielsen, B.O., Mahler, V. and Rasmussen, P. 2000. An arthropod assemblage and the ecological conditions in a byre at the Neolithic settlement of Weier, Switzerland. *Journal of Archaeological Science* 27: 209-18.

Oudemans, T.M. and Boon, J.J. 1991. Molecular archaeology - analysis of charred food remains from prehistoric pottery by pyrolysis-gas chromatography mass-spectrometry. *Journal of Analytical and Applied Pyrolysis* 20: 197-227.

Parker Pearson M., Sharples, N. and Mulville, J. 1996. Brochs and Iron Age society: a reappraisal. *Antiquity* 70: 57-67.

Parker Pearson M., Sharples, N. and Mulville, J. 1999. Excavations at Dun Vulan: a correction. *Antiquity* 73: 149-52.

Payne, S. 1973. Kill-off patterns in sheep and goats. *Anatolian Studies* 23: 281-303.

Rasmussen, P. 1993. Analysis of goat/sheep feces from Egolzwil-3, switzerland - evidence for branch and twig foddering of livestock in the Neolithic. *Journal of Archaeological Science* 20: 479-502.

Regert, M., Bland, H.A., Dudd, S.N., van Bergen, P.F. and Evershed, R.P. 1998. Free and bound fatty acid oxidation products in archaeological ceramic vessels. *Proceedings of the Royal Society of London Series B - Biological Sciences* 265: 2027-32.

Reid, A. 1996. Cattle herds and the redistribution of cattle resources. *World Archaeology* 28: 43-57.

Rice, P.M. 1987. *Pottery Analysis: a sourcebook*. Chicago: University of Chicago Press.

Rowley-Conwy, P. 2000. Milking caprines, hunting pigs: the Neolithic economy of Arene Candide in its west Mediterranean context. In P. Rowley-Conwy (ed.) *Animal Bones, Human Societies*. Oxford: Oxbow Books. 124-32.

Sato, I. 1992. Detection of alpha s1-casein in vomit from bottle-fed babies by enzyme-linked-immunosorbent-assay. *International Journal of Legal Medicine* 105:127-31.

Sherratt, A.G. 1981. Plough and pastoralism: aspects of the secondary products revolution. In I. Hodder, G Isaac and N. Hammond (eds) *Patterns of the Past: studies in honour of David Clarke*. Cambridge: Cambridge University Press. 261-305.

Sherratt, A.G. 1983. The secondary exploitation of animals in the Old World. *World Archaeology* 15: 90-104.

Sherratt, A.G. 1997. *Economy and Society in Prehistoric Europe: changing perspectives*. Edinburgh: Edinburgh University Press.

Smith, P.R. and Wilson, M.T. 1992. Blood residues on ancient tool surfaces - a cautionary note. *Journal of Archaeological Science* 19: 237-41.

Tuross, N., Barnes, I., and Potts, R. 1996. Protein identification of blood residues on experimental stone tools. *Journal of Archaeological Science* 23: 289-96.

Wells, H.G. 1908. Thermo-stability of milk proteins. *Journal of Infectious Disease* 5: 449.

Whittle, A. 1996. *Europe in the Neolithic: the creation of new worlds*. Cambridge: Cambridge University Press.

Neolithic and Early Bronze Age 'food' from northern Greece: the archaeobotanical evidence

Soultana-Maria Valamoti

Department of Archaeology, University of Thessaloniki

Introduction

It was approximately a decade ago that the need to link palaeodietary remains to food was pointed out by Andrew Sherratt as 'People don't eat species, they eat meals' (Sherratt 1991: 221). A few years later, in a discussion of palaeodiet in the Aegean (Vaughan and Coulson 2000), two important issues were raised: the need for a thorough and systematic retrieval of food-related archaeological remains (Sarpaki 2000), and the need for an approach capable of treating 'consumption as a social practice and a signification code which maintains life and social relationships, produces, reproduces and signifies inequalities, constructs and deconstructs power and hierarchy' (Hamilakis 2000: 55). As far as the retrieval of archaeobotanical remains from Greek archaeological sites is concerned, some progress has since been made. A considerable body of archaeobotanical data has been generated together with an increased awareness for the usefulness of this kind of data, albeit locally confined depending on the availability of native specialists.

At the interpretive level, the emphasis has largely been placed on subsistence rather than on the social meaning of food. Approaching food within the social context of a given prehistoric society or culture is a difficult task involving the decipherment of whole sets of rules and behaviours through their material remains. Food *per se* is what is consumed, transformed into energy and feelings associated with its consumption, most of which are not likely to leave material traces. Plants, among other things, constitute an important element of food (including drinks, medicines and aromatic substances) not just as 'diet' but also as an essential part of the underlying structure and the communication system operating within a society. To what extent can archaeobotanical remains be used to approach prehistoric food within the framework defined by Hamilakis (2000: 55)?

This paper brings together the available archaeobotanical evidence from northern Greek Neolithic (seventh-fifth millennia B.C.) and Early Bronze Age sites (fourth-third millennia BC). It attempts to investigate the plant component of food (rather than resources) and the cultural messages this food might have conveyed. What follows is only a partial glimpse into the realm of food, in particular its plant component as revealed by charred plant remains. The remains of other foodstuffs (meat, fish, shellfish), food preparation facilities and food consumption equipment are other related categories of material culture which are only briefly considered here. Inevitably, the approach starts from the material, easily discernible, parameters of food and gradually moves to those less tangible but equally interesting aspects.

Available and used ingredients

A first step in our approach to Neolithic and Early Bronze Age food in northern Greece is the recognition of the ingredients available to those in charge of preparing and cooking meals. The long species lists found in published archaeobotanical reports suggest an impressive range of potential ingredients ranging from cereals, pulses, fruits, nuts and oil plants to medicinal and aromatic plants, and wild edible herbs (Kroll 1983; Valamoti 2001a). These, however, do not necessarily represent used foodstuffs as 'not everything edible constitutes food' (Hamilakis 2000: 57). This definition can be illustrated by, for example, the food practices of the aLuund of Zaire: 'foodstuffs become food only insofar as they are the result of reciprocity and exchange' (de Boeck 1994: 267). Thus, among the archaeobotanist's tasks is to discern which of those species that left their traces in the archaeological record were actually consumed as food, and which found their way into the archaeological deposits *via* other routes (*e.g.* dung, thatching) so as to separate potential from actual ingredients.

As we lack the 'reciprocated and exchanged' dimension of plant remains we resort to other lines of evidence, materialistic and utilitarian, yet informative. This can be done by identifying which species have been deliberately processed and used (*cf* Dennell 1976; Halstead 1994: 200). Needless to say, these species preserved by charring are only a fraction of those that either might have been eaten or used untreated in their vegetative state, or were unlikely to survive charring, or needed no contact with fire for their preparation (Dennell 1976: 231). Because we study material culture, our approaches are confined within the limits set by the archaeobotanical record: what has been deposited in areas other than those excavated, what has not been preserved, and what - because of the way it was prepared and consumed - is bound to evade our reconstructions.

For Neolithic and Early Bronze Age Greece a wide range of plant species appears to have been used, involving cereals, pulses, fruits, nuts, oil, medicinal and aromatic plants judging from archaeobotanical evidence from Neolithic and Bronze Age sites (Halstead 1994: 204-5; Valamoti 2001a). Most of the evidence originates from Late and Final Neolithic sites (5400-4500 and 4700-3100 BC), a temporal distribution reflecting, partly at least, the lack of a universal retrieval of plant remains from Greek prehistoric sites. Cereals and pulses dominate the archaeobotanical assemblages from northern Greece. A wide range of cereals was used; various species of wheat and barley include einkorn, emmer and bread/macaroni wheat, two and six-row barley, hulled and naked. In the Late Bronze Age (1700-1100 BC), the range of cereals found in store is broadened by the addition of two more species, spelt and millet.

Pulses are the constant companions of cereals throughout Neolithic, Bronze Age and historical times in Greece and many pulse species have been identified as crops at several Late Neolithic sites in northern Greece (Valamoti 2001a) including lentils, peas, bitter vetch and grass pea, while Celtic bean appears in the Early Bronze Age. The significance of the co-existence and complementary dietary role of cereals and pulses during prehistoric times has been pointed out by many scholars (Renfrew 1972: 300-1; Halstead 1981: 317; Sarpaki 1992: 74). Cereals are the starch providers that, together with the protein providers (pulses) 'and a few drops of olive oil', have formed the basis of Mediterranean cuisine over at least the last 3000 years. Although Neolithic and Early Bronze Age kitchens of northern Greece do not appear to have been honoured with the presence of olive oil (replaced perhaps by linseed oil or animal fat), the pairing of pulses and legumes emerges as the major component of Neolithic and Bronze Age diet in the region.

The range of Neolithic and Bronze Age ingredients is not exhausted by cereals and pulses. Flax is recorded as a crop from the Late Neolithic in northern Greece and it could well have been used for its edible seeds, oil and medicinal properties besides its use for linen production. Fruits and nuts add to the variety of the ingredients. Grapes, figs, blackberries, wild pears, acorns and terebinth nuts have been identified as separate concentrations, in some cases in storage contexts. New ingredients such as strawberries, opium poppy seed, gold of pleasure, medicago (Kroll 1983; Halstead 1994: 204-5), coriander (Renfrew 1973: 171, pl. 44) and an unidentified, probable oil plant (Valamoti 1999: 173) make their appearance during the Bronze Age.

The ingredients list for Neolithic and Bronze Age Greece thus points to a considerable variety, which seems to have increased during the Late Bronze Age. Although variety is the overriding pattern in terms of the available ingredients, a clear pattern emerges from the available evidence from Late Neolithic sites: glume wheats (emmer, einkorn) predominate over free-threshing cereals (bread/macaroni wheat, barley). Among the glume wheats, einkorn is the most commonly and most abundantly occurring species, compared to emmer and the new type recently identified (Jones *et al.* 2000), during the Late Neolithic (Valamoti 2001a). This predominance of einkorn over the other glume wheats appear to have continued until at least the Late Bronze Age (Kroll 1983: 116-17; Halstead 1994: 209). Compared to the emphasis on one cereal (einkorn), pulses appear to be stored in great variety even within single households. At the Dikili Tash houses (Koukouli *et al.* 1997), einkorn or barley were each stored together with at least two pulse crops while at nearby Arkadikos, a room contained only stored pulses. For the Late Neolithic, variety in the available 'grain ingredients' appears to have been emphasised through pulses rather than cereals.

From ingredients to recipes: reconstructing Neolithic and Bronze Age cuisine

A recipe is 'a statement of the ingredients and the procedure required for preparing cooked food' (Oxford English Dictionary). Do such recipes exist for the period we are investigating? For most of the ingredients listed in the previous section any reconstruction of recipes rests on the properties of the different species, the traditional ways of their preparation for consumption and a dash of imagination. Categories of foodstuffs and rules concerning their cooking and mixing are culture-specific, closely interwoven with the classification system of a given society (*cf* Bourdieu 1977: 129-30, 142-6, 157). The general categories (cereals, pulses *etc.*) into which ingredients are grouped and discussed correspond to our modern classification of botanical remains and foodstuffs, not necessarily perceived in the same way by our prehistoric counterparts, but necessary as a starting point for our analyses. Moreover, what among these ingredients was actually consumed as food or was rather intended as fodder is also culturally determined (*cf* Jones 1998: 96; Halstead 1990:152-3), and as ethnographic observations (Jones and Halstead 1995; Jones 1998: 96) and folklore (Megas 1988: 65, 68, 86) suggest, the boundary between food and fodder can be quite flexible, depending on availability of foodstuffs and the specific purposes of their consumption (*i.e.* ordinary or ceremonial). Acorns are a good example of a foodstuff largely despised and considered suitable for fodder nowadays, while it appears to have been widely appreciated as a food source in the recent and distant past (Mason 1995).

The mixing and food preparation possibilities of the ingredients presented above could be several and we are unable, at present, to associate cooking methods and prepared foods with Neolithic and Bronze Age worldviews. Cereals and pulses could have been used on their own or in combination, roasted, ground into coarse or fine meal, mixed with water, boiled or baked. Oil-rich seeds and nuts such as linseed, poppy seed, gold of pleasure, and terebinth nuts could have been used for their edible oil, which might have been an essential component of recipes. Fruits or aromatic/medicinal plants (*e.g.* grapes, terebinth, opium poppy) could have been added to food mixtures or used on their own. Fruits could be consumed dried or transformed into fermented beverages. Various toxin/tannin rich ingredients like grass pea, bitter vetch and acorns could have been processed in order to remove these undesirable substances through roasting, soaking or burying in the ground (Renfrew 1973: 154; Mason 1995: 19-22; Valamoti 1998a: 252-3; Hansen 2000: 19-23). Again, although these species are generally associated with animal fodder (*e.g.* Halstead and Jones 1989; Mason 1995) there are historic and ethnographic records of their consumption as food (Hodkinson 1988: 43; Renfrew 1973: 154; Mason 1995: 12-13).

The archaeobotanical contribution to the investigation of prehistoric food leaves out other foodstuffs such as meat, fish and sea food; these were also consumed but how often and in what combinations is at present hard to tell. In a recent isotopic analysis of skeletal remains from northern Greece, dating from Late Neolithic to Classical times, it is suggested that Neolithic populations relied heavily on plant food resources compared to populations of the Late Bronze Age (Triantafyllou 2001). This observation highlights the importance of archaeobotanical remains in reconstructions of Neolithic meals. The same study suggests that marine resources (fish and shellfish) made an insignificant contribution to the diet, despite abundant remains of marine molluscs and more restricted finds of fish bone.

Meat in Greece until recent times was a rare component of meals, its consumption marking special occasions in the annual cycle, related to religious festivals, celebrations such as weddings, or Sundays for those who could afford meat on a more regular basis. In the Classical period meat was consumed less frequently than other foodstuffs and the 'fresh meat of domestic animals formed a sacrifice, butchered with appropriate religious ritual' (Dalby 1996: 22-3). Whether human flesh, a 'food' attributed by Greeks to barbarians and to the 'wild' state of humanity (Vidal-Naquet 1983: 26-7, 384-6) - paralleling pork in taste (Dalby 1996: 61) - was consumed within the context of Neolithic and Bronze Age cultures of northern Greece, remains to be investigated in future studies of skeletal remains; such a practice would appear though, as the antipode of a Neolithic, plant-based diet.

Despite the lack of recipes for the majority of the available ingredients, some have survived the passage of time by sheer luck, brought to us through their charred remains. As will be seen, some of these recipes are more straightforward than others while some still require a little imagination to bring the ingredients together. Each of these recipes involves one of the main foodstuff categories mentioned above - cereals, pulses, fruits and aromatic/medicinal plants - and are presented in some detail below.

'Pligouri' (bulgur)

Pligouri - the Greek term for bulgur - represents a recipe and an ingredient at the same time. Bulgur-making is a recipe in itself, as it represents 'a statement of the ingredients and the procedure required for preparing cooked food'; for its preparation wheat grain (preferably emmer or durum wheat) is boiled, subsequently dried and, with the bran removed, is finally ground (Abdalla 1990: 30; Bayram 2000: 81-2). Despite the boiling of the grain (*i.e.* cooking) involved in bulgur preparation, the end product is dry ground grain, forming an ingredient for various meals: through simple soaking (mixing with water) or boiling, it is ready for consumption. Nowadays it is a widely used ingredient in Mediterranean cuisine as has been shown by a survey of traditional Mediterranean cooking (Rivera-Nuñez and Obon de Castro 1989). It is used instead of rice and also for the preparation of many dishes (Abdalla 1990: 33-5; Scott 1983: 120).

Pligouri can be regarded as one of the few direct pieces of evidence as regards prehistoric recipes from northern Greece as it has been identified from the Early Bronze Age deposits of the tell settlement of Mesimeriani (Grammenos and Kotsos in press) in the region of central Macedonia. Experimental charring of wheat grain treated with water, including bulgur fragments, helped identify the Bronze Age remains of wheat grain fragments as 'grain fragmented prior to charring and probably treated with hot water prior to charring' (Valamoti in press). Ground barley grain has also been identified from Archondiko (Papaefthymiou-Papanthimou and Pilali-Papasteriou 1995; Papaefthymiou-Papanthimou *et al.* 2002), a tell site in the region of western Macedonia, originating from Early Bronze Age deposits, though it is not at present possible to determine whether the fragments were originally loose or in the form of cakes (Valamoti in press).

Although grinding of cereals or other crops for feeding to animals is reported both ethnographically (Peña-Chocarro 1999: 36; Halstead and Jones 1989; P. Halstead pers. comm.) and in animal feeding handbooks (Morisson 1957), these archaeological finds have been interpreted as human food (Valamoti in press). This interpretation is based on:

a) the storage context of the finds and their high density suggesting stored processed cereals,
b) the timing of boiling (prior to storage) characterising the Mesimeriani find,
c) the absence of chaff from the mixture, unnecessary when glume-wheat grain is fed to animals, and
d) the laborious and time-consuming process of cracking grain with Bronze Age facilities (grinding stones and perhaps wooden mortars leaving no trace).

This last argument is based on the assumption that notions of time and effort in preparing animal feed were experienced in the same way in Early Bronze Age northern Greece as they are in our day, something which need not necessarily be the case. Whether Bronze Age bulgur was cooked/served with pulses or meat (as is done nowadays in Greece and elsewhere) is impossible to tell at present, though chemical analysis of pottery may, in the future, provide some clues (*cf* Tzedakis and Martlew 1999).

Cooking grass peas

Bitter vetch and grass pea are two pulses rich in toxins dangerous to humans (Cohn and Kislev 1987; Dalianis 1993). Although their repeated consumption as famine foods, attested both in the recent and Classical past, has degrading effects for human well-being (Halstead 1990; Hodkinson 1988), their use as food during the Neolithic cannot be ruled out (*cf* van Zeist 1988: 56). Bitter vetch is particularly toxic: Dioskourides, the ancientl Greek doctor, refers to the toxicity of bitter vetch when consumed by humans causing headaches, disturbance of the bowels and drawing of blood in the urine (Scarborough 1991: 158), while lathyrism, a condition caused by the consumption of grass pea, results in paralysis (Cohn and Kislev 1987). Leaching of the toxins by soaking or boiling and discarding the water prior to cooking (a method practiced for grass pea in modern Greece), however, renders these pulses suitable for consumption (bitter vetch: van Zeist and Bottema 1971; 1988; grass pea: Dalianis 1993). Removal of the *testa* (*i.e.* the outer coat of the seed where most of the toxins are present) is perhaps even more efficient and, nowadays, grass pea is sold in Greek supermarkets in this form (without *testa* and split). Cooking of grass pea in modern Greece involves a combination of the two steps recommended in scientific literature for the removal of toxins: removal of *testa* (*i.e.* the seed coat), brief boiling in water, disposal of the first water and subsequent cooking in water until it forms a smooth paste like pease pudding.

A fragment of such a recipe is reported from Middle Neolithic Servia, where finds of grass pea seeds in a collar-necked jar were interpreted as 'soaking in water to rid them of the poisonous component when the building set fire' (Hubbard 1979: 227). This interpretation is based on the even arrangement of seeds around the broken jar, considered unlikely to be generated when the dry contents of a falling or overturned pot are scattered. However, this 'arrangement' of seeds could have also been generated if the pot had been crushed *in situ* irrespective of whether water was present in the pot. Thus, this appears more as a potential rather than a confirmed recipe, and other ways to recognise soaking of pulse seeds need to be developed, for example by experimentally charring soaked and unsoaked pulse seeds and observing surface structure (Valamoti in preparation; *cf* experiments for bulgur identification: Valamoti in press). *Testa* removal and splitting of pulse seed, a practice associated with Late Bronze Age finds of Celtic bean (*Vicia faba*) from Thebes in central Greece (Jones and Halstead 1993), has not been recognised in any archaeobotanical finds from the north. The absence of *testa* from most charred remains of pulse seeds from Greece could be due to their deliberate removal in order to remove toxins (Hansen 2000: 21) or could be an outcome of charring.

Must or 'wine'

The extraction of juice from grapes is another recipe suggested from the remains of 'wine-pressings' found at Final Neolithic Dikili Tash and demonstrated to be such through experimental charring of fresh grapes, raisins and wine-pressings (Mangafa 1990; Mangafa *et al.* 1998: 166-8). Whether this juice was consumed as such, was reduced to some form of syrup through boiling, or was allowed to ferment into an alcoholic beverage like wine, is at present impossible to tell. 'Wine pressings' at Dikili Tash could represent stored by-products of grape juice extraction, or the remains of fermentation of must together with the 'wine-pressings' (*tsipoura* in Greek), a process that gives wine its red colour (Renfrew 1995: 263). Pottery analysis from the areas yielding the charred 'wine-pressings' may facilitate interpretation of the 'recipe' that generated the archaeobotanical find.

Flavourings and medicinal potions

Are we to imagine that the Neolithic and Early Bronze Age inhabitants of northern Greece 'enjoyed' a bland 'cuisine' whose meals lacked any flavourings from spices (including aromatic herbs)? Unfortunately little evidence is available in this respect, especially for the Neolithic. At the same time it should be borne in mind that aromatic plants of which the flowers or leaves are used are not likely to be preserved by charring, except perhaps for their stems if they are used in a dried form. Terebinth is so far the only aromatic/medicinal plant the use of which can be attributed to the Neolithic. The edible oil extracted from its nuts is aromatic and has medicinal properties (Gennadios 1997: 783) and its resin, tapped from its trunk was widely used in the Bronze Age (Knapp 1991:28; Artzy 1994; Haldane 1993) and in historical times. It might have been used as flavouring, as is nowadays done with mastic (resin of the lentisk tree, *Pistacia lentiscus*) in various parts of rural Greece where it is added to bread (Lambraki 1997: 255). Coriander seeds are reported from Early Bronze Age Sitagroi and they could have been used as a flavouring in meals and drinks (Renfrew 1973: 171, pl. 44). Opium poppy seeds are reported from Early Bronze Age Mandalo (Valamoti 2001a) but in numbers too small to indicate likely use of the species. Later on, during the Late Bronze Age, they make their appearance at Kastanas as significant concentrations of more or less pure seeds (Kroll 1983). Thus, terebinth nuts, coriander and opium poppy seeds could have been sprinkled on dishes or used to flavour alcoholic or other drinks.

The same species are also known for their medicinal properties (Bown 1995, Gennadios 1997). Terebinth essence and oil are used, for example, in treating bronchial, urinary and renal infections (Bown 1995: 331), and coriander seeds can be taken for minor digestive problems (*ibid.*: 267), while opium poppy is widely known for its analgesic properties (*ibid.*: 322). They could therefore have been used in medicinal preparations, together with a wide range of available wild plants with aromatic/medicinal properties, visible in our species lists (Valamoti 2001b) but in numbers too small or in 'inappropriate' contexts to assign them such a medicinal use. Besides plants that have a more or less clear medicinal use, foodstuffs also have medicinal properties and, although in our analytical classifications they widely feature as food, in our interpretations we need also consider their medicinal properties. Flax is one such example (linseed is used, for example, for curing constipation, pharyngitis and bronchial complaints; Gennadios 1997: 615-16; Bown 1995: 304-5) but the list could also include barley, figs, blackberries and wine.

In this respect, could the peculiar finds of a Neolithic pit at Makriyalos represent remains of the ingredients of medicinal potions? A characteristic example is represented in the finds from a pit associated with an oblong building with an apsidal end at Late Neolithic Makriyalos phase II (Pappa and Besios 1999). The pit contained concentrations of linseed, terebinth, fig, lentils and a wide range of wild plants including *Urtica* sp.,

Solanaceae, *Anagallis arvensis*, *Trigonella* sp., indet. Compositae, *Brassica rapa/nigra*, *Camelina* sp., *Medicago* sp. and *Hypericum* sp. The first four plants are edible and they could represent food remains thrown into the pit as refuse or cooking accidents. The wild plants could be interpreted as weeds. All the plant species identified - except the lentils -have, however, medicinal properties and have been widely used in traditional herbal medicine. *Anagallis arvensis* (scarlet pimpernel) was used as a diuretic and a treatment for fever and various other health problems (Bown 1995: 237; Gennadios 1997: 111); a mixture of flax seeds and *Brassica/Sinapis* seeds is recommended for chest complaints (Bown 1995: 305). Thus an alternative interpretation for some of the contents of this pit could be that they represent the remains of medicinal preparations related perhaps to a special (ritual?) function for this particular building (Valamoti 2001a).

Food for thought: deciphering the 'vegetative' component of cuisine

As stated in the introduction to this paper, food is much more than the material substance that enters a human body in order to satisfy its energy requirements. Food reflects the classification system operating in a society and is loaded with a set of meanings intrinsic to a society or culture. Its procurement, preparation and consumption reflect systems through which the world is perceived. Different kinds of food are eaten during different times of the year; food carries symbolic meaning and is used in this context (Bourdieu 1977: 129-30, 142-6, 157). Food forms a code of communication among members of a society and between different societies. It is not only good to eat but good to think (Lévi-Strauss 1962). These microscopic, often invisible plant remains incorporated in the excavated soil are the remains of daily human interaction and buried within them lie fragments of prehistoric thoughts, meanings and feelings which constitute worldviews of people who lived during Neolithic and Bronze Age times.

Deciphering what the specific patterns of food preparation, consumption and deposition, as reflected in the archaeobotanical record, may represent in terms of the Neolithic/Bronze Age worldview is a difficult task. Plant remains are only one of many material components related to food, thus their contribution to the study of food is only partial. Furthermore, and as is the case with all archaeological finds, when interpreting material culture we risk the danger of projecting into the past our own beliefs and perceptions. What follows is a first attempt to approach the 'good to think' component of food through the archaeobotanical remains available from northern Greece. It is based on the surviving ingredients and recipes which are used to explore cultural choices and identity as expressed through food; it is partial, speculative and probably subjective but necessary in order to move archaeobotany towards enquiry into this aspect of past life in the region.

From crop choices to food choices: archaeobotanical remains as signifiers of culture and identity

Cereal staples of the north

The predominance of glume wheats (and einkorn in particular) appears as one of the characteristics of Late Neolithic settlements, continuing, albeit less strongly, into the Bronze Age in northern Greece. Before this predominance of einkorn is interpreted in terms of food choice, taphonomic factors such as differential preservation, treatment and deposition of the various wheat species need to be considered. Processing by-products of free-threshing cereals (*e.g.* bread/macaroni wheat, barley) are more vulnerable to charring conditions (Boardman and Jones 1990), so these species could be under-represented; the same is not true, however, among stored products from which free-threshing wheat is absent. Likewise, it is unlikely that taphonomic biases are involved in the predominance of einkorn over emmer and new-type wheat, as all three are glume wheats and therefore require the same type of processing when intended for human consumption. Glume wheats differ from free-threshing cereals in that, in order to remove the grain from the husks surrounding them, an extra set of operations is required in addition to threshing, winnowing and sieving. These operations involve pounding of the spikelets and the removal of the husks either by sieving, winnowing or both (Hillman 1984; Peña Chocharro 1999).

The next question to ask is whether einkorn was used for human consumption. Chaff-rich deposits of glume wheat from Greek sites have been interpreted as refuse deposits, containing the by-products of dehusking glume wheat spikelets, a process unnecessary when these are used as fodder, and thus as the remains of food for people rather than animals (Valamoti 1995; 1998a). Recent experimental work has shown that whole spikelets (*i.e.* glume wheat grain with the glumes still on), when fed to goats, come out in dung pellets as glume bases (Charles and Valamoti in prep.). Thus finds of glume wheat chaff cannot be directly associated with human food. The possibility that einkorn was being fed selectively to animals, thus ending up in dung and subsequently used as fuel, needs to be considered.

Refuse and storage contexts may contain remains of food, fodder, or both. At Makriyalos, where a large part of the site has been excavated and where both food and fodder-rich refuse deposits would be likely to occur, einkorn dominates the assemblage. At Dikili Tash, the only wheat species found in store was einkorn. Also, glume wheats and einkorn in particular continue to be the dominant cereals in Bronze Age storage deposits in the region (Kroll 1983: tab. 19, 114, 117; Jones 1983; Halstead 1994: 209). It therefore seems unlikely that the majority of Neolithic and Bronze Age deposits, both refuse and storage, would be selectively containing remains of einkorn fodder; it is more likely that einkorn was indeed the cereal staple during the Late Neolithic and through to the Bronze Age in northern Greece.

Food can be used to unite and divide people within the same community, within communities of the same region and within different regions (Goody 1982): 'sharing the same food habits, eating the same staple, results in social integration and cohesion' (Hamilakis 2000: 58). Different staples have characterised different peoples throughout history. In antiquity, barley is associated with ancient Greek diet, and emmer wheat with Roman and Egyptian diet (Braun 1995). In our minds rice is the staple associated with people from east Asia, corn with America, and the combination of wheat, wine and olive oil with Mediterranean cultures. In this respect, cultures inhabiting northern Greece in the Late Neolithic and Bronze Age appear to have been einkorn growers. The Late Neolithic predominance of einkorn continued into the Bronze Age, suggesting a strong local tradition. Einkorn can be considered as the crop *par excellence* through which a cultural identity based on einkorn consumption might have been experienced in northern Greece during the Neolithic and Bronze Age. We do not know whether a particular einkorn food was also in general use alongside any variability in local culinary traditions; the finds of Early Bronze Age ground wheat and barley at two different settlements could suggest differences in cereal food preparation between different sites or could simply reflect different snapshots of daily life captured by the destructive effect of fire. For the Bronze Age settlements of Kastanas the predominance of einkorn is interpreted as a commodity exchanged with southerners bringing Mycenaean 'goodies' (Kroll 1983: 116-17); at Assiros einkorn dominated surplus, represented provision against famine and provided an agricultural surplus produced for the acquisition of exotic craft goods (Halstead 1994: 209, 211).

If einkorn was the dominant cereal staple in northern Greece, as the evidence seems to suggest, playing the role of a 'group identity signifier' (Hamilakis 2000: 58) for the Neolithic and Bronze Age inhabitants of the region, then emmer probably represented something else. We need to ask is how this 'otherness' of emmer was perceived within the context of Neolithic and Bronze Age consumption. The site of Mandalo (Kotsakis *et al.* 1989) clearly has more emmer in store than einkorn whereas at Makriyalos and Makri, emmer represents a small or negligible component of chaff dominated refuse, with only a few, spatially isolated exceptions. As we are talking of the same region, it is clearly not a case of different environments favouring different species and, in any case, both einkorn and emmer have a wide range of tolerances in terms of growth requirements and are very hardy species (Christidis 1963). It seems more likely therefore that emmer held a special position as a cereal among northern Greek settlements of the Late Neolithic.

Some tentative suggestions can be put forward, as working hypotheses awaiting further archaeobotanical work from a larger number of sites in the region: emmer may have been particular to certain settlements in northern Greece (*e.g.* Mandalo) perhaps belonging to emmer-growing cultures of the north, not yet recognisable in the archaeobotanical record. Such a regional distribution of the two species has been recognised for Neolithic Bulgaria, where einkorn predominates in the Strymon valley and north Bulgaria and emmer in the Thracian plain (E. Marinova pers. comm.). Alternatively, einkorn could represent the 'traditional' cereal of the region, initially grown by the first farmers in the area and continuing to be cultivated for millennia, whereas emmer was a crop of 'others', an 'exotic' crop accumulated or exchanged for status reasons, perhaps consumed only on certain occasions or by certain people.

Hansen argues that for the whole of Greece 'emmer wheat and two-row barley may be the more common cereal species' (Hansen 2000: 18) throughout the Neolithic and Bronze Ages. This picture is probably the result of an overrepresentation of sites located south of Macedonia and Thrace (47 out of a total of 54 sites; Hansen 2000: 17). Could this suggest the existence of einkorn growers in the north and emmer growers in the south, of groups culturally differentiated through the staples they relied upon? Is it then possible that the 'north-south divide' (Halstead 1994: 195, 209) as regards settlement pattern and subsistence, may also be visible in allegiance to different cereal staples? The evidence at present is limited to only a few sites systematically and extensively investigated in the north for archaeobotanical remains, so generalisations are dangerous for the time being. In order to evaluate the 'otherness' of emmer we need adequate data from areas further north, south and east with which mainland Greece had cultural contacts. At present any answer to the question would be premature but it is important to address the issue and carefully collect and examine the necessary evidence. Such a task falls beyond the scope of this contribution but it is certainly interesting and should be examined together with other remains of material culture - pottery in particular - which are closely related to all aspects of food preparation and consumption.

Despite the long tradition of einkorn in northern Greece, newcomers became established during the course of the Bronze Age. The new cereal species that made their appearance were spelt wheat and millet. Spelt, being a glume wheat like einkorn and emmer, was probably not as 'foreign' as millet, which is different from the other cereals both in its appearance (small and round) and in its growth cycle (short). Both crops have been identified from archaeological deposits north and northeast of Greece and dated earlier than the Greek finds: spelt is found in Transcaucasia during the fifth millennium BC (Zohary and Hopf 1994: 52) while millet occurs in east and central Europe also during the fifth millennium BC (Zohary and Hopf 1994: 80). Millet is reported only once from a Late Neolithic context at the site of Olynthos (Mylonas 1929). Interestingly enough, it was found as the contents of a miniature pot, implying a special, perhaps non-utilitarian function, or the precious little seeds of an unknown crop brought from elsewhere in order to try and grow for the first time. Isotope analysis of human bone assemblages from northern Greece suggests a heavier reliance on C_4 plants with millet being 'the only obvious candidate' (Triantafyllou 2001). Bread wheat, a free-threshing cereal whose presence is very limited during the Neolithic, is found stored at Late Bronze Age Assiros.

These crops could signify contacts with new cultures already using these species, the adoption of new production and consumption practices, and the use of new recipes and cooking methods during the Bronze Age. Storage at Assiros is attributed a redistributive character, provisioning against famine (Jones *et al.* 1986; Halstead 1994: 209), and the presence of these new crops could reinforce differences between an 'elite' in charge of redistribution and 'ordinary' people, the former controlling access to the new crops and possessing the know-how for their cultivation and preparation. To test this hypothesis archaeobotanical evidence from a large number of Bronze Age settlements, of different status in regional hierarchies, is obviously needed.

The role of pulses

Pulses, as mentioned earlier, are considered as the companion of cereals for Neolithic and Bronze Age Greece. According to some authorities, however, a temporal trend towards an increase in the importance of pulses is discernable in the Late Neolithic or Early Bronze Age (Hansen 2000: 18; Demoule and Perlès 1993; Blouet 1986: 138). This opinion has been taken a step further and clay cooking vessels (*i.e.* with evidence that they had been placed on the fire), which appeared during the Middle Neolithic and became more common during the Late Neolithic (Perlès and Vitelli 1999: 98), have been associated with the cooking of pulses (Vitelli 2000: 11). This is an interesting suggestion but the archaeobotanical evidence is inadequate, at present, for such a connection between pottery and charred seeds to be made. At least for northern Greece, very little is known archaeobotanically from sites earlier than the Late Neolithic (one is Early Neolithic - Nea Nikomedeia [van Zeist and Bottema 1971] - and the other is a Middle Neolithic site - Servia [Hubbard 1979]). Thus the variety in stored pulses observed for the Late Neolithic might or might not have had parallels in the earlier periods; we simply cannot tell. Besides, when comparisons of the archaeobotanical material are attempted between periods, care should be taken to compare similar types of contexts.

A comparison of refuse-context, chaff-rich deposits with storage-context, grain-rich deposits would probably yield crop differences that might have had little to do with crop choices, reflecting differences in use and discard patterns of different crops. As far as food preparation methods are concerned, there is no real necessity for cooking pots to have been preferentially associated with pulses. As discussed earlier, some pulses need processing in the form of boiling or soaking for the removal of toxins, but soaking does not require cooking vessels that rest on the fire and boiling also applies to cereals as these too need cooking before eating. Thus the association of pulses and cooking vessels is not self-explanatory and, as far as the archaeobotanical evidence is concerned, it is also premature.

Vicia faba is probably a Bronze Age newcomer, as it has not been found as a crop earlier than the Bronze Age (Halstead 1994: 204-5). In northern Greece it occurs as a crop only at the site of Mandalo in Early Bronze Age deposits (Valamoti 2001a). The consumption of *Vicia faba* seeds can act against malaria (Katz 1987: 142-3; Hansen 2000: 23) and might have been used for this purpose, though at present there is no bioarchaeological evidence to indicate that malaria was a disease in northern Greece during these times.

The great variety in pulses observed during the Late Neolithic (*cf* Dikili Tash where each household stored a variety of pulses) seems to decrease in the Bronze Age where only bitter vetch at Kastanas and bitter vetch with lentils at Assiros have been identified as separate crops. A possible interpretation for the limited representation of pulses on the Bronze Age sites may be related to the function of the storerooms at Assiros and the purposes that the stored produce served. At Assiros storage might have focused on staples like glume wheats, emergency crops like millet and a fodder crop like bitter vetch; the other pulse species may have represented perhaps minor crops cultivated by individual households and thus are not included in the crop repertoire of the large storerooms at Assiros. Such a range of crops probably re-enforces the 'redistributive' character attributed to Assiros (Jones *et al.* 1986): it resembles the situation in the Bronze Age palaces of southern Greece where the archives are concerned with staples and do not mention pulses which are, however, found in the non-palatial sector (Halstead 1992). Kastanas, however, without the Assiros storerooms, with a different spatial organisation and probably occupying a different position in the regional hierarchy (Andreou *et al.* 1996: 581), also demonstrates the same pattern in terms of pulses. At present, the evidence is too limited and it would be unwise to generalise from only two Bronze Age sites, extensively sampled for archaeobotanical remains, and argue for changes in the position of pulses, as crops and consumed foodstuffs, during the course of the Bronze Age.

Wine

If einkorn played the unifying role in northern Greece among the cereal staples, was there an equivalent among the drinks consumed? Ancient Egyptians drank beer, ancient Greeks drank wine and nowadays beer is associated with northwestern and central Europe and wine with Mediterranean cultures. Such regional differences may have existed in Neolithic and Bronze Age times, yet the evidence needed to recognise them in the archaeological record is inadequate at present. Drinking of wine in southern Greece during the Bronze Age (Hamilakis 1996; 1999) is suggested from a range of botanical, artefactual and epigraphic evidence. In northern Greece, drinking is mainly inferred from pottery finds: handled vessels for dispensing and consuming liquids, probably inspired by metal originals and interpreted as 'drinking sets' (Sherratt 1987: 85-6) appear at Sitagroi IV (3500-3100 BC; Renfrew 1986); other 'drinking sets' characterise the Late Bronze Age (Kiriatzi *et al.* in press; Wardle *et al.* in press).

What was drunk out of these 'drinking sets' in the north? The possible candidates that come to mind are wine, beer and mead (*cf* Sherratt 1987). Mead, which is based on honey, has been identified from chemical analysis of residues in Minoan and Mycenaean pottery (Tzedakis and Martlew 1999) no such remains have been reported from northern Greece. Wine is generally associated with the grape and beer with barley, though other fermented drinks could have been produced from other fruits (*e.g.* figs; Gorny 1995: 155) and cereals (*e.g.* wheat and millet; Samuel 2000: 547, 557). Thus the ingredients necessary for beer production in Neolithic and Bronze Age Greece were widely available. Whether the taste for such a product was there is impossible to tell in the absence of finds of sprouted barley grain or residues of beer-making like those found in Egypt (Samuel 2000). The beverage consumed in the Late Bronze Age drinking sets is thought to have been wine (Kiriatzi *et al.* in press; Wardle *et al.* in press), a suggestion further supported by the archaeobotanical evidence: charred wine pressings from morphologically cultivated grapes were recovered at the site of Toumba Thessalonikis (Mangafa and Kotsakis 1996).

This Late Bronze Age tradition seems to have its roots at least in the Final Neolithic of northern Greece as the evidence from Dikili Tash suggests: the juice of grapes was definitely being extracted during the Final Neolithic (end of the fifth millennium BC) at Dikili Tash and the next step to the production of a fermented beverage, although not directly documented, can safely be inferred. How widespread its production was is hard to tell, and it may or may not have been a well-kept secret known only to certain communities, such as that of Dikili Tash.

Although drinking sets of the Late Bronze Age have been studied in detail (Kiriatzi *et al.* in press; Wardle *et al.* in press), very little or nothing is known about the Early Bronze Age and the Late Neolithic. Some liquid containers were, from a technological point of view, used for the storage of liquids other than water, as pottery analysis at Late Neolithic Makriyalos suggests, but whether their contents had been some form of wine is far from clear.

The particulars of wine consumption in northern Greece during the Late Neolithic and Early Bronze Age may be as yet unknown, but for the Late Bronze Age wine consumption and related pottery forms have been approached from different perspectives: a local-style symposium during which alcohol is consumed from a communal bowl is gradually replaced by a Mycenaean-style symposium during which alcohol is consumed from individual cups (Wardle *et al.* in press). According to Kiriatzi *et al.* (in press), the consumption of wine was performed by a small part of the population or on special occasions of consumption only while later on, when Mycenaean pottery dominates, wine and food 'consumption episodes' were aimed towards maintaining group cohesion. Wine consumption seems to have dominated the drinking scene during the Late Bronze Age in northern and southern Greece alike, albeit in different drinking sets and perhaps in different contexts. In this case, wine is a foodstuff that unifies different cultures of the Aegean (and beyond) but the particulars of its consumption may well have differed from region to region and through time, acting as signifiers of intra- and intercultural differences.

The wild and domesticated boundary of foodstuffs

In ancient Greek thought a polarity between the cultivated and the wild is first observed in the most ancient literary texts: in Hesiod and Homer, civilised man is a cultivator, a bread-eater, as opposed to the cannibal and the wild (Vidal-Naquet 1983: 26-7). In Homer, the opposition between the civilised and the wild refers to an opposition between the cultivated and wild countryside (*ibid.*: 32). According to Vidal-Naquet, hunting in ancient Greek myth is the first level of the rupture between humanity and the wild condition, but hunting is also the wild side of man (*ibid.*: 28). Beyond humanity lies the realm of gods, cannibals, and non-cultivators which Ulysses abandons when he returns to Ithaca, to the land of cereals (*ibid.*: 62). Thus, in certain societies of cultivators, the distinction between wild and domesticated food may be culturally significant.

Wild fruits such as terebinth, acorns and wild pears were the diet of young male Persians in antiquity when they were sent out to tend the flocks, probably as a rite of passage; the diet was different to that which they received at home which comprised bread, barley cakes, cardamom and roasted or boiled meat (Sancisi-Weerdenburg 1995: 287-8). For the Neolithic we place our hopes for approaching such social dimensions through the study of material culture. Let us examine in what ways plant remains can be instructive in this respect.

For Greece as a whole, it has been argued that the increasing frequency of wild resources, both plant and animal, towards the end of the Neolithic and during the Bronze Age, suggests the gradual isolation of the household 'and the concomitant erosion of obligations to share in favour of rights to store' (Halstead 1999: 86). The archaeobotanical record from northern Greece, at least for the Late Neolithic (as earlier periods are not adequately represented), indicates the use of wild fruits and nuts (Valamoti 2001a). An emphasis on wild animals, greater than usual for the period, is also observed in the area (*e.g.* the sites of Makri [Curci and Tagliacozzo 1998] and Kryoneri [Mylona 1999]). This suggests the lack of an overriding pattern in terms of food choices and may indicate a variability in the ways different groups of people incorporate the wild and the domesticated in their worldview and consumption habits.

Approaching perceptions of the 'wild' and the ways it is incorporated by 'domesticated' societies is extremely difficult as they cannot be directly inferred from the material remains that constitute the archaeological record. Wild plant foodstuffs were used by Neolithic, and Early Bronze Age communities living in northern Greece, and they might or might not have been perceived as such. This would depend on where the line between wild and domesticated was drawn during these periods. Acorn-producing oak trees, blackberry bushes, pear and fig trees, as wild plants (not systematically husbanded/cultivated) might have represented part of the wild woodland vegetation. Alternatively they could have corresponded to the 'domesticated' landscape, as selected oak trees in the vicinity of fields, as bushes and trees along field fences, and therefore would not have been perceived as wild. Whether considered as part of the wild or domesticated world, harvests of acorns, figs, pears and blackberries are found together with cereals and pulses in storage contexts, suggesting that the boundary (if it existed at all) was removed once these foodstuffs entered the house.

The use of, and close contact with, one at least of these wild plants, however, had the power to gradually transform it, leading to its incorporation into the domain of the 'domesticated'. This is the case of the vine, an interesting and instructive one in this respect. Grapes were presumably initially gathered from the wild as the common (but in small numbers) occurrence of morphologically wild grape pips suggests (Renfrew 1995, Valamoti 1998b). Then, in the Late Neolithic, we are faced not only with large concentrations of charred grape pips but, at the site of Dikili Tash, also with what has experimentally been shown to represent wine pressings (Mangafa *et al.* 1998). A Neolithic beverage from grapes, perhaps wine, was in all likelihood being produced at the site but, as for the ingredients, did they originate from the domain of the wild or the domesticated? The archaeobotanical evidence (grape-pip measurements) suggests wild grapes (Mangafa and Kotsakis 1996) clearly distinguishable from domesticated examples dating to the Bronze Age. Although morphology of grape pips and type of vine exploitation are nowadays clearly distinguishable in most cases (but not always as the pinot noir grapes suggest; Nunez and Walker 1989: 214) this need not have been the case in the Neolithic.

The Neolithic pips, although morphologically wild according to our perceptions, may have been derived from plants that were treated as part of the 'domesticated' world, husbanded in a way that might not have immediately resulted in morphologically domesticated pips. Alternatively, the morphologically domesticated grapes and wine produced from them, attested from the Late Bronze Age, could still embody a meaning related to the original wild condition of the plant. Such a process of negotiation between the wild and the domesticated aspects of nature, invisible from our botanical classifications of wild and domesticated plants, may be reflected in ancient Greek mythology: Dionysus, god of trees, wild vegetation, wine and ecstatic behaviour (including consumption of raw animal and human flesh), becomes, according to some versions of the myth, the son of Demeter, goddess of cereals and cultivation (Kakridis 1986: 207).

Food in hospitality and ritual

Consideration of modern Greek folklore, away from the 'westernised' lifestyle of the cities (of which the author is a 'victim'), shows that life is enriched by a series of symbolic and ritual activities closely linked to food production and consumption. Although nowadays this ritual is closely interwoven with Greek Orthodox religion it seems to date to pre-Christian rituals as old as Classical times at least (Megas 1988, Nilsson 1961). Specific foods prepared for religious celebrations marking certain times of the year are consumed in feasts involving the whole village community (Megas 1988: 20, 93, 243). It would therefore be naïve to think of Neolithic and Bronze Age life as lacking in this respect.

The cultural significance of food consumption has been emphasised by many scholars and attempts to link material culture with consumption 'habits' and cultural change have been made (*e.g.* Sherratt 1987; 1991; 1995; Hamilakis 1996; 1999). Some categories of material culture related to food, such as pottery, may be more amenable to a decipherment of their symbolic meaning as they are more likely to preserve it in a durable form through time. Although food is the main constituent of feasts and ceremonies, food remains *per se* are often approached in terms of subsistence in the sense of calorific value, storage life and risk-buffering potential. This may partly result from the fact that food acquires its symbolic meaning through its manipulation within a specific context, a meaning that may be lost in the refuse or storage context of the archaeological find.

Neolithic and Bronze Age food consumed in this part of the world has been associated with a certain sophistication in relation to preparation, serving and consumption (Sherratt 1991: 226-8). Pots are flat-based and have elaborate decoration during the Neolithic, while miniature tables indicate the presence of such furniture, and these two aspects together are regarded as suggestive of tableware (Sherratt 1991: 226; Halstead 1999: 80). Changes in pottery decoration and the location of hearths (indoors or outdoors) have been linked to changing patterns of food consumption. The shift, according to Halstead (1999: 80-1), is:

a) from 'public' sharing of cooked food during the Early and Middle Neolithic, when cultural values were placed on hospitality, to

b) a restriction of sharing to immediate neighbours and the less inclusive nature of Late Neolithic hospitality, followed by

c) a further restricted sharing between households during the Final Neolithic/Early Bronze Age.

The evidence used for this interpretation focuses on the location of cooking facilities and on pottery characteristics. The shift is from cooking facilities in open spaces (Early-Middle Neolithic) to cooking facilities in enclosed domestic areas (Late Neolithic) followed by cooking facilities in enclosed yards (Final Neolithic/Early Bronze Age); and from fine, skilfully decorated ceramic vessels with continuous design fields (Early/Middle Neolithic) to pottery decoration with bounded panels (Late Neolithic) to a rarity of fine decorated table ware (Final Neolithic/Early Bronze Age; Halstead 1999: 80).

Are we then to imagine Early and Middle Neolithic hospitable villages, each with its tableware, giving way to less hospitable neighbourhoods which, in turn, evolved to even less hospitable households? Of course hospitality and the parties involved are socially defined and can vary between different societies of different periods; needless to say, the organisation of food preparation, presentation and consumption is closely linked to this. On the other hand, indoor and outdoor cooking facilities are present throughout the Neolithic at least and, although settlement segmentation may be indicated by ditches and yard walls, hospitality and sharing might still have been there in the same degree but within a different framework. At Early Bronze Age Archondiko, food was prepared within individual households, as the numerous ovens and hearths found in houses suggest (Papaefthymiou-Papanthimou *et al.* 2002). This is not equivalent to not sharing food or sharing among close family only. Whether food is prepared inside or outside may depend on the organisation of a given society and the meaning assigned to the inside and the outside; this can be closely related to various annual cycles (*e.g.* cooking, women's activities, daily rhythm) defined within a framework of 'pertinent oppositions' (Bourdieu 1977: 144, 147, 149, 157).

One could argue that the more people become aware of boundaries (*e.g.* through ditches and walls), within or between households or neighbourhoods, the more people become aware of sharing, and the more emphasis hospitality/sharing acquires as an institutionalised social interaction: while food is being cooked in an open yard it is still attractive to the curious neighbour through its smell, but there is a wall to cross in order to approach this food and probably rules regulating who crosses which boundaries. Thus, 'hospitality' should be more clearly recognised in the archaeological record as 'household isolation' (Halstead 1999: 86) evolves during the course of the Neolithic.

The problem lies in what can be defined as the remains of hospitality. Pottery is one obvious candidate but different meanings can be extracted from it depending on the approach: the transformation of pottery from Neolithic to Bronze Age times from fine, decorated and non-utilitarian into utilitarian, non-cooking vessels does not necessarily imply the 'end' of hospitality; it may reflect its replacement by impressive newcomers: metal vessels (*cf* Sherratt 1987; Halstead 1994; Perlès and Vitelli 1999). Assuming that decorated pottery places the emphasis on the container and food presentation, and that elaborate decorated pottery may signify formal occasions of commensality (Halstead 1999: 80), the question of what meanings were being conveyed through food consumption (equal access to food and food sharing or social hierarchy and inequality) remains open. Presence and absence of decoration on pottery may not necessarily correspond respectively to the antithetic couples defined by Halstead (1999): a) communal and hospitable; b) private and inhospitable. Such antithetic couples are likely to have characterised communities both of the Neolithic and the Bronze Age. The understanding of how they were experienced by people and expressed through material culture still requires more data and the close parallel investigation of archaeological context, spatial organisation, distribution of finds, use of space, *etc.*

As was mentioned in a previous section, the Bronze Age saw a greater variety of foodstuffs and the introduction of new species, at least from the Late Bronze Age onwards. This could imply the adoption of new recipes and new cooking methods, or an incorporation of the new ingredients to old ways of food preparation, a shift of emphasis from the traditional einkorn to other, more 'exotic' cereals such as spelt, free-threshing wheat and millet. New ways of cooking food have, in fact, been suggested on the basis of skeletal studies in the region (Triantafyllou 2001). This increased variety of ingredients and possible recipes and dishes contradicts interpretations of an introverted 'within the household' pattern of consumption inferred from the absence of decorated pottery. This contradiction is further underlined by the clear evidence for drinking, probably of wine, from the Final Neolithic onwards, since alcohol consumption probably accompanied occasions of feasting or hospitality. Whether this consumption was meant to bring the community together or emphasise social differences or do both, is not possible to say.

Likewise, the idea of tables and tableware (Sherratt 1991: 226) and their association with an emphasis on hospitality (Halstead 1999: 80) should be considered with some caution; pots need to have a base in order to balance securely, be it standing on earthen floors, wooden surfaces or tables. The ethnographic account concerning the introduction of the table in recent peasant communities is very instructive in this respect. Greek peasant communities of the nineteenth and early twentieth centuries ate from a communal food container, placed on the earthen floor, or on mats, or wooden surfaces placed on the floor; for a long time tables, in the form we know them nowadays, were kept, if kept at all, as a decorative object (Skouteri-Didaskalou 1999). Neolithic miniature representations of tables therefore do not necessarily reflect daily food consumption - or even communal festive consumption - or occasions requiring the expression of hospitality. They might have performed a special function similar to that attributed to Early Neolithic pots which, with their probably non-domestic utilitarian use and low levels of production, have been regarded as the products of 'ritual healers', used for medicinal and other ritual potions (Perlès and Vitelli 1999: 103).

Technological and stylistic changes in pottery taking place from the Middle Neolithic onwards are seen as reflecting changes in the role of 'potter/healers' and in the contexts within which rituals relating to pottery and potion making were performed (Perlès and Vitelli 1999: 104). Thus consumption of pots and their contents could have been highly ritualised but gradually lost its ''mystique' and ritual power' (Perlès and Vitelli 1999: 104) and, in the Final Neolithic, 'the role of pottery is... in the ordinary domestic sphere' (Perlès and Vitelli 1999: 105). Food consumption must have been an essential part of both the ritual and secular spheres of Neolithic and Bronze Age societies, though we cannot tell where the boundary between the two was drawn.

It is now time to return to plants and the archaeobotanical record and see how they relate to theories about ritual 'consumption' of potions. This is deep water as nearly all archaeobotanical evidence from the area - and from Greece as a whole - has largely been approached from a utilitarian point of view: species represent crops good to cultivate, good to store, and good to eat; burnt remains are burnt fuel, cooking accidents or rubbish. A quick examination of Greek folklore involving the use of foodstuffs that constitute elements of archaeobotanical assemblages is very instructive, revealing the symbolic dimension of plants, food and their ritual use.

'Three days before Christmas, on Saint Anastasia's day, they prepared certain sweets made of wheat grains, a little flour, raisins, sugar, walnuts, figs and cinnamon which they distributed to the households. Before doing that they would throw upwards three spoonfuls under the chimney (for the evil Christmas spirits called 'Skarkantzali')' (Megas 1988: 43).

On Saint Varvara's day, a mixture of boiled grains is taken to church for the evening mass. This is done for protection against *evlogia* (smallpox). From the blessed grains they take up to 40 items and pierce them to form a rosary (*komboloi*), which they place above the head of Saint Varvara (on the household's icon) hanging it from the same nail as the icon. There it stays all year as an amulet, so that the disease does not enter the house (Megas 1988: 37). In many places offerings of olive oil and/or wine are made to the fireplace, remnants of the ancient worship of the fireplace (*ibid.*: 55). Wheat grain is involved in divination in Greek folklore: people take wheat grains and throw them on the hot surface of the *parastia* (hearthside). From the way these grains jump or the way they char it is concluded whether a person will live or emigrate. They then throw these grains into the village fountain or well (*ibid.*: 58). A symbolic act which aims to increase the prosperity of the household during the forthcoming year is carried out on New Year's Day and involves throwing grain in the yard, onto the front doorstep (*katofli*) or inside the house (*ibid.*: 60, 70, 75).

Various species are attributed special meanings and are involved in certain rituals. The pomegranate, for example, symbolises fertility and abundance and is thus broken open on the plough before sowing, or on the doorstep on New Year's Day (Megas 1988: 30, 76). Pulses in general (Megas 1988: 34), and *Vicia faba* (Katz 1987: 150) in particular, have a close connection to souls and death, a situation also observed in many circum-Mediterranean cultures (ancient Greek, Roman and Egyptian); in ancient Greece they were avoided by certain groups of people (*e.g.* the Pythagoreans; Katz 1987: 150). Similar symbolism might have been attributed to *Vicia faba* or other plants during Neolithic and Bronze Age times, though at present such a function cannot be inferred from the archaeobotanical remains.

Hardly any archaeobotanical evidence has been presented in relation to ritual activities and symbolic meaning, but this does not imply that these aspects were not experienced. Rather, it reflects our tendency to interpret the archaeobotanical record in utilitarian terms. Of course it would be wrong to overlook this major function of plant food, *i.e.* the satisfaction of daily energy requirements. This other, symbolic dimension of food, however, should not be overlooked and we need to become open-minded enough to allow space for other interpretations of the archaeobotanical record.

A characteristic example is represented by the finds from a pit at Makriyalos, for which a possible medicinal/ritual function may be assigned (see above). A few metres away from this pit, another was found containing, together with einkorn chaff, seeds of *Hypericum* and *Juncus*. These species could have been grazed, included in dung, burnt as fuel and discarded in the pit. There is also, however, the 'ritual alternative': Greeks of Delvino in southern Albania light hearths (or open fires) on which they throw a flower they call St John's wort (probably *Hypericum perforatum*) and, in the evening, when goats return to the village, they make them jump over the 'hearth' in the same way as themselves, as fire cleanses and expels all evil from people and their animals alike (Megas 1988: 223). Could then these wild seeds found in only very few of the numerous Makriyalos pits represent the remains of such activities? And what about the numerous cases of edible seeds that are found among the remains of ashes interpreted usually as grain accidentally charred during cooking? Could they have been charred not so accidentally, as the ethnographic example of burning grain in divination suggests? These archaeological examples are cited not to suggest that archaeobotanical remains should be seen solely as part of ritual activities; they only aim to bring out the possibility that such activities may also be involved in the formation of the archaeobotanical record and therefore should be considered alongside daily, ordinary activities.

Concluding remarks

Food, ways of preparing it, tastes, medicinal and other potions are likely to have changed through time, alongside changes in contexts of consumption, settlement patterns, pottery shapes, decoration *etc.* Despite the relatively recent interest in the retrieval of archaeobotanical material and the increase in the number of archaeobotanists working on Greek material, the available data are still unevenly distributed in space and time for meaningful comparisons to be made. Nevertheless, some patterns are now discernible and can cautiously be outlined, more as working hypotheses rather than certainties.

Einkorn appears to be the preferred cereal staple in the north (although not at every single site), perhaps playing a culturally unifying role in the region from the Neolithic though to the Bronze Age. In the Bronze Age, tradition goes hand in hand with the adoption of new cereal species. Ground cereals and the know-how for the preparation of some type of bulgur were available at least from the Early Bronze Age onwards, and dishes that contained a wheat equivalent to modern parboiled rice might well have been prepared. Fruits, nuts and aromatic and medicinal plants were used and wine was probably produced from the Final Neolithic onwards, marking a close relationship with the vine, though the exact timing of cultivation replacing collection from the wild remains unknown.

A Neolithic meal might have included an einkorn 'dish', accompanied by a legume soup or 'fava'-type meal, a little 'wine' perhaps and some fruit, though whether all were eaten together or separately, at different times of the day or different times of the year, is impossible to tell. By comparison, Bronze Age cuisine might have been more varied in the range of possible ingredients and foodstuffs prepared from them. These are general trends attributed to the whole of northern Greece, but it is very likely that variability might have existed in the food habits among different communities inhabiting the region, despite the apparently similar crop range. This possibility is reflected in the diverging patterns of plant/animal exploitation demonstrated from archaeobotanical assemblages of five Late Neolithic sites from northern Greece (Valamoti 2001a). Clearly a lot more work needs to be done before patterns of food consumption can be reconstructed in detail.

Archaeobotanical data provide just one more line of evidence in the investigation. They are essential, however, as they represent the direct remains of what was consumed and therefore potentially carry within them valuable evidence for our approaches and interpretations of food consumption. For their meanings to be extracted it is imperative that archaeobotany become a mainstream practice - rather than a fancy accessory whose selection depends on the good will of excavators – and for sampling strategies to be as detailed and context-related as possible. It is also essential that practitioners of this discipline become aware of the multiple levels of meaning that our objects of study convey.

Acknowledgements

I wish to thank Mike Parker Pearson for offering me the opportunity to write this paper and approach my data from a slightly different perspective; it may be significant that the offer occurred in a drinking consumption context in a place called the 'Revolution'. The paper is largely based on data included in my PhD thesis and I should therefore like to express my gratitude to Glynis Jones and Mike Charles, my supervisors, and also to Paul Halstead, for their constant support, ideas and critical comments. I am deeply grateful to Eleonora Skouteri-Didaskalou, for guiding me through the relevant anthropological literature on food and generously providing articles as well as her university teaching notes; and to Ken Wardle and Vangelio Kiriatzi for kindly providing articles on Late Bronze Age drinking.

References

Abdalla, M. 1990. Bulgur - an important wheat product in the cuisine of contemporary Assyrians in the Middle East. In H. Walker (ed.) *Oxford Symposium on Food and Cookery 1989: staple foods*. London: Prospect Books. 27-37.

Andreou, S., Fotiadis, M. and Kotsakis, K. 1996. Review of Aegean prehistory V: the Neolithic and Bronze Age of northern Greece. *American Journal of Archaeology* 100: 537-97.

Artzy, M. 1994. Incense, camels and collared rim jars: desert trade routes and maritime outlets in the second millennium. *Oxford Journal of Archaeology* 13: 121-47.

Bayram, M. 2000. Bulgur around the world. *Cereal Food World* 45: 80-2.

Blouet, B. 1986. Development of the settlement pattern. In C. Renfrew, M. Gimbutas and E.S. Elster (eds) *Excavations at Sitagroi: a prehistoric village in northeast Greece, 1*. Monumenta Archaeologica 13. Los Angeles: University of California Press. 133-43.

Boardman, S. and Jones, G. 1990. Experiments on the effects of charring cereal plant components. *Journal of Archaeological Science* 17: 1-11.

Boeck, F. de 1994. 'When hunger goes around the land': hunger and food among the aLuund of Zaire. *Man* 29: 257-82.

Bourdieu, P. 1977. *Outline of a Theory of Practice*. Cambridge: Cambridge University Press.

Bown, D. 1995. *The Royal Horticultural Society Encyclopedia of Herbs and their Uses*. London: Dorling Kindersley.

Braun, T. 1995. Barley cakes and emmer bread. In J. Wilkins, D. Harvey and M. Dobson (eds) *Food in Antiquity*. Exeter: University of Exeter Press. 25-37.

Christidis, V. 1963. *Χειμωνιάτικα σιτηρά* (Winter cereals). Thessaloniki.

Cohn, D.F. and Kislev, M.E. 1987. Human neurolathyrism. *Adler Museum Bulletin* 13: 5-8.

Curci, A. and Tagliacozzo, A. 1998. Preliminary analysis of the fauna of Makri. In N. Efstratiou *et al*, Excavations at the Neolithic settlement of Makri, Thrace, Greece (1988-1996): a preliminary report. *Saguntum* 31: 45-7.

Dalby, A. 1996. *Siren Feasts: a history of food and gastronomy in Greece*. London: Routledge.

Dalianis, K. 1993. *Ψυχανθή για Καρπό και Σανό* (Leguminosae for seed and hay). Athens: Stamoulis Publications.

Demoule, J.P. and Perlès, C. 1993. The Greek Neolithic: a new review. *Journal of World Prehistory* 7: 355-416.

Dennell, R. 1976. The economic importance of plant resources represented on archaeological sites. *Journal of Archaeological Science* 3: 229-47.

Gennadios, P. 1997 [1914]. *Lexikon Phytologikon* (Phytological Lexicon). Athens: Trochalia Publications.

Goody, J. 1982. *Cooking, Cuisine and Class: a study in comparative sociology*. Cambridge: Cambridge University Press.

Gorny, R.L. 1995. Viticulture in ancient Anatolia. In P.E. McGovern, S.J. Flemming and S.H. Katz (eds) *The Origins and Ancient History of Wine*. Luxembourg: Gordon and Breach. 133-74.

Grammenos, D. and Kotsos, S. In press. Ο προϊστορικός

οικισμός της Μεσημεριανής Τούμπας (The prehistoric settlement of Mesimeriani Toumba). *Μακεδονικά*.

Haldane, C. 1993. Direct evidence for organic cargoes in the Late Bronze Age. *World Archaeology* 24: 348-60.

Halstead, P. 1981. Counting sheep in Neolithic and Bronze Age Greece. In I. Hodder, G. Isaac and N. Hammond (eds) *Pattern of the Past: studies in honour of David Clarke*. Cambridge: Cambridge University Press. 307-39.

Halstead, P. 1990. Waste not, want not: traditional responses to crop failure in Greece. *Rural History* 1 (2): 147-64.

Halstead, P. 1992. Agriculture in the Bronze Age Aegean: towards a model of palatial economy. In B. Wells (ed.) *Agriculture in Ancient Greece*. Stockholm: Swedish Institute at Athens. 105-17.

Halstead, P. 1994. The north-south divide: regional paths to complexity in prehistoric Greece. In C. Mathers and S. Stoddart (eds) *Development and Decline in the Mediterranean Bronze Age*. Sheffield Archaeological Monograph 8. Sheffield: J.R. Collis Publications. 195-219.

Halstead, P. 1999. Neighbours from hell? The household in Neolithic Greece. In P. Halstead (ed.) *Neolithic Society in Greece*. Sheffield: Sheffield Academic Press: 77-95.

Halstead, P. and Jones, G. 1989. Agrarian ecology in the Greek islands: time stress, scale and risk. *Journal of Hellenic Studies* 109: 41-55.

Hamilakis, Y. 1996. Wine, oil and the dialectics of power in Bronze Age Crete: a review of the evidence. *Oxford Journal of Archaeology* 15: 1-32.

Hamilakis, Y. 1999. Food technologies/technologies of the body: the social context of wine and oil production and consumption in Bronze Age Crete. *World Archaeology* 31: 38-54.

Hamilakis, Y. 2000. The anthropology of food and drink consumption and Aegean archaeology. In S.J. Vaughan and W.D.E. Coulson (eds) *Palaeodiet in the Aegean*. Oxford: Oxbow Books: 55-63.

Hansen, J.M. 2000. Palaeoethnobotany and palaeodiet in the Aegean region: notes on legume toxicity and related pathologies. In S.J. Vaughan and W.D.E.Coulson (eds) *Palaeodiet in the Aegean*. Oxford: Oxbow Books: 13-27.

Hillman, G. 1984. Traditional husbandry and processing of archaic cereals in recent times. The operations, products, and equipment which might feature in Sumerian texts, part I: the glume wheats. *Bulletin on Sumerian Agriculture* 1: 114-52.

Hodkinson, S. 1988. Animal husbandry in the Greek polis. In C.R. Whittaker (ed.), Pastoral economies in Classical antiquity. *Cambridge Philological Society Supplementary Volume* 14: 35-74.

Hubbard, R.N.L.B. 1979. Ancient agriculture and ecology at Servia. *Annual of the British School at Athens* 74: 226-8.

Jones, G. 1983. The use of ethnographic and ecological models in the interpretation of archaeological plant remains: case studies from Greece. Unpublished PhD thesis, University of Cambridge.

Jones, G. 1998. Distinguishing food from fodder in the archaeobotanical record. In M. Charles, P. Halstead and G. Jones (eds), Fodder: archaeological, historical and ethnographic studies. *Environmental Archaeology* 1: 95-8.

Jones, G. and Halstead, P. 1993. An early find of 'fava' from Thebes. *Journal of the British School at Athens* 88: 103-4.

Jones, G. and Halstead, P. 1995. Maslins, mixtures and monocrops: on the interpretation of archaeobotanical crop samples of heterogeneous composition. *Journal of Archaeological Science* 22: 103-14.

Jones, G., Valamoti, S. and Charles, M. 2000. Early crop diversity: a 'new' glume wheat from northern Greece. *Vegetation History and Archaeobotany* 9: 133-46.

Jones, G., Wardle, K., Halstead, P. and Wardle, D. 1986. Crop storage at Assiros. *Scientific American* 254 (3): 96-103.

Kakridis, I. (ed.) 1986. *Helliniki Mythologia, Hoi Theoi* (Greek Mythology, The Gods), *Volume 2*. Athens: Ekdotiki Athinon.

Katz, S.H. 1987. Fava bean consumption: a case for the co-evolution of genes and culture. In M. Harris and E.B. Ross (eds) *Food and Evolution: towards a theory of food habits*. Philadelphia: Temple University Press. 133-59.

Kiriatzi, E., Andreou, S. and Psaraki, K. In press. 'Old wine in new cups?' Pottery and the politics of drinking in LBA central Macedonia. Paper presented at the Sheffield Aegean Round Table on Food, January 2001.

Knapp, A.B. 1991. Spice, drugs, grain and grog: organic goods in Bronze Age east Mediterranean trade. In N.H. Gale (ed), Bronze Age Trade in the Mediterranean. *Studies in Mediterranean Archaeology* 90: 21-68

Kotsakis, K., Papanthimou, A., Pilali-Papasteriou, A., Savopoulou, P., Maniatis, Y. and Kromer, B. 1989. Carbon 14 dates from Mandalo, W. Macedonia. In Y. Maniatis (ed.) *Archaeometry: proceedings of the 25th International Symposium*. Amsterdam: Elsevier. 679-85.

Koukouli-Chrysanthaki, Ch., Treuil, R. and Malamidou, D. 1997. Προϊστορικός οικισμός Φιλίππων «Ντικιλί Ταş»: δέκα χρόνια ανασκαφικής έρευνας (Prehistoric settlement of Philippoi 'Dikili Tash': ten years of excavation research). *ΑΕΜΘ* 10β: 681-704.

Kroll, H.J. 1983. *Die Pflanzenfunde. Kastanas: Ausgrabungen in einem Siedlungshugel der Bronze und Eisenzeit Makedoniens 1975-1979*. Berlin: Spiess.

Lambraki, M. 1997. *Τα Χόρτα* (Greens). Athens: Trochalia Publications.

Lévi-Strauss, C. 1962. *La Pensée Sauvage*. Paris: Plon.

Mangafa, M. 1990. The plant remains from the Late Neolithic/Early Bronze Age site of Dikili Tash, Macedonia, Greece. Unpublished MSc dissertation, University of Sheffield.

Mangafa, M. and Kotsakis, K. 1996. A new method for

the identification of wild and cultivated charred grape seeds. *Journal of Archaeological Science* 23: 409-18.

Mangafa, M., Kotsakis, K. and Andreou, S. 1998. Αμπελοκαλλιέργεια στην προϊστορική Μακεδονία: τα δεδομένα της προϊστορικής Τούμπας Θεσσαλονίκης (Viticulture in prehistoric Macedonia: the evidence from Toumba Thessalonikis). In *Αμπελοοινική Ιστορία στο Χώρο της Μακεδονίας και της Θράκης* (Viticulture in Macedonia and Thrace). Athens: ETBA Publications. 158-69.

Mason, S. 1995. Acornutopia? Determining the role of acorns in past human subsistence. In J. Wilkins, D. Harvey and M. Dobson (eds) *Food in Antiquity*. Exeter: University of Exeter Press. 12-23.

Megas, G.A. 1988. *Ελληνικές Γιορτές και Έθιμα της Λαϊκής Λατρείας* (Greek Festivals and Customs of Folk Worship). Athens: Odysseas Publications.

Mylonas, G.E. 1929. *Excavations at Olynthus. I, The Neolithic Settlement*. Baltimore: Johns Hopkins University Press.

Nilsson, M. 1961. *Greek Folk Religion*. New York: Harper.

Nunez, D.R. and Walker, M. 1989. A review of palaeobotanical findings of early Vitis in the Mediterranean and of the origins of cultivated grapevines, with special reference to new pointers to prehistoric exploitation in the western Mediterranean. *Review of Palaeobotany and Palynology* 61: 205-37.

Pappa, M. and Besios, M. 1999. The Neolithic settlement at Makriyalos, northern Greece: preliminary report on the 1993-1995 excavations. *Journal of Field Archaeology* 26: 177-95.

Papaefthymiou-Papanthimou, A. and Pilali-Papasteriou, A. 1995. Ανασκαφή στο Αρχοντικό Γιαννιτσών (Excavations at Archondiko Giannitson). *AEMΘ* 6: 151-6.

Papaefthymiou-Papanthimou, A., Pilali-Papasteriou, A., Basogianni, D., Papadopoulou, E., Tsagaraki, E. and Fappas, I. 2002. Αρχοντικό 2000: τυπολογική παρουσίαση και ερμηνευτικά προβλήματα των πηλόκτιστων κατασκευών (Archondiko 2000: typological presentation and interpretive problems concerning the clay structures. *AEMΘ* 14: 421-33.

Peña-Chocarro, L. 1999. *Prehistoric Agriculture in Spain: the application of ethnographic models*. Oxford: British Archaeological Reports (International Series) 818.

Perlès, C. and Vitelli, K.D. 1999. Craft specialisation in the Neolithic of Greece. In P. Halstead (ed) *Neolithic Society in Greece*. Sheffield: Sheffield Academic Press. 96-107.

Renfrew, C. 1972. *The Emergence of Civilisation: the Cyclades and the Aegean in the third millennium BC*. London: Methuen.

Renfrew, C. 1986. The Sitagroi sequence. In C. Renfrew, M. Gimbutas and E.S. Elster (eds) *Excavations at Sitagroi: a prehistoric village in northeast Greece, 1*. Monumenta Archaeologica 13. Los Angeles: University of California Press. 147-74.

Renfrew, J.M. 1973. *Palaeoethnobotany: the prehistoric food plants of the Near East and Europe*. New York: Columbia University Press.

Renfrew, J.M. 1995. Palaeoethnobotanical finds of *Vitis* from Greece. In P.E. McGovern, S.J. Flemming and S.H. Katz (eds) *The Origins and Ancient History of Wine*. Luxembourg: Gordon and Breach. 255-67.

Rivera-Nuñez, D. and Obon de Castro, C. 1989. La dieta cereal prehistorica y su supevivencia en el area mediterranea. *Trabajos de Prehistoria* 46: 247-54.

Samuel, D. 2000. Brewing and baking. In P.T. Nicholson and I. Shaw (eds) *Ancient Egyptian Materials and Technology*. Cambridge: Cambridge University Press. 537-76.

Sancisi-Weerdenburg, H. 1995. Persian food stereotypes and political identity. In J. Wilkins, D. Harvey and M. Dobson (eds) *Food in Antiquity*. Exeter: University of Exeter Press. 286-302.

Sarpaki, A. 1992. The palaeoethnobotanical approach: the Mediterranean triad or is it a quartet? In B. Wells (ed.) *Agriculture in Ancient Greece*. Stockholm: Swedish Institute at Athens. 62-76.

Sarpaki, A. 2000. The study of palaeodiet in the Aegean: food for thought. In S.J. Vaughan and W.D.E. Coulson (eds) *Palaeodiet in the Aegean*. Oxford: Oxbow Books. 115-21.

Scarborough, J. 1991. The pharmacology of sacred plants, herbs, and roots. In Ch.A. Faraone and D. Obbink (eds) *Magika Hiera: ancient Greek magic and religion*. Oxford: Oxford University Press. 138-74.

Scott, D. 1983. *Traditional Arab Cookery*. London: Rider.

Sherratt, A. 1987. Cups that cheered. In W.H. Waldren and R.C. Kennard (eds) *Bell Beakers of the Western Mediterranean*. Oxford: British Archaeological Reports (International Series) 331(i). 81-114

Sherratt, A. 1991. Palaeoethnobotany: from crops to cuisine. *Paleoecologia & Arquelogia* 2: 221-36.

Sherratt, A. 1995. Sacred and profane substances: the ritual use of narcotics in later Neolithic Europe. In P. Garwood, D. Jennings, R. Skeates and J. Toms (eds) *Sacred and Profane: proceedings of a Conference on Archaeology, Ritual and Religion*. Oxford: University Committee for Archaeology. 50-64.

Skouteri-Didaskalou, E. 1999. Από το σοφρά στο τραπέζι (From 'sofra' to table). In *The Greek World between East and West, 1453-1981: proceedings of the 1st European Conference on Modern Greek Studies, Berlin, October 1998*. Athens: Hellinika Grammata. 533-59.

Triantafyllou, S. 2001. *A Bioarchaeological Approach to Prehistoric Cemetery Populations from Western and Central Greek Macedonia*. Oxford: British Archaeological Reports (International Series) 976.

Tzedakis, Y. and Martlew, H. (eds) 1999. *Minoans and Myceneans: flavours of their time*. Athens: Kapon Editions.

Valamoti, S.M. 1995. Γεωργικά προϊόντα από το νεολιθικό οικισμό Γιαννιτσά Β: μία προκαταρκτική προσέγγιση μέσω των αρχαιοβοτανικών δεδομένων. (Agricultural products from the Neolithic settlement of Yiannitsa B: a preliminary approach by means of archaeobotanical data). *AEMΘ* 6: 177-84.

Valamoti, S.M. 1998a. Η επεξεργασία της τροφής κατά τη Νεολιθική και την Εποχή του Χαλκού: τα αρχαιοβοτανικά δεδομένα. (The processing of foods during the Neolithic and Bronze Age: the archaeobotanical data). In *Ancient Greek Technology: proceedings of the 1st International Conference, Thessaloniki, 1997.* Thessaloniki: Society for Macedonian Studies and Museum of Thessaloniki. 249-57.

Valamoti, S.M. 1998b. Η εκμετάλλευση της αμπέλου στο χώρο της Μακεδονίας και της Θράκης κατά τη Νεολιθική και την Πρώιμη Εποχή του Χαλκού. (The exploitation of the vine in Macedonia and Thrace during the Neolithic and Early Bronze Age). In *Αμπελοοινική Ιστορία στο Χώρο της Μακεδονίας και της Θράκης (*Viticulture in Macedonia and Thrace*).* Athens: ETBA Publications: 137-49.

Valamoti, S.M. 1999. Αρχοντικό 1997: προκαταρκτικά δεδομένα από την αρχαιοβοτανική έρευνα στον οικισμό (Archondiko 1997: preliminary results from the archaeobotanical research at the settlement). *AEMΘ* 11: 172-75.

Valamoti, S.M. 2001a. Archaeobotanical investigation of Late Neolithic and Early Bronze Age agriculture and plant exploitation in northern Greece. Unpublished PhD thesis, University of Sheffield.

Valamoti, S.M. 2001b. Φαρμακευτικά και αρωματικά φυτά στη Μακεδονία και τη Θράκη κατά τη Νεολιθική και την Εποχή του Χαλκού: τα αρχαιοβοτανικά δεδομένα (Medicinal and aromatic plants in Macedonia and Thrace in the Neolithic and Bronze Age: the archaeobotanical evidence). In *Φαρμακευτικά και Αρωματικά Φυτά* (Medicinal and Aromatic Plants). Athens: ETBA Publications. 23-38.

Valamoti, S.M. In press. Food remains from Bronze Age Archondiko and Mesimeriani Toumba in northern Greece. *Vegetation History and Archaeobotany.*

Vaughan, S.J. and Coulson, W.D.E. (eds) 2000. *Palaeodiet in the Aegean.* Oxford: Oxbow Books.

Vidal-Naquet, P. 1983. *Le Chasseur Noir* (Greek translation). Athens: Livanis-Nea Synora Publications.

Vitelli, K.D. 2000. Observations on palaeodietary research in the Aegean. In S.J. Vaughan and W.D.E. Coulson (eds) *Palaeodiet in the Aegean.* Oxford: Oxbow Books. 10-12.

Wardle, K.A., Wardle, D. and Wardle, N.M.H. In press. The symposium in Macedonia: a prehistoric perspective. *AEMΘ.*

Zeist, W. van 1988. Some aspects of early Neolithic plant husbandry in the Near East. *Anatolica* 15: 49-67.

Zeist, W. van and Bottema, S. 1971. Plant husbandry in Early Neolithic Nea Nikomedeia, Greece. *Acta Botanica Neerlandica* 20: 524-38.

Zohary, D. and Hopf, M. 1994. *Domestication of Plants in the Old World.* Oxford: Oxford University Press.

Changing paradigms: food as a metaphor for cultural identity among prehistoric fisher-gatherer-hunter communities of northern Europe

Liliana Janik

Department of Archaeology, University of Cambridge

Introduction

Food is explored in this paper as a metaphor in the definition of cultural identities among prehistoric fisher-gather-hunters of northeastern Europe. It has often been assumed within the adaptive ecological paradigm of prehistoric food procurement that choices of what to eat depended on what was provided by 'nature' rather than on any cultural preferences in food consumption. An underlying assumption of such interpretations is that natural environment catastrophe leads to food shortage. In this light, prehistoric communities are seen to adapt by fishing, gathering and hunting all available food sources. I shall argue, however, that we should shift our paradigm for fisher-gatherer-hunter societies to look at possibilities other than economic utilisation of animal or plant resources.

Animals, fish and plants as food can be seen as components of wider cultural categorisations of what was collected, hunted and fished. Various types of plants, fish and animals provide the principal categories of food for fisher-gatherer-hunters as well as for farming communities and represent an explicit set of remains recoverable as archaeological data. Are, for example, all types of edible nuts present in the natural environment gathered and incorporated in the diet or do culturally governed choices direct what is gathered and whether its residues are deposited in recoverable archaeological contexts?

To answer these questions of food as metaphor in defining cultural identities I propose to look at the archaeological data from various perspectives. The issue of scale - general models *versus* a finer-grained contextual approach - is used here to explore, firstly, the overall use of animals and plants as categories of food. Secondly, I shall compare three separate regions to illustrate the various scales of contextual analysis of different archaeological cultures and economies. Thirdly, I shall concentrate on the way in which different interpretive paradigms may permit us to add a cultural dimension to food procurement strategies in mid-Holocene Europe.

Contemporary categorisations of food are based on the ways in which we classify our surrounding world. Being part of one or another cultural tradition means that, if we are Chinese, we might not eat carnivores whereas if we are European, we might favour the food of carnivores or omnivorous animals; what we eat and how we eat it is an obvious expression of cultural tradition. As we move to the smaller scale, different groups, communities and nations celebrate the important events of their lives with food which becomes an independent entity in marking the passage of time (through seasonal food and associated food events), in cultural events and religious activities, as well as at very private moments in our lives. The social life of every community is mapped through interactions between individuals with food taking the form of an independent mediator in signifying such relationships. Food as a cultural category transcends labelling as economy or mode of production. Within any single economic formation it is the food that distinguishes people. French and Germans or Dutch and Italians share the same economy but their food is different. In the case of Poland, for example, economic changes since the socialist era have made Poland today one of the most successful economies in Europe and yet the cherished food preferences of former days are still the same. I would propose that it is not so much the type of economy that is important but cultural preferences; therefore, when looking at the past one should keep this in mind when studying food procuring, producing or consuming.

To state that food is a cultural category sounds almost trivial but, when we put it into the context of prehistoric fisher-gatherers-hunters in northern Europe, we find little evidence for such a statement from the existing literature unless utilised in ethnographic analogies. On the other hand, calorific food value is a concept often used by archaeologists interpreting these prehistoric communities; when we think about ourselves, however, we do not postulate that we might eat maggots or insects as a highly calorific alternative to other foods. It is culturally unacceptable for Europeans and yet, if we go to other parts of the world, we may be offered such delicacies. In other words, our economic evaluations of food value in prehistoric non-farming communities may overshadow the issue of food as a cultural category.

Chronology

Archaeological periods are treated here merely as a means of periodisation rather than as evolutionary phases relating to changes in food acquisition systems. The change from food procuring to food producing occurred late in the southeastern Baltic Sea region, taking place in the Iron Age or later. There was a short introduction and abandonment of domesticated animals and plants in the Early Bronze Age as one of several strategies of food-getting (Janik 1998). The domesticates, however, did not play an important part in terms of the amounts of food produced in this way (*ibid.* 1998). The data presented here derive from a period spanning three climatic periods, namely the Boreal, Atlantic and Sub-Boreal, from *c.* 9000 BC to 2000 BC (Table 11.1).

Archaeological period	Uncalibrated dates BC	Climatic period
Early Mesolithic	c.7600 – 5800	Boreal
Middle Mesolithic	c.5800 – 5400	Boreal
Late Mesolithic	5400 – 4600	Atlantic
Early Neolithic	c.4600 - c.3300	Atlantic
Middle Neolithic	c.3300 - c.2200/2100	Sub-Boreal
Late Neolithic	c.2200/2100 - c.1800/1600	Sub-Boreal
Early Bronze Age	c.1800/1600	Sub-Boreal

Table 11.1. *Relationship between archaeological periods, uncalibrated dates and climatic periods in northeastern Europe.*

Symbol	Archaeological Period
EM	Early Mesolithic
MM	Middle Mesolithic
LM	Late Mesolithic
EN	Early Neolithic
MN	Middle Neolithic
LN	Late Neolithic
EB	Early Bronze Age
/	indicates interval between two particular periods, *e.g.* LM/EN period between Late Mesolithic and Early Neolithic

Table 11.2. *Abbreviations for archaeological periods used in the tables.*

The southeastern Baltic Sea region: its cultural materials, archaeological record and locational features

Human occupation of the southeastern Baltic Sea region (Figure 11.1) began after the northern retreat of the ice in the early Holocene. The archaeological remains from the Mesolithic and Neolithic period in this region form part of the Kunda and Niemen traditions, which are based on the extensive use of bone and antler tools as well as microlithic flint tools.

The inhabitants of Mesolithic and Neolithic settlements primarily exploited inland resources despite living close to the coast. Although marine resources were exploited, occupation was concentrated around inland lakes, streams and rivers. This is illustrated by the Kunda or Narva-town settlements. Other settlements were located further inland (Figure 11.2) and their inhabitants were not able to exploit the marine environment. This pattern is apparent throughout the period discussed in this paper. The extensive exploitation of inland resources is one of the features particular to the southeastern Baltic Sea region in comparison to other Baltic or North Sea regions such as Denmark or Scania.

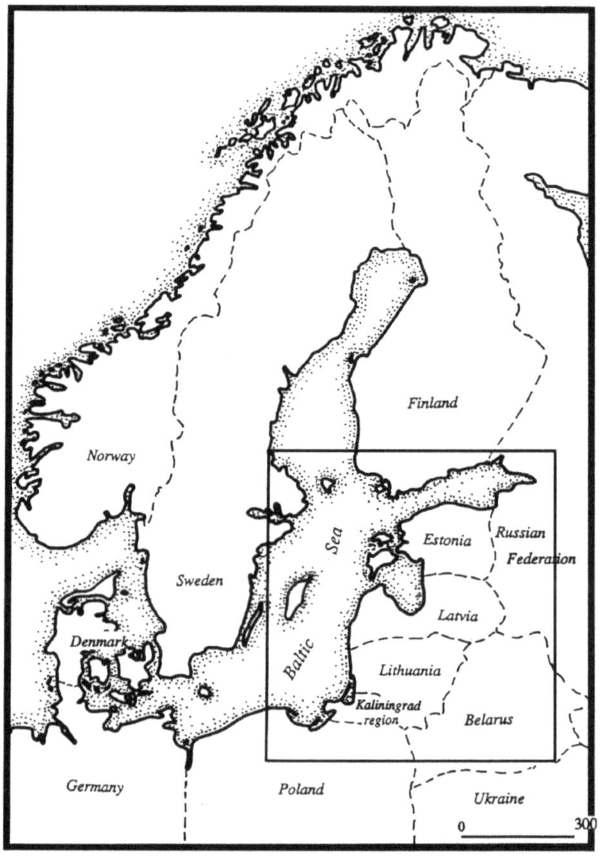

Figure 11.1. *Northern Europe, showing the outline of the southeastern Baltic Sea region.*

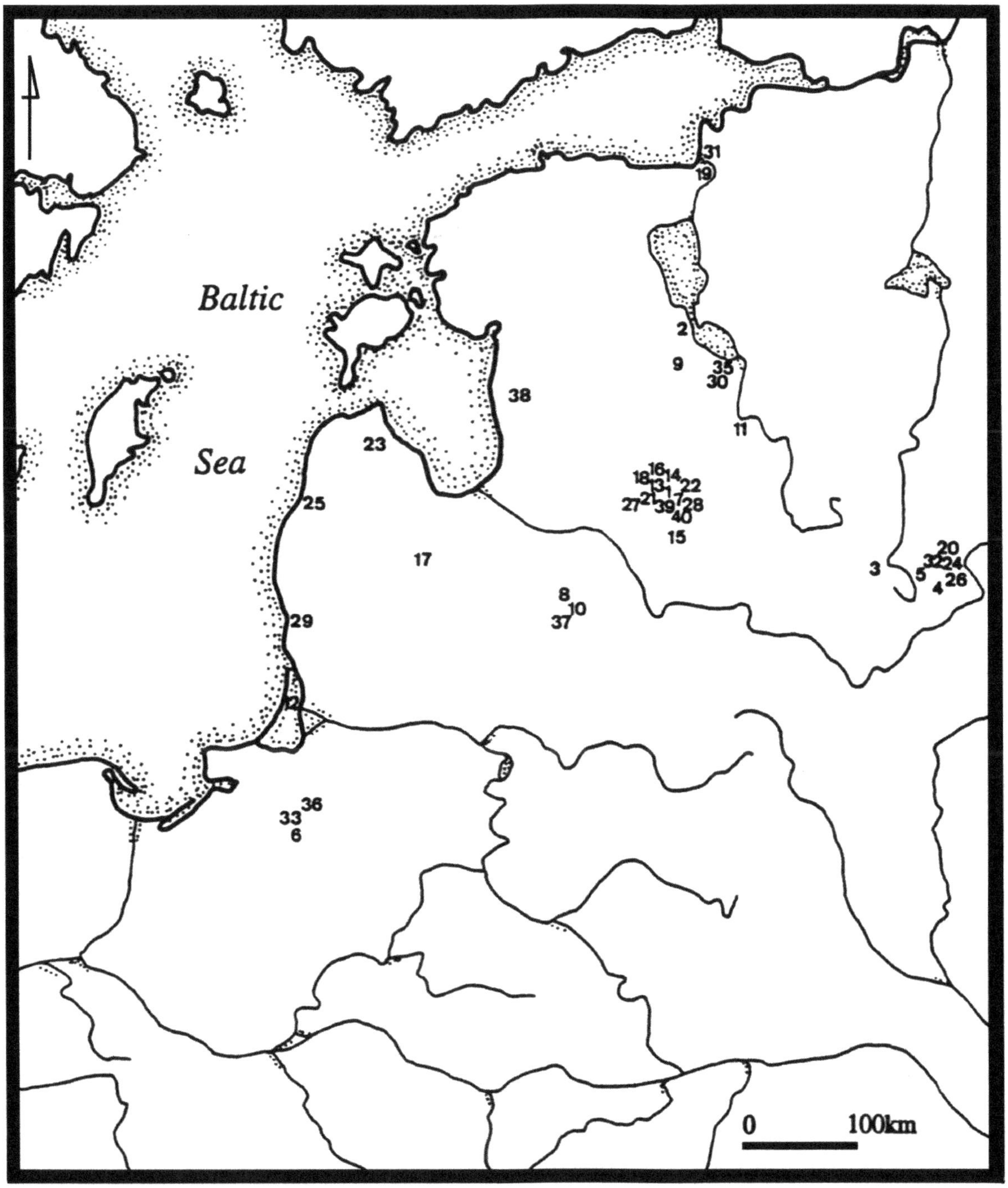

Figure 11.2. *Locations of the major Mesolithic and Neolithic sites in the southeastern Baltic Sea region.*
1, Abora 1. 2, Budianka. 3, Dyazditsa. 4, Dubokrai V. 6, Dudka. 7, Eini. 8, Jara. 9, Kiaeta. 10, Kretuonas 1B and 1C. 11, Kreichi. 12, Kurskaya Kosa. 13, Kvapani. 14, Lagaza. 15, Leimanishki. 16, Malmuta. 17, Leias-Ciskas. 18, Nainekste. 19, Narva I and II, Narva-town layers I, II and III. 20, Naumovo B, B/A and A. 21, Osa. 22, Piestinia. 23, Puclems. 24, Rudnya Serteiskaya. 25, Sarnate. 26, Serteia. 27, Sulagalas. 28, Sulka. 29, Sventoji. 30, Tamula. 31, Tervala. 32, Usvyaty IVB, B/A and A. 33, Utinoye Baloto. 34, Valma. 35, Villia. 36, Zedmar A and D. 37, Zemaitiske 1 and 2. 38, Zvejnieki. 39, Zvejsalas. 40, Zvidze.

The introduction of pottery and domestic plants and animals emphasises the complexity of the introduction of new techniques and modes of food production. Whereas the appearance of cultivated plants and domesticated animals took about five hundred years in western Europe, it took about two thousand years in the southeastern Baltic Sea region and was eventually established in the Iron Age (Janik 1998). The introduction of pottery took place in the southeastern Baltic Sea region at about the same time as pottery appeared in the Danish Ertebølle culture. The process of its introduction into the southeastern Baltic Sea region, however, is not fully understood. In the Ertebølle zone in southern Scandinavia, pottery was adopted from nearby farming communities by fisher-gatherer-hunters. In the southeastern Baltic Sea region at this time, however, these farming communities were occupying areas far away from the fisher-gatherer-hunters. It therefore seems that the knowledge required for pottery production was passed on to the fisher-gatherer-hunter communities of the southeastern Baltic Sea region by other food-procuring communities closer than the farmers. This suggests that the process of adoption of technological innovations in the southeastern Baltic Sea region was more complicated than previously understood (Zvelebil 1985; 1986; 1994; 1996; Zvelebil and Dolukhanov 1991). Despite these differences between southern Scandinavia and the southeastern Baltic Sea region in the appearance of pottery, both regions have been treated in the archaeological literature as undergoing the same process of development, the only difference being that it took longer in the southeastern Baltic Sea region than in southern Scandinavia. The particular situation in the southeastern Baltic Sea region, however, emphasises the importance of a contextual understanding, and prevents our making any generalising statements about the introduction of pottery and food production into these northern regions.

Mesolithic and Neolithic settlements in the southeastern Baltic Sea region can be divided into two types. The first are long-term occupations and the second are those that were occupied only on a seasonal basis. Settlements are located along water features such as lakes, rivers and streams, in a postglacial landscape made up of moraine hills cut by river plains with river terraces and streams. Networks of rivers and streams join postglacial lakes in a landscape of lowland plains interspersed with moraine hills. The locations of these settlements close to the water, as well as the high groundwater levels, have resulted in very good preservation of organic materials. Locational factors also play a key role in the extent to which these settlements have survived to the present day. They vary from sites located on sand, whose remains survive only as a few burnt bones, flint scatters, stones and sand features, to the most spectacular waterlogged sites where rubbish areas, houses, fish traps, tools, vessels, and figurines have all survived.

The archaeological record from the southeastern Baltic Sea region is thus very diverse and rich, including the highest number of inland and coastal waterlogged settlements in Europe. There is little good evidence for the actual sizes of settlements. What evidence there is, however, suggests that many sites were large. The area of Naumovo is estimated to be 6,000m², of which 316m² have been uncovered. Abora 1 covers an estimated area of 5,000m², of which 1311m² were uncovered. The areas excavated at other settlements differ from site to site: 890m² at Zvidze, 225m² at Sulagalas, and 76m² at Sulka.

The amount of archaeological material recovered varies from settlement to settlement. Between 10 and 12 pottery vessels were recovered at Rudnya Serteiskaya, while the number of pottery sherds varied from 56,229 fragments at Zvidze, *c.* 25,000 at Naumovo and *c.* 120,000 at Kretuonas 1C. The number of bone and antler tools also varies from site to site: 905 at Abora 1, 157 at Eini and 139 at Lagaza. Some settlements contained just a few wooden or amber objects, whilst at Zvidze there were 289 wooden objects and 924 amber artefacts. Fish nets, traps, pine bark floats and stones wrapped in birch bark (used as fishing-weights) have been found on a number of sites.

The largest known fisher-gatherer-hunter cemetery in northern Europe is located in the southeastern Baltic Sea region. Zvejnieki lies on a moraine hill on the northern bank of Lake Burnieku. The chronology of the cemetery stretches from the Middle Mesolithic to the Late Neolithic and Early Bronze Age (from the early part of the sixth millennium BC to the second half of the second millennium BC). Two graves dating to the Medieval period have been also uncovered but are understood to be intrusive amongst the fisher-gatherer-hunter burials (Zagorskis 1987). The continuous use of the cemetery dates from the Late Mesolithic (late fifth millennium BC) to the second half of the second millennium BC. The excavation covered 4200m² and uncovered 317 graves. The earlier graves were situated within the cultural layer of a settlement, known as Zvejnieki 2, while the later graves have no relation to this settlement site (Zagorskis 1987). The graves at Zvejnieki contained numerous bone and antler tools, pendants, figurines, stone tools, pendants, amber and pottery (Zagorska and Zagorskis 1985). In addition, the use of powdered ochre and stone was recorded.

Categories of food as a metaphor for cultural identity

I suggest three general categories of food - plant, animal and fish - are possible representations of cultural identity. Enquiring into the use of food as a metaphor of how identities of particular groups were defined brings us to the issue of scale in archaeological discourse. I shall postulate that large-scale geographical models cannot reveal variety or diversity within prehistoric communities or indicate the choices made by local communities. Archaeology in this region is still wedded to the so-called 'grand narrative' of linear development (Zvelebil and Dolukhanov 1991; Zvelebil 1994; 1996) whose models cover all spheres of life from economic to social and symbolic, drawing on relationships between various categories of data through time and space as if the cultural/symbolic identities of particular prehistoric communities were non-existent or irrelevant for interpreting the lives of these prehistoric communities. The cultural context of each particular regional group is, in contrast, the focus of this study of food as a metaphor of cultural identity.

Distinctions between plants, animals and fish as drawn in this paper are ultimately culturally arbitrary; in other words, these distinctions reflect our present-day categorisation, rather than that of past societies. However, we can use these modern categories as a heuristic device to allow us to look at possible similarities and differences between past communities.

When time is taken as an additional factor we can observe further differences between regions as well as within regions in terms of the types of food found in archaeological contexts (Tables 11.3-11.15). This allows us to develop a high-resolution analysis which puts particular communities in the 'spotlight' as well as adopting a long-term perspective that allows us to trace the ways in which particular groups of north European fisher-gatherer-hunters identified themselves through food categorisation.

Plants

The main plant remains that can be recognised incontrovertibly as food are the different varieties of nuts present on coastal and inland sites. The preservation of shells by charring, whether or not in waterlogged conditions, makes them the most common and easily recognisable component of the prehistoric fisher-gatherer-hunter diet. Remains of water chestnut, hazel nut and acorn are commonly present in archaeological remains (Girininkas 1990; pers. comm.; Levkovskaya 1987; Loze 1988; Miklyayev 1970; 1979; pers. comm.; Miklyayev and Dolukhanov 1986; Rimantiene 1979; Vankina 1970).

Nut remains can be studied from sites in three regions: the Lubana Lake region in Latvia, northwestern Russia, and the Kretuonas Lake region (Girininkas 1990; pers. comm.; Levkovskaya 1987; Loze 1988; Miklyayev 1970; 1979; pers. comm..; Miklyayev and Dolukhanov 1986).

Nut type	Northwestern Russia	Lubana Lake lowlands	Kretuonas Lake region
water chestnut	X	X	X
hazel nut	X	X	X
acorn	X	-	-

Table 11.3. *Regional differences in nut procurement in the southeastern Baltic Sea region.*

A difference is apparent between northwestern Russia and the other two regions. Oak (*Quercus*) is present in the pollen diagrams from all regions for all periods of human occupation. The consumption of acorns is, however, restricted only to northwestern Russia and the acorn is absent from plant remains in settlements from Latvia and Lithuania. It appears that, in these other two regions, this nut was rejected perhaps because of the labour required for tannin extraction. This could indicate the operation of cultural choice in deciding what types of nuts were procured. It is also interesting that other *Quercus* macrofossils are absent from waterlogged sites in these two regions (Levkovskaya 1987: tab. 4; Guman 1978).

These different types of nuts were gathered in various archaeological and climatic periods suggesting that these were long-term cultural choices made by inhabitants of particular regions. The palynological evidence (Girininkas 1990) indicates that acorns were growing in all regions and so the choice of whether to collect them cannot be explained by their absence from the surrounding natural environment. There was no difference between regions in terms of climate or the availability of oak woodlands and so it seems that the choice to exploit acorns was made not on the basis of their availability but according to priorities specific to the community making this choice.

Water chestnuts (*Trapa natans*) were the most commonly gathered and - and we assume - consumed nut in the Lubana Lake region. Hazel nuts provided other possibilities for nut consumption but have been found from only two archaeological periods and two archaeological sites, Zvidze (LM) and Abora 1 (LN), although hazel pollen is present in all archaeological periods from the region (Levkovskaya 1987; Loze 1988). When considered in the long-term, there were stronger differentiations between these regions and within them (Tables 11.4-11.6).

Nut type	LM	EN	MN	LN	EB
water chestnut	X	X	X	X	X
hazel nut	X	-	-	X	-
acorn	-	-	-	-	-

Table 11.4. *Presence of different types of nut in various archaeological periods from the Lubana Lake region.*

Nut type	MN	LN/EB	EB
water chestnut	-	X	X
hazel nut	X	X	-
acorn	-	-	-

Table 11.5. *Presence of different types of nut in various archaeological periods from the Kretuonas Lake region.*

In the Kretuonas Lake region, nut remains were found on the Late Neolithic and Early Bronze Age sites of Zemaitiske 1 and 2 and Kretuonas 1B and 1C (Table 11.5). Although there was a general collecting strategy of gathering water chestnuts and hazel nuts, closer analysis of individual sites indicates different choices being made by gatherers from particular settlements. Hazel nuts were found in Middle Neolithic layers at Kretuonas 1B and water chestnuts were found at Early Bronze Age Zemaitiske 2, while both water chestnuts and hazel nuts were present in the period between the Late Neolithic and Early Bronze Age at Zemaitiske 1. There are no nuts from Early Bronze Age Kretuonas 1C (Girininkas 1990; pers. comm.). This indicates that there may have been considerable intra-regional variation in nut collection.

A different pattern of plant use is seen in northwestern Russia (Table 11.6). Plant remains collected from this area from the Early Neolithic to the Early Bronze Age show a consistency in nut types collected. Each settlement produced hazel nuts, acorns and water chestnuts (Miklyayev 1970; 1979; pers. comm.; Miklyayev and Dolukhanov 1986). Thus in northwestern

Russia all species of available nuts were procured for food, indicating a different conceptualisation of their edibility in comparison to the Lubana Lake and Kretuonas Lake regions.

Nut type	EN	MN	MN/LN	LN	EB
water chestnut	X	X	X	X	X
hazel nut	X	X	X	X	X
acorn	X	X	X	X	X

Table 11.6. *Presence of different types of nut in various archaeological periods from northwestern Russia.*

Plants as a metaphor for cultural identity

The cultural categorisation of various types of nuts is specific to particular regions and time periods. When interpreting patterns of plant remains in various regions, local preferences based on cultural rather than ecological factors have to be taken into account. Variability in the nuts collected shows shared features between the Kretuonas Lake region and the Lubana Lake lowlands. The availability of particular species of nuts within the natural environment provided fisher-gatherer-hunter communities with opportunities for exercising choices based on cultural preferences. There were also differences over time. Only in northwestern Russia was there a long-term, consistent exploitation of water chestnuts, hazel nuts and acorns.

Animals

Choices made by fisher-gatherer-hunter communities in the southeastern Baltic Sea region provide different consumption patterns depending on our scale of analysis. Consideration of animal food sources in this paper is restricted to mammals. The overall picture, regardless of regionality or period (Table 11.7), provides a general pattern against which choices within regions or periods can be identified (Tables 11.8-11.11).

Mammal type	EM	LM	EN	MN	MN/LN	LN	LN/EB	EB
elk	85.8	31	28.5	35.5	71.8	9.7	45.1	30
red deer	0	36	14.7	8	0	1.8	5	12.5
roe deer	0.8	6.3	2.8	1.2	0	0.8	0.8	3.7
wild pig	5.8	11	31.0	20.2	4.8	19.8	12.8	11
wild cattle	0	0	2.4	6.7	0	10.9	7.2	3.4
beaver	0	0	2.2	17.7	4.8	21.6	20.4	15.3
bear	0.5	0	3.4	3.4	18.5	1.4	0.4	3
Total no. of bones	400	64	1736	23490	124	25498	3058	5530

Table 11.7 *Composition (by percentage) of bone assemblages in various archaeological periods from the inland part of the southeastern Baltic Sea region.*

Sites from which data on bone assemblages are derived:

Boreal: Dudka (EM and MM), Sulagalas (EM)

Atlantic (LM): Osa*, Dudka and Zvidze* (3 sites representing 5 phases).

Atlantic (EN): Dubokrai V*, Dudka, Osa, Rudnya Serteiskaya*, Zedmar A, Zvidze.

Sub-Boreal (MN): Dudka (2 phases), Kiaeta, Kretounas 1B, Kreichi (early phase), Osa*, Piestinia, Usvyaty IVB, Utinoye Baloto West, Utinoye Baloto East, Valma, Zvidze*.

Sub-Boreal (MN/LN): Naumovo B.

Sub-Boreal (LN): Abora 1, Budianka, Eini, Kreichi (late phase), Leias-Ciskas, Malmuta, Naumovo A/B, Tamula, Usvyaty IVA.

Sub-Boreal (LN/EBA): Zemaitiske 1, Lagaza, Dudka and Villia.

Sub-Boreal (EBA): Naumovo A, Kretounas 1C, Zemaitiske 2.

(*: only presence/absence data available and therefore not included in Table 11.7)

(Daugnora 1994; Girininkas 1990; pers. comm.; Loze 1979; 1988; 1992; Miklyayev pers. comm.; Miklyayev and Dolukhanov 1986; Miklyayev *et al.* 1984; Paaver 1960; Rimantiene 1992; Timofeev 1979; 1983; 1991; Vereshchagin *et al.* 1979).

Site	Dudka	Dudka	Dudka	Dudka	Dudka
Mammal type	**LM**	**EN**	**MN**	**LN**	**LN/EBA**
elk	3.1	2	1.4	2.9	14
red deer	36	25.4	35.4	36.5	36.7
roe deer	6.3	5.4	8.8	9.4	7.7
wild pig	11	0	14.7	10.4	14
wild cattle	0	0	0.2	1.1	0
beaver	0	3.4	0	0	2
bear	0	0	0	0	0
Total no. of bones	64	205	1056	556	207

Table 11.8. *Composition (by percentage) of bone assemblages in various archaeological periods from Dudka.*

Site	Sulagalas	Zvidze	Osa	Piestinia	Abora 1	Eini	Lagaza
Mammal type	**EM**	**EN**	**EN**	**MN**	**LN**	**LN**	**LN/EB**
elk	85.8	36.9	13.5	20	36.7	40.7	52.7
red deer	0	14	1.7	5	4	3.6	0.8
roe deer	0.8	1.4	1.1	?	0.6	0.9	0.1
wild pig	5.8	35.8	50.6	36.3	26.1	22.8	17
wild cattle	0	2.4	1.7	4.1	10.8	14.7	7
beaver	5	1.7	4.5	6.3	13.8	10.8	14.3
bear	0.5	4	4.5	1.7	?	?	?
Total no. of bones	400	1250	178	416	3503	334	1487

Table 11.9. *Composition (by percentage) of bone assemblages in various archaeological periods from the Lubana Lake region.*

Site	Naumovo B	Naumovo B/A	Naumovo A	Usvyaty IVB	Usvyaty IVA
Mammal type	**MN/LN**	**LN**	**EB**	**MN**	**LN**
elk	71.8	45.4	39.2	50.6	31.4
red deer	0	0	0	0	0
roe deer	0	0	0	1.1	0
wild pig	4.8	0	14	16.9	12.7
wild cattle	0	0	0.5	3.4	3
beaver	4.8	0	9	2.2	0
bear	18.5	23	20.4	16.9	35.3
Total no. of bones	124	196	186	89	102

Table 11.10. *Composition (by percentage) of bone assemblages in various archaeological periods from northwestern Russia.*

Site	**Kretuonas 1B**	**Zemaitiske 1**	**Kretounas 1C**	**Zemaitiske 2**
Mammal type	**MN**	**LN/EB**	**EB**	**EB**
elk	19.7	21	31.1	6.8
red deer	50.5	60	12	27.7
roe deer	1.2	3	4.1	0.3
wild pig	11.1	10	11.3	4.5
wild cattle	3.6	0	3.7	0
beaver	4.9	3	14	41.5
bear	2.9	0	1.7	12.5
Total no. of bones	1936	100	5024	311

Table 11.11. *Composition (by percentage) of bone assemblages in various archaeological periods from the Kretuonas Lake region.*

Regional differentiation

When comparing different periods, we can observe differences between regions as well as within them in terms of the species of animals found in archaeological contexts. The same three regions are discussed here but I shall broaden the study by adding one more site, Dudka, which lies in northeastern Poland, the southernmost part of the region under consideration (Figure 11.2). It is essential to stress here that, despite geographical variation, there are no environmental reasons which would account for varying proportions of elk and deer in different areas of the southeastern Baltic Sea region.

The occupation of the settlement at Dudka lasted for thousands of years, covering several archaeological and climatic periods. Despite this long period of occupation, however, the importance of deer in hunting practices continued and in this area there was a culturally governed preference for deer over elk. This preference is clearly visible in that, despite significant variations in numbers of bone recovered in each period, almost the same percentage of deer bones are present in each period's assemblage. Over a thousand bones recovered from a number of sites in the Lubana Lake lowland region (see Figure 11.2; Table 11.9) suggest a different pattern to that at Dudka: all show an overwhelming preference for elk. I therefore suggest that the Lubana Lake lowland region is a zone of elk preference.

In the northwestern Russia region we can see a similar pattern in which elk are favoured over deer (Table 11.10), suggesting that this region belongs to the area of elk preference. The Kretuonas Lake region (see Figure 11.2) lies in between the two zones of different preference - elk to the north and deer to the south - and its bone assemblages similarly reflect this mix of deer and elk. This mixed pattern, which continued for several thousand years, indicates that both types of animal were valued although deer were of greater importance.

Animals as a metaphor for cultural identity

The cultural categorisation of various types of mammals is specific to particular regions and time periods. When interpreting the pattern of animal remains in different regions, local preferences based on cultural rather than ecological factors have to be taken in to account. Variability in the mammals hunted shows shared features between the Kretuonas Lake region and Dudka, while the Lubana Lake lowland is associated with northwestern Russia in its preference for hunting elk. Availability of a wide variety of animal species provided fisher-gatherer-hunter communities with opportunities for selecting particular animals, the choice of which was continued in subsequent archaeological periods.

Fish and molluscs

The fisher-gatherer-hunters chose fish on the basis of principles of culturally governed selection similar to those with which they chose animals, rather than merely eating what was available. Despite the same lack of emphasis on fish as on birds in archaeological publications, enough evidence has been recovered to show that fish were one of the major food sources available to these fisher-gatherer-hunter communities on an everyday basis. Similarly, recent research into isotope ratios in human bones has revealed the significant role of fish in the diet of central Eurasian groups (O'Connell*et al.* 2000). In the southeastern Baltic two classes of archaeological evidence indicate the importance of fish: the fish bones themselves and the material culture of fishing.

Many sites that have no reported fish bones do, however, have artefacts related to fishing, including harpoons, hooks, net weights and net floats. Such artefacts indicate that fishing activities took place even at sites where no fish bones have been recovered (Girininkas 1978; Jaanits 1991). Fish traps have been found at three settlements: Zvidze (Early Neolithic), Abora 1 (Late Neolithic) and Kretuonas 1C (Early Bronze Age) (Girininkas pers. comm.; Loze 1979; 1988). Fish traps are also known from the southeastern Baltic Sea's coast from the Late Neolithic layers at Sventoji 1A (Rimantiene

Fish type	Kretuonas Lake region	Northwestern Russia	Lubana Lake region
pike (*Exus lucius*)	X	X	X
catfish (*Siluris glanis*)	X	-	X
zander (*Stizostedion lucioperca*)	-	X	X
perch (*Pruca fluviatilis*)	X	X	X
tench (*Tinca tinca*)	X	-	X
roach (*Rutilus rutilus*)	-	X	X
common bream (*Abramis brama)*	-	X	X
crucian carp (*Carassius carassius*)	-	X	X
asp (*Aspius aspius*)	-	-	X
chub (*Leuciscus cephalus*)	-	-	X
silver bream (*Blicca bjoerkna*)	-	-	X
ide (*Leuciscus idus)*	-	-	X
rudd (*Scardinius erythrophthalmus*)	-	-	X
blue bream (*Abramis ballarus*)	-	-	X
ruff (*Gymnocephalus cernuus*)	-	-	X
molluscs (*Anodonta cellenis* and *Unio pictorum*)	-	X	X

Table 11.12. *Presence of fish bones from three areas in the southeastern Baltic Sea region from the Late Mesolithic to the Early Bronze Age.*

1992). Fish traps are also known from Scandinavia (Agerod V; Larsson 1983) and other parts of Europe, suggesting their widespread use in the procurement of fish (Gramsch 1992; Mordant and Mordant 1992). Near the fish trap on the Zvidze settlement the remains of 16 pike were found, ranging from 65cm to 136cm in length, suggesting that pike was a readily available food resource.

Fish bone assemblages have been studied at 22 settlements. Although some of these analyses are only on the basis of presence/absence rather than by percentages of different species, it is possible to argue for culturally governed choice and local diversity of fishing strategies. Fifteen types of fish and two types of mollusc have been identified from these sites (Table 11.12).

Regional differentiation

The study of regional differences in fishing is constrained by the fact that a coherent database of fish remains is not available. Moreover the quality of material and published references varies greatly between each of the regions discussed below. Variation in fish bone assemblages from the three areas of the southeastern Baltic Sea region indicates the exercise of local culturally governed choices in fish procurement (Table 11.12). There is no evidence to suggest any differences in the fish populations in the natural environment in these three areas and thus there is no environmental explanation as to why particular species were fished and others not. In the following discussion of assemblages from particular settlements the preferences of particular communities become even more apparent.

Many settlements located in the Kretuonas Lake region contain fish remains. However, analyses of these bones are still being carried out at present and therefore it is difficult to discuss the presence or absence of particular species of fish (Girininkas pers. comm.). However, pike, catfish, perch and tench bones are present in Middle Neolithic and Late Neolithic/Early Bronze Age occupation layers (Girininkas 1990) (Table 11.13).

Fish type	MN	LN/EB	EB
pike	X	X	X
catfish	X	X	X
perch	X	X	X
tench	X	X	X
molluscs	-	X	X

Table 11.13. *Presence of fish bones from the Kretuonas Lake region from the Middle Neolithic to the Early Bronze Age.*

The most significant finding for this region, however, is not the numbers or species of fish caught but rather the presence of two types of freshwater molluscs in the cultural layers of Zemaitiske 1 and 2 (Girininkas 1990). The same molluscs were also found on two settlements in Estonia (Narva I and Akali) and in northwestern Russia, at Usvyaty IVB (Miklyayev 1972; Jaanits 1957).

Therefore it was not only fish that were being exploited from the waters of this region. Shellfish-gathering introduces a further element to the available foods and to the complexity of choice in exploiting the ecosystem. The presence of mollusc shells in the settlements of this particular region emphasises regional flexibility in local preferences.

In northwestern Russia (Table 11.14), perch bones have been recovered from Early Neolithic Dubokrai V, and fishing was an important subsistence activity in the same period at Rudnya Serteiskaya (Miklyayev and Dolukhanov 1986). For the Middle Neolithic and Late Neolithic, bones of pike, zander, perch, roach, common bream and crucian carp were recovered from the occupational layers of Usvyaty IV and Naumovo; the last two were present at Usvyaty IV only. Fish were also part of the food supply at Dyazditsa, Dubokrai 1, Serteia 1 and Serteia 2 but, once again, the species of fish are not described (Miklyayev and Dolukhanov 1986). The same pattern of fishing continued into the Early Bronze Age. Four species have been identified in the Early Bronze Age layers at Naumovo A: pike, zander, perch and roach (Miklyayev *et al.* 1984). Molluscs were also found on the settlement of Usvyaty IVB (Miklyayev 1972).

Fish type	EN	MN	MN/LN	LN	EB
pike	-	X	X	X	X
zander	-	X	X	X	-
perch	X	X	X	X	X
roach	-	X	X	X	X
common bream	-	X	-	X	-
crucian carp	-	X	-	X	-
molluscs	-	X	-	-	-

Table 11.14. *Presence of fish bones from northwestern Russia from the Early Neolithic to the Early Bronze Age.*

Continuity and uniformity throughout successive archaeological and climatic periods characterised fishing strategies in the northwestern Russian region. This coherence reflects a flexible but persistent pattern of exploitation. The inhabitants of this region were selecting from the fish that were available, rather than fishing for whatever was there in the surrounding lakes, rivers and streams.

Analyses of osteological remains from the Lubana Lake lowlands (Table 11.15) are the most specific of all three regions, although the publication of the results of these analyses is not uniform (Loze 1979; 1988). Continuity of fishing strategies characterises this region as much as northwestern Russia. Pike, catfish, perch and common bream are present in all archaeological periods and zander and crucian carp are present in three of the four periods.

Fish type	LM	EN	MN	LN	LN/EB
pike	X	X	X	X	X
catfish	X	X	X	X	?
zander	-	X	X	X	?
perch	X	X	X	X	?
tench	-	-	X	X	?
roach	-	-	-	X	?
common bream	X	X	X	X	?
crucian carp	X	X	-	X	?
asp	-	-	X	-	?
chub	-		-	X	?
gustjera	-	-	-	X	?
ide	-	-	-	X	?
rudd	-	-	-	X	?
blue bream	-	-	-	X	?
ruff	-	-	-	X	?

Table 11.15. *Presence of fish bones from the Lubana Lake region from the Late Mesolithic to the Early Bronze Age.*

Fish as a metaphor for cultural identity

Despite the lack of consistency in the published records of fish species, I interpret regional differentiation in the choice of fish procured as evidence that the fisher-gatherer-hunters of the southeastern Baltic Sea region were consciously selecting preferred species rather than merely taking what was available to them. In purely economic terms fish were a common element of the food resources in all three regions. On the other hand, only a few major species were consistently fished from the Mesolithic to the Early Bronze Age.

Conclusion

I have attempted in this paper to shift our perspective of how we look at food in fisher-gatherer-hunter societies away from the practical rationality of economic utilisation of animal and plant resources to looking at food as embedded in the cultural categories of prehistoric foragers. Furthermore, examples used in this paper support the view that the cultural perceptions of food may, in the long term, transcend typologies of economic forms or modes of production.

References

Daugnora, L. 1994. Laukiniu gyvunu galuniu kaulu, rastu Kretuono 1 C gyvenvieteje, matavimai. In A. Girininkas (ed.) *Gyven Vieciu ir Keramikos Raida Baltu Zemese*. Vilnius: Savastis. 29-33.

Girininkas, A. 1978. Siaures rytu Lietuvos akmens amziaus paminklai. *Trudy Akademi Nauk Litovskoi SSR* 3 (64) s.A: 63-72.

Girininkas, A. 1990. *Kryatuonas: sredny i pozdny neolit.* Archeologia Litvy 7. Vilnius: Mokslas.

Gramsch, B. 1992. Friesack Mesolithic wetlands. In B. Coles (ed.) *The Wetland Revolution in Prehistory.* Exeter: The Prehistoric Society and WARP. 65-72.

Guman, M.A. 1978. Antropogennye izmeneniya rastitelnosti yuga Pskovskoi oblasti v golotsene. *Botanichesky Zhurnal* 63: 1415-29.

Jaanits, L. 1957. Neue Graberfunde auf dem spateneolitischen Wohnplatz Tamula in Estland. *Soumen Muinaismuistoyhdistyksen Aikakauskirja* 58: 80-100.

Jaanits, L. 1991. Rybolostvo i morskoi promysel na territorii Estnonskoi SSR. In N.N. Gurina, *Rybolostvo i Morskoi Promysel v Epokhu Mezolita - Rannego Metala v Lesnoi ZoneVvostochnoi Evropy.* Leningrad: Nayka. 25-38.

Janik, L. 1998. The appearance of food producing societies in the southeastern Baltic Sea region. In M. Zvelebil, L. Domanska and R. Dennell (eds) *Harvesting the Sea, Farming the Forest.* Sheffield: Sheffield Academic Press. 237-44.

Larsson, L. 1983. Agerod V: an Atlantic bog site in central Scania. *Acta Archaeologica Lundensia* 8.12.

Levkovskaya, G.M. 1987. *Priroda i Chelovek v Srednem Golotsene Lubanskoi Niziny.* Riga: Zinatne.

Loze, I.A. 1979. *Pozdny Neolit i Rannyaya Bronza Lubanskoi Ravniny.* Riga: Zinatne.

Loze, I.A. 1988. *Poseleniya Kamennogo Veka Lubanskoi Niziny.* Riga: Zinatne.

Loze, I.A. 1992. The early Neolithic in Latvia. *Acta Archaeologica* 63: 119-40.

Miklyayev, A.M. 1970. Pamyatniki usvyatskogo mikroraiona. Pskovskaya oblast'. *Arkheologichesky sbornik* 11: 18-40.

Miklyayev, A.M. 1972. Neoliticheskoe svainoe poselenie na Usvyatskom ozepe. *Arkheologichesky sbornik* 13 8-29.

Miklyayev, A.M. 1979. O svainykh poseleniyakh tret'ego - vtorogo tysyachletiya do nashei ery v baseine zapadnoi Dviny. *Arkheologichesky sbornik* 18: 10-14.

Miklyayev, A.M. and Dolukhanov, P.M. 1986. Iz istorii razvitiya khozyaistva drevnego naseleniya na pravoberezhe zapadnoi Dviny i v verkhov'yakh Lovati. *Arkheologichesky sbornik* 27: 1-7.

Miklyayev, A.M., Dolukhanov, P.M. and Guman, M.A. 1984. Usvyati 1 U, Naumovo - ozerniye poseleniya epoch neolita i bronzy v verkhov'yakh zapadnoi Dviny. In *Archelogiya i Paleogeografiya Mezolita y Neolita Ruskoy Ravniny.* Moskva: Nauka. 67-81.

Mordant, D. and Mordant, C. 1992. Noyen-sur-Seine: a Mesolithic waterside settlement. In B. Coles (ed.) *The Wetland Revolution in Prehistory.* Exeter: The Prehistoric Society and WARP. 55-64.

O'Connell, T.C., Levine, M. and Hedges, R.E.M. 2000. The importance of fish in the diet of central Eurasian peoples from Mesolithic the Early Iron Age. In K. Jenabert, L. Larsson, R. Petré and B. Wyszomerska-Werbart (eds) *Late Prehistoric Exploitation of the Eurasian Steppe.* Lund: Almqvist and Wiksell International.

Paaver, K. 1965. *Formirovaniye teriofauny i izmenchivost' mlekopitayushchikh Pribaltiki v Golotsene.* Tartu: AN Estonia.

Rimantiene, R. 1979. *Sventoji. Narvos Kultoros Gyvenvietes.* Vilnius.

Rimantiene, R. 1992. Neolithic hunter-gatherers at Sventoji in Lithuania. *Antiquity*, 251: 367-76.

Timofeev, V.I. 1979. Pamyatniki kamennogo veka vosochnoi chasti kaliningradskoi oblasti. *Kratkiye Soobshcheniya* 157: 61-72.

Timofeev, V.I. 1983. Keramika neoliticheskogo poseleniya Utinoe Boloto I. *Kratkiye Soobshcheniya* 173: 103-8.

Timofeev, V.I. 1991. Neolithic sites of the Zedmar type in the southeast Baltic area. In K. Jenabert, L. Larsson, R. Petré and B. Wyszomierska-Werbart (eds) *Regions and Reflections: in honour of Marta Stromberg.* Lund: Almqvist and Wiksell International. 15-26.

Vankina, L.V. 1970. *Torfyanikovaya stoyanka Sarnate.* Riga.

Vereshchagin, N.K., Dolukhanov P.M. and Miklyayev, A.M. 1979. Khozaistvo i ekologiya svainogo poseleniya Naumovo v Pskovskoi oblasti. *Izvestiya Vsesoyuznogo Geograficheskogo Obshchestva* 111 (4): 363-8.

Zagorska, I and Zagorskis, F. 1985. The bone and antler inventory from Zvejnieki 2, Latvian SSR. In C. Bonsall (ed.) *The Mesolithic of Europe.* Edinburgh: John Donald Publishers Ltd. 414-23.

Zagorskis, F. 1987. *Zvejnieku Akmens Laikmeta Kapulauks.* Riga: Zinatne.

Zvelebil, M. 1985. Iron Age transformation in northern Russia and the northeast Baltic. In G. Barker and C. Gamble (eds) *Beyond Domestication in Prehistoric Europe.* Oxford: Basil Blackwell. 147-80.

Zvelebil, M. 1986. Mesolithic prelude and Neolithic revolution. In M. Zvelebil (ed.) *Hunters in Transition.* Cambridge: Cambridge University Press. 5-15.

Zvelebil, M. 1994. Plant use in the Mesolithic and its role in the transition to farming. *Proceedings of the Prehistoric Society* 60: 35-74.

Zvelebil, M. 1996. The agricultural frontier and the transition to farming in the circum-Baltic region. In D.R. Harris (ed.) *The Origins and Spread of Agriculture and Pastoralism in Eurasia.* London: University College London Press Limited. 323-45.

Zvelebil, M. and Dolukhanov, P. 1991. The transition to farming in eastern and northern Europe. *Journal of World Prehistory* 5 (3): 233-78.

Mead, chiefs and feasts in later prehistoric Europe

Eva Koch

National Museum, Copenhagen

Introduction

Within Europe north of the Alps, prehistoric finds with evidence for the use of honey are primarily rich graves from *c.* 1700 BC to *c.* 400 BC. Such burials date from the Early Bronze Age to the earliest centuries of the Iron Age and are found in Britain, Denmark and Germany. These elaborately equipped burials are interpreted as the products of some of the number of small chiefdoms which are thought to have existed at this time (Jensen 1982: 168). In societies of this kind the chief typically has a tripartite role (Service 1962: 145; Sahlins 1968: 86, 110; Feinman and Neitzel 1984):

1. He has to collect and redistribute economic surplus in the best possible ways - not only for himself, but also for his people.
2. He has to handle diplomatic connections with neighbouring chiefdoms, with regard both to trade and to alliances in times of war.
3. He is charged with reducing strife between the people in his area.

The chief does not have any formalised instruments of power, but has to obtain consensus for most of his decisions and, consequently, he has to exert his power through diplomacy rather than by armed force. For this reason, large-scale feasting was one of the axes of political life; at great feasts disputes were solved and the surplus was handed out. Guests and invited neighbours could be impressed by the chief's largesse, influence, and economic and religious abilities. A number of finds from the Bronze Age and Early Iron Age seem to indicate that honey, most probably in the form of mead, played a central role in preparing and staging these feasts.

Honey, mead and a mixed honey-drink

Until cane sugar and beet sugar became available at a reasonable price, honey was one of the few kinds of sweet food available in northern Europe. It was probably used for different culinary purposes, as well as being eaten directly from the comb. Honey's preservative and medicinal qualities may also have been known from quite early on. It is mentioned, for example, as a medicinal ingredient on an Egyptian papyrus from about 1900 BC (Crane 1999: 502).

Yet the few certain finds of honey from the Bronze Age and Early Iron Age of northern Europe all derive from residues in containers that seem to have held some kind of beverage made with honey. Most likely, this was an alcoholic drink, even though we have no conclusive proof of fermentation. Perhaps new methods of genetic analysis will one day allow us to ascertain whether these residues indeed contain traces of yeast cells. Until then, our evidence suggests that there were two kinds of drink made with honey. One is mead proper, made from honey and water, perhaps with spices but without any other ingredients. The other is a kind of honey-beer, brewed from a mixture of cereals, honey and water, and occasionally with berries added.

Mead

Mead is a fermented drink made of honey. It is made from a honey-water solution, in which one part of honey is mixed with three or four parts of water. In the written recipes for making mead - both those from Medieval times and those that can be downloaded from the internet today - the solution is boiled in order to destroy bacteria and to facilitate the fermentation process. Yeast is then added when the solution is lukewarm. Yet it is not absolutely necessary to boil the solution and add yeast - a honey-water solution of the right strength will ferment naturally if it is placed at room temperature or a little warmer. Different herbs such as meadowsweet, St. John's wort or bog myrtle can be added as spices, but basically it is quite a simple brew (Kjersgaard 1978: 99; see also internet websites such as http://www.pastforward.co.uk/vikings/mead.html).

The strength of the mead depends on the amount of honey and on the kind of yeast. It is possible to make mead very strong, close to the strength of modern fortified wines such as port and sherry, but that requires special types of yeast which are able to stand these high levels of alcohol. When fermented with common types of wild yeast, of the sort that occur naturally in honey or on the surfaces of berries, mead probably had a strength akin to that of modern red wine.

Honey-beer and other mixed honey-drinks

Honey-beer is recorded in various written sources from Medieval Sweden (Husberg 1994: 333), from peasant cultures in Germany several centuries ago (Armbruster 1931:2) and is still brewed in many African societies (Hertz 1996: 25).

Mixed drinks with honey are made in two fundamentally different ways. In some versions, the honey is added before fermentation. With this method, the purpose is to obtain a beverage with an alcohol content higher than ordinary beer. In other cases the honey is added after fermentation, just before the drink is served. In this case the purpose is to sweeten the drink, which would have an alcohol content similar to beer or fruit wine. We have no possibility of knowing which of these two recipes was the more common in prehistoric times. All we can know, on the basis of certain discoveries, is that there was a mixed drink made with honey as well as a mead drink.

Bronze Age contexts with residues of honey

From the European Bronze Age contexts that have produced evidence of honey, nine are graves with containers whose residues include large amounts of pollen from bee-pollinated plants. A tenth find is from a well shaft of votive offerings at Berlin-Lichterfelde, where pots filled with a liquid containing honey were ritually deposited.

Rich male graves with mead

Six of these nine graves are well-equipped male burials, which were most likely the graves of chiefs. In all of them, residues inside the pot or container indicate that the dead man had received a grave gift in the form of a pot or metal vessel which, at the moment of burial, had contained a drink made of honey and water, most probably the strongest form of mead. Three of these graves are from the first half of the Bronze Age, from the centuries between 1700 BC and 1200 BC. In the second half of the Bronze Age, the rite of cremation burial became common on the Continent and also in Britain and no honey/mead residues have been found in connection with any of these cremation graves. Inhumation graves from the earlier Iron Age include another three rich male graves with containers whose residues indicate contents of honey or mead.

The Ashgrove grave

The earliest of these chiefly graves is the stone cist grave from Ashgrove in Scotland (Henshall 1964; Dickson 1978; Dickson and Dickson 2000: 79-81). It was sealed so well with clay that the grave chamber had remained dry and free of earth until the moment of excavation. The grave contained the skeleton of a man, aged about 55, placed in a sleeping position (Figure 12.1) and provided with a flat bronze dagger with a triangular blade (Figure 12.2) and a Beaker vessel (Figure 12.3). Between them, the vessel and the dagger date the grave to about 1700 BC, so the radiocarbon determination on plant material of 1250 BC, mentioned by Dickson, must be some hundreds of years too late.

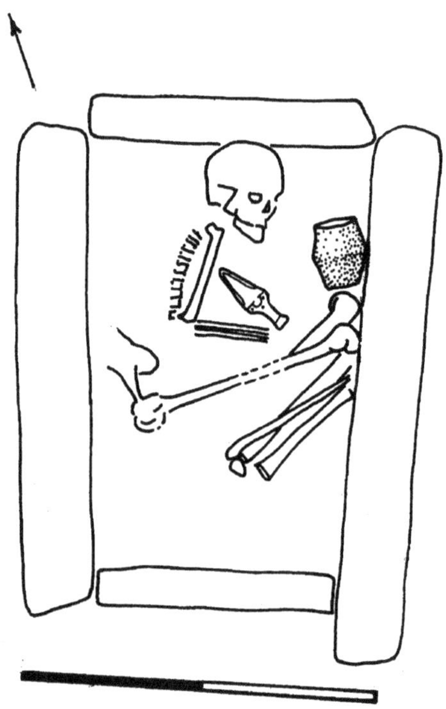

Figure 12.1. *Excavation plan of the man's grave at Ashgrove in Scotland. Scale 1:25. The length of the scale at the bottom of the drawing is 1m. After Henshall 1966: fig.3.*

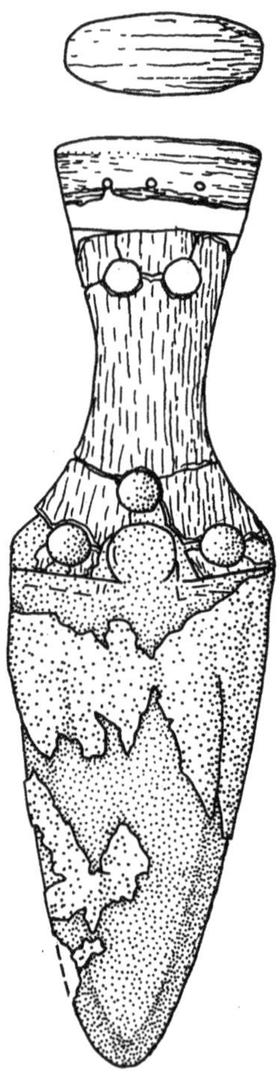

Figure 12.2. *Dagger found in the man's grave at Ashgrove in Scotland. The blade is made of bronze, the grip is made of horn, and the pommel is made of whale-tooth. This dagger is mentioned by S. Gerloff (1975: 42 f., no.27) as a dagger of the type 'flat rivetted blade, type Butterwick'. Scale 1:2. After Henshall 1966: fig.5.*

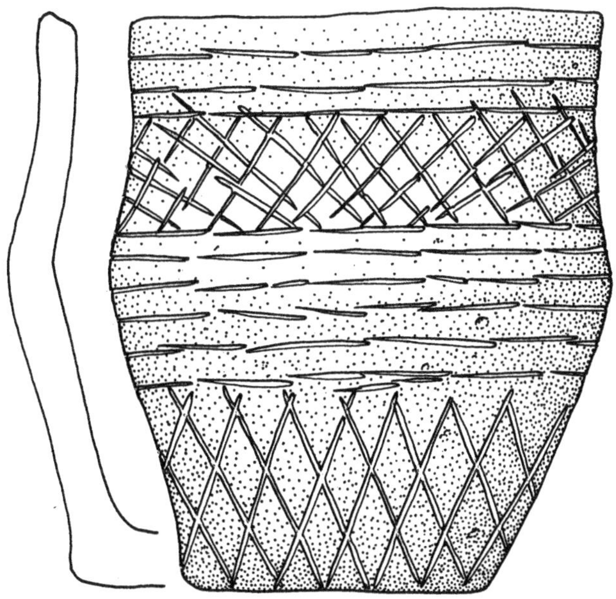

Figure 12.3. *Clay pot from the man's grave at Ashgrove in Scotland. The volume of the vessel can be calculated to about 0.93 litres, so if it were about 3/4 full it would have contained c. 7 decilitres of honey or mead. This vessel is mentioned by D.L. Clarke (1970: 517, no.1662 I) as a Beaker of his type S4. Scale 1:2. After Henshall 1966: fig.4.*

Figure 12.4. *Clay pot from a man's grave at Nandrup on the island of Mors. Scale 1:3. After Aner and Kersten in press: Ke5282, Taf.70. Drawn by M. Hansen.*

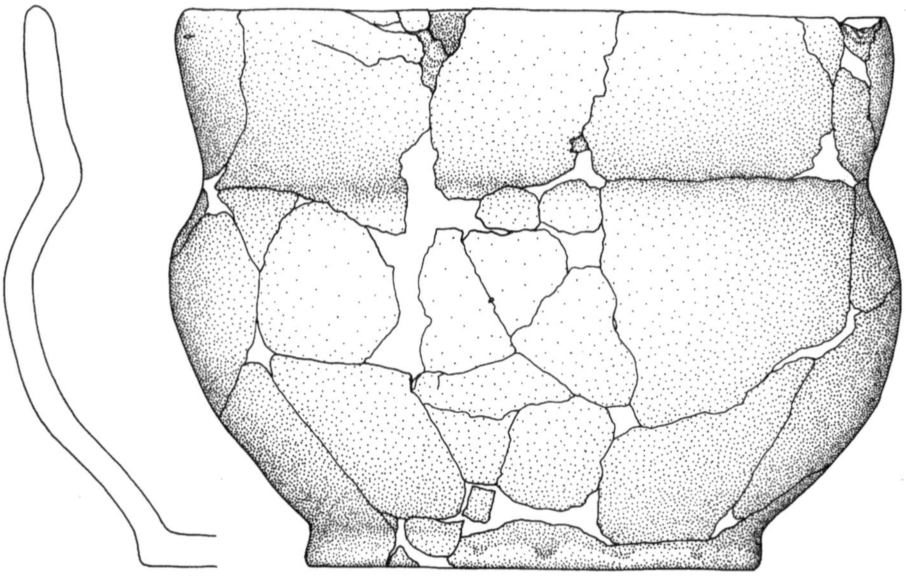

Figure 12.5. *Clay pot from a man's grave at Bregninge in the northwestern part of the island of Zealand. Residue in the bottom of the vessel showed that it had contained honey or mead. The volume of the vessel can be calculated to about 0.8 litres - so if the vessel were about 3/4 full, it would have contained about 6 decilitres of honey or mead. Scale 1:2. Drawn by E. Koch.*

There was a black, porous mass of mostly sphagnum moss on top of the man's breast and arms. Pollen analysis of this moss revealed high proportions of lime pollen (54%) and pollen from flowers such as meadowsweet, heather, ribwort plantain and the thyme/mint family. This result was surprising because neither today nor in the Bronze Age could lime trees grow in Scotland. This particular composition of pollen could only be interpreted as the remains of honey or mead. The cushion of moss had been saturated with a liquid of honey or mead, either of which must have been imported into the area. An alternative interpretation is that the honey in the moss could have been used as a treatment for wounds but this can be dismissed because the Beaker vessel - with a volume of about 0.9 litres - was the evident source of the pollen, lying beside the moss and tipped over on its side. It was possible to sample the Beaker's interior and to confirm that it contained the same mixture of pollen as the samples from the moss. Dickson (1978) concluded that the Beaker had contained a beverage made with honey, probably mead, and that it had tipped over during the moment of burial.

The Nandrup grave

The grave at Nandrup in Denmark was found in 1936 by a farmer ploughing a barrow on the island of Mors in the Limfjord. In the absence of a drawn plan, the dead man appears to have been laid in a stone cist on his back and in an extended position. He was provided with a long sword with a massive bronze hilt, and with a large dagger. On his clothes he was wearing a worn, long fibula of bronze. A large pot (Figure 12.4) stood in the western end of the grave and, together with its thick crust of black residue, it contained a small, flat, amber button, perhaps the fitting from a vanished lid made of organic material. The styles of the bronze weapons date this grave to Montelius' Period II of the Nordic Bronze Age (*c.* 1500-1300 BC).

As early as 1938 the palynologist Johannes Iversen examined this black residue. He found a large amount of pollen, mostly from lime and meadowsweet with some white clover. He concluded that the pot had contained honey or mead (Broholm and Hald 1939: 102, n. 3). Later on, this crust was re-examined by J.J. Troels-Smith who also mentioned that the black discoloration on the interior reached as high as 14.5cm up the wall of the pot. This makes it possible to calculate that the vessel had contained about 6.2 litres of liquid at the time of the burial. Troels-Smith also confirmed Iversen's identification of the contents as being the remains of honey or mead.

The Bregninge grave

Another Danish grave, from Bregninge in the northwestern part of the island of Zealand, was excavated in 1976 (Nielsen 1977; 1978; 1988). This grave was also well equipped, being the central burial within a barrow of 16m-17m diameter. The buried man had been placed in an extended position within an oak coffin. His grave goods consisted of a sword with a massive bronze hilt and a bronze battle-axe, together with a toilet set consisting of tweezers, an awl and a razor. He also wore a bronze fibula on his clothes. This burial is also dated to Period II (*c.* 1500-1300 BC) from the styles of the bronzes.

A medium-sized pot, 10cm tall with a volume of about 0.8 litres (Figure 12.5), was placed towards the foot end of the coffin. This vessel was removed within a single lump of earth and later analysed by Troels-Smith. He found large amounts of pollen from lime and from flowers such as meadowsweet, white clover, various Compositae and knotgrass. We may conclude that this vessel had also contained honey or, more likely, mead.

Graves with a mixed honey drink

Two of the north European Bronze Age graves with residues indicate the presence of grave gifts of some kind of mixed drink with honey, and it may be significant that both of these are of women whereas the Bronze Age male graves (see above) contained residues most probably of mead. A recent find from northwestern Anatolia, the grave of King Midas, has provided evidence of a similar mixed drink brewed from cereals, grapes and honey (McGovern 1999). The two female graves from the first half of the Bronze Age are considered here as containing residues of a mixed honey drink. The third is a male grave from the La Tène Iron Age (see below) and, interestingly, is not quite as rich as near-contemporary Iron Age princely graves provided with the mead.

The Egtved grave

The famous Danish Bronze Age burial at Egtved, from the central part of the Jutland peninsula at Vejle, contained the body of a young woman of 16-18 years buried in an oak coffin under a barrow (Thomsen 1929; Hvass 1981). The styles of the bronzes date the grave to Period II of the Nordic Bronze Age (*c.* 1500-1300 BC) and the tree-trunk coffin has also provided a dendrochronological date for the burial of 1370 BC.

This grave is special because of its exceptional preservation, especially with regard to the fine dress that the girl was wearing. She was placed, fully dressed and with a number of grave goods, on a cow hide that was then folded over her. On top of this hide, at the foot of the coffin, was placed a bucket made of birch-bark (Figure 12.6), with a volume of about 2.2 litres. It had contained a drink that was still preserved as a brown, porous residue on the bottom. As early as 1929 the botanist Bille Gram described this as containing cowberries (or cranberries), grains of wheat, glandular hair from bog myrtle, and an extensive collection of pollen grains from lime trees and other plants. He interpreted the drink as a combination of beer and fruit-wine, with the addition of honey to make it stronger. The pollen was later re-examined by J.J. Troels-Smith, and he was able to confirm the contents of the residue as deriving from bee-pollinated flowers (Figure 12.7; Christensen and Jensen 1991: 16).

The North Mains grave

Some years ago, another woman's grave, at North Mains in Scotland, provided potential evidence of a honeyed drink (Barclay 1983). Within a stone cist was placed the body of a woman of 20-25 years, radiocarbon dated to about 1750 BC (GU-1381), approximately contemporary with the Ashgrove burial mentioned above.

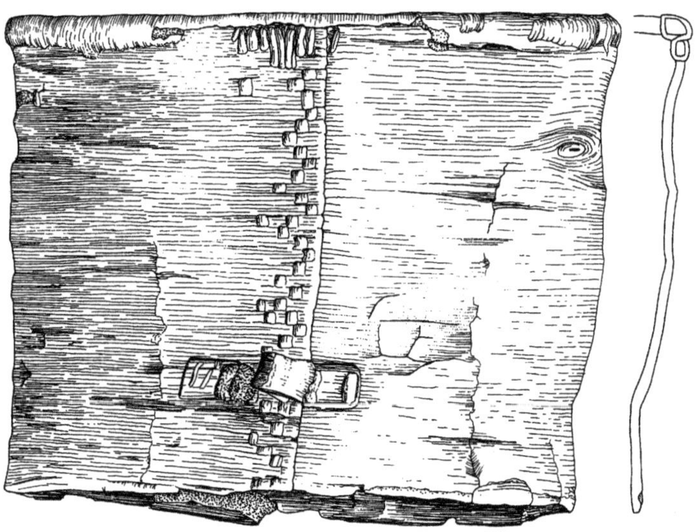

Figure 12.6. *Birch-bark bucket found in the woman's grave at Egtved. The bucket contained residue from a drink fermented from honey, wheat and cranberries or cowberries. If the bucket were about 3/4 full at the time of deposition, it would have contained 1.5 litres of fermented beverage. Scale 1:2. After Aner and Kersten 1990: Ke 4357 A, Taf.15. Drawn by E. Koch.*

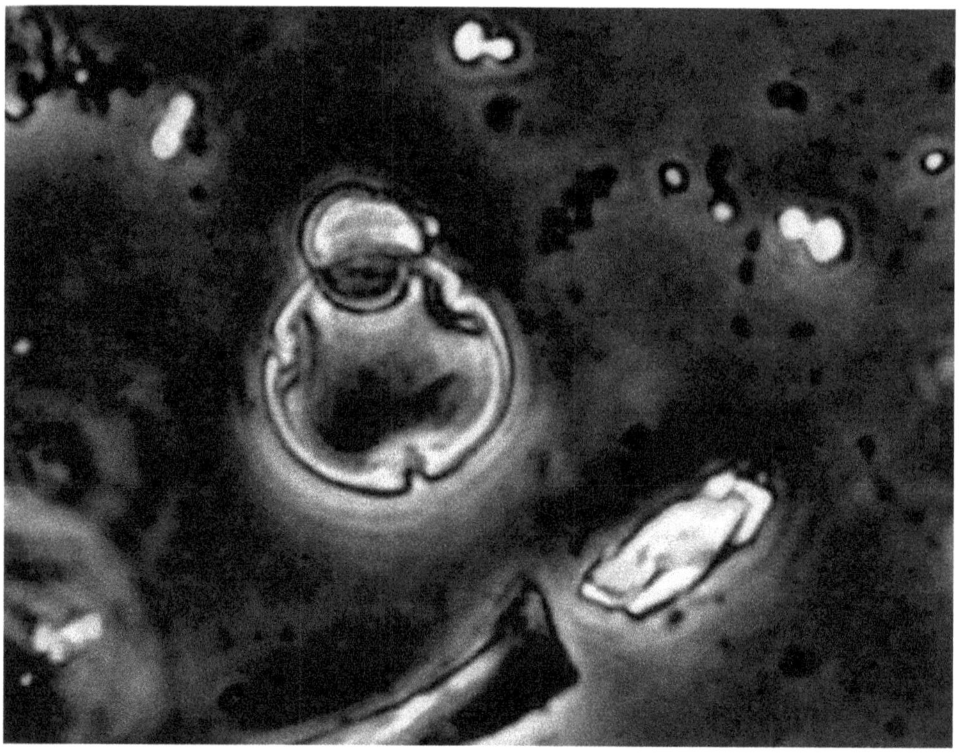

Figure 12.7. *Pollen grains and yeast cells from the bucket found at Egtved. The large grain in the centre is lime pollen; the yeast cells are visible at the top and top right of the photograph. By permission of M. Kielland-Brandt. Copyright Carlsberg Research Centre.*

Like the man in the Ashgrove grave, the woman was placed in a sleeping position (Figure 12.8). Next to her head was her only grave good, a richly decorated Food Vessel (Figure 12.9). Samples from the black, greasy matter found inside the vessel, and another sample scraped from the inside of the pot, yielded pollen for which one interpretation was that it formed the remains of a drink brewed from cereals and spiced with flowers of meadowsweet (the meadowsweet pollen constituted about 80% of the samples; Bohncke 1983). Bohncke considered the rest of the pollen to derive from background pollen-rain and interpreted the contents as either porridge or a fermented ale, sweetened with meadowsweet flowers. Yet about half of the remaining 20% of the pollen is from bee-pollinated flowers and, in my opinion, this indicates the addition of honey to a drink. The meadowsweet pollen may have come from adding meadowsweet flowers to the drink as a spice, but meadowsweet is also a bee-fodder plant. In an analysis of 556 samples of Danish honey from the years 1943-1945, meadowsweet pollen was found in 39% of the samples (Hammer *et al.* 1948: table 1). In one case, the content of meadowsweet pollen was as high as 58% of the total pollen in the sample (Hammer *et al.* 1948: 325). So I think it is possible to argue that the drink in the vessel was a mixed drink made of cereals and honey, a kind of honey-beer.

Iron Age graves with honey drinks

The Eberdingen-Hochdorf grave

The earliest of the Iron Age graves with honey is that at Eberdingen-Hochdorf. It is an overwhelmingly rich grave and is also both well preserved and well excavated and analysed, since it was found only some 20 years ago (Biel *et al.* 1985). The grave was beneath a barrow within a timber-built chamber lined with textiles and made for a strong middle-aged man (Figure 12.10). He was lying in an extended position on a luxurious bronze couch padded with cushions, dressed in fur and textiles, and wearing a cone-shaped birch-bark hat and peaked shoes. Opposite the couch was a four-wheeled wagon, without horses but with harnesses for two horses. On this wagon were placed flat bowls and plates laid for a dinner with nine participants.

But most interesting for us is a large bronze cauldron placed at the foot of the couch. It was made in one of the Greek provinces and could hold about 500 litres. There was a 8mm-10mm thick black residue at the bottom of it. To enjoy the contents of the cauldron, the prince had with him eight bronze-studded horns of aurochsen, and also one even larger drinking-horn made of iron. This iron horn could hold 5.5 litres: merely lifting it to the mouth would have been something of a feat, let alone emptying it.

The black residue proved to be of the same kind as the

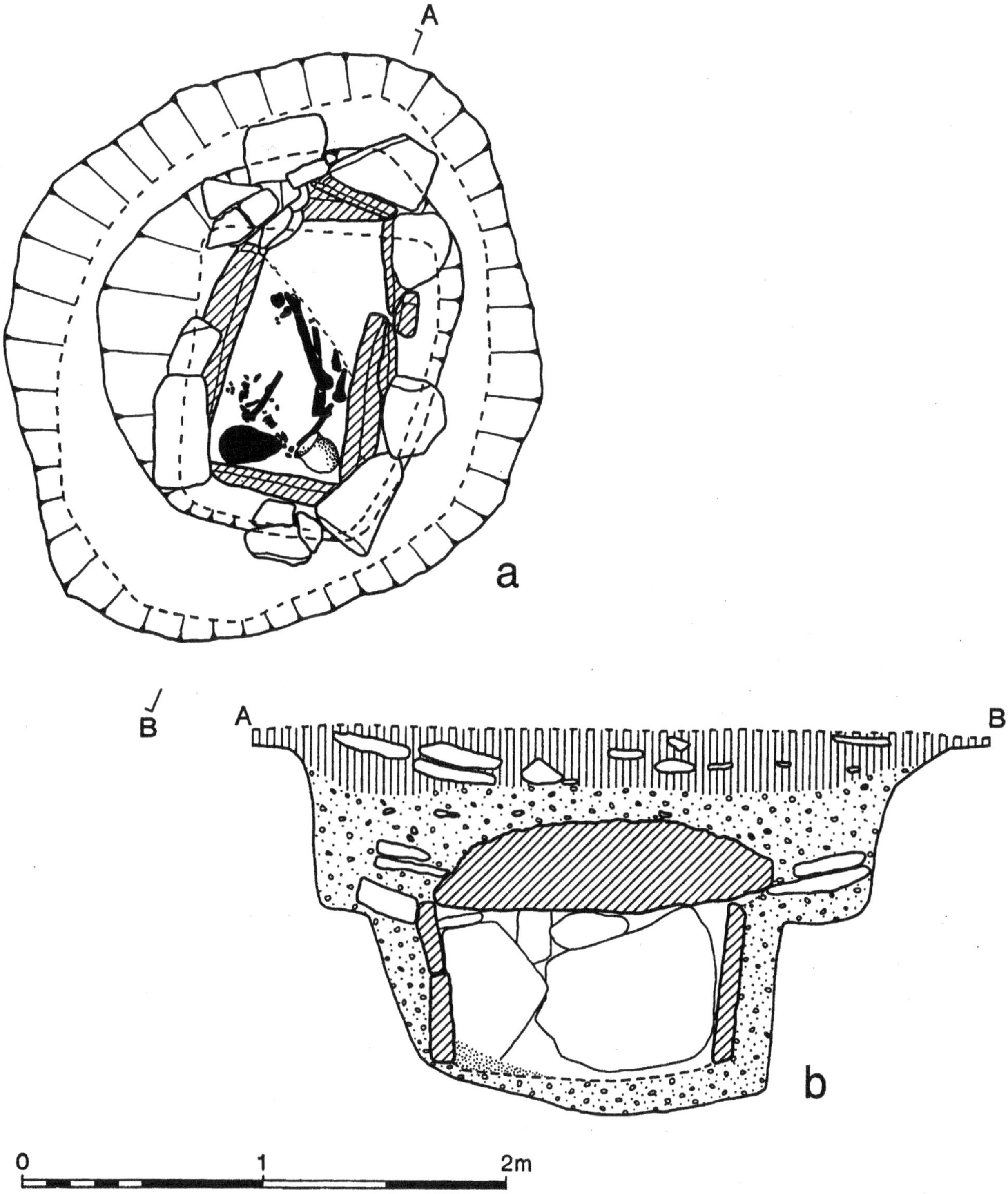

Figure 12.8. *The woman's grave at North Mains. Scale 1:25. After Barclay 1983: fig.15.*

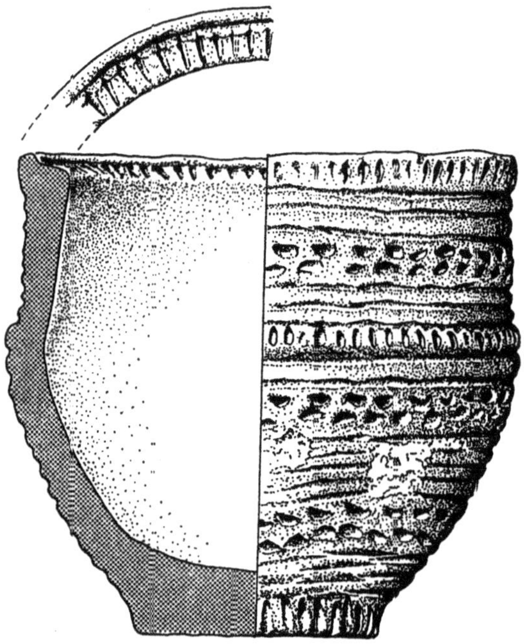

Figure 12.9. *The vessel from the woman's grave at North Mains. Scale 1:2. After Barclay 1983: fig.29a.*

above-mentioned residues from the Bronze Age male graves. It contained large quantities of pollen from a number of flowering herbs, mostly thyme and other plants of open countryside. This amount of pollen can be calculated as having derived from about 100 litres of honey. Discoloration on the inside of the cauldron indicates that it had originally been about three-quarters full; in other words, it had contained about 350 litres of liquid. This quantity, together with the calculated amount of honey, indicates that the liquid had indeed been mead (Körber-Grohne 1985: 98; Vorwohl 1985: 126).

The grave was obviously laid out as a spectacle during the burial ceremony. Many of the items were covered with thin gold foil which must have been provided specially for the burial. The deliberate posture and position of the dead 'prince' portrayed him as a host, ready to welcome his guests to a party where copious amounts of mead were to be consumed. His descendants must have deliberately wished to show him in this role, in which his status found its clearest expression: as the great giver and host. This role accords well with our perception of the Iron Age aristocracy of southern Germany around 525 BC; the Hochdorf burial is of a somewhat grander style than those encountered in the Bronze Age and presents the deceased incontrovertibly as the host rather than as the recipient of the alcoholic gift. This burial illustrates some of the political mechanisms and ideals of the chief, as the great host and giver of gifts, which must have been important in Iron Age society; similar practices and ideals may have been valid for the chiefdoms of the Bronze Age but were not given such clear material form.

The Glauberg burials

The princely grave from the very large mound of Melberg, close to the contemporaneous fortified hilltop settlement known as the Glauberg, also shows a similar development to that of Hochdorf. The grave dates to La Tène A, that is about 450-400 BC. Apart from the barrow itself, a number of large ditches were constructed in the area, highlighting the ceremonial quality of the monument. One of them encircles the mound, which has a ceremonial road leading up to it. Two large sandstone figures of princes or heroes were found in connection with this mound. The monument is a fitting tribute to an early Celtic prince, founding his power - and the power of his descendants - in part on a religious and ideological base.

Two graves were found within the mound and both were very well preserved. They were excavated and removed in block and examined under laboratory conditions. Although analysis has not yet been completed, initial results have been published recently (Bartel *et al.*1998; Rösch 1999).

Grave 1 was constructed as a large wooden chamber (Figure 12.11). The man buried within was about 30 years of age. He was placed on his back, with his head to the southeast, perhaps so that he could look up at his stronghold from where he lay. He was buried wearing his jewellery and his weapons, but the only real grave gift was a 50cm tall bronze jug with a volume of about four litres. At the moment it was placed in the grave, the jug was filled with fluid which is preserved as 275 grams of dark brown organic residue at its bottom. Before deposition, it had been sealed very carefully and wrapped up in a large cloth fastened with blue-striped ribbons (Figure 12.12). The dark brown organic layer at the bottom of the jug was analysed, both for macroscopic remains and for pollen. The only macrofossil found was a seed of flax but there was an overwhelming amount of pollen, approximately 100,000 pollen grains per gram. Most of these were from plants with nectar-yielding flowers that thrive within open, cultivated conditions. The amount of honey that this residue represents has recently been calculated as about 1.4 litres - so the fluid in the jug was indeed mead of a good strength.

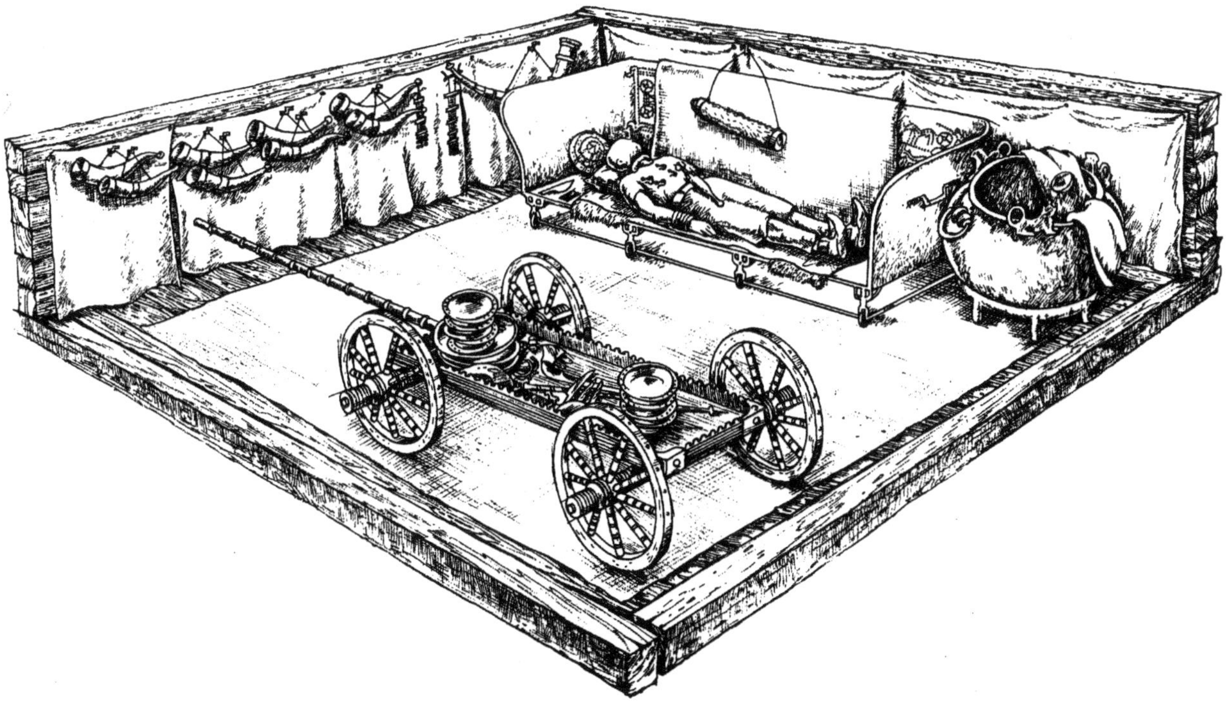

Figure 12.10. *Reconstruction of the grave at Eberdingen-Hochdorf. After Biel et al. 1985: Abb.119.*

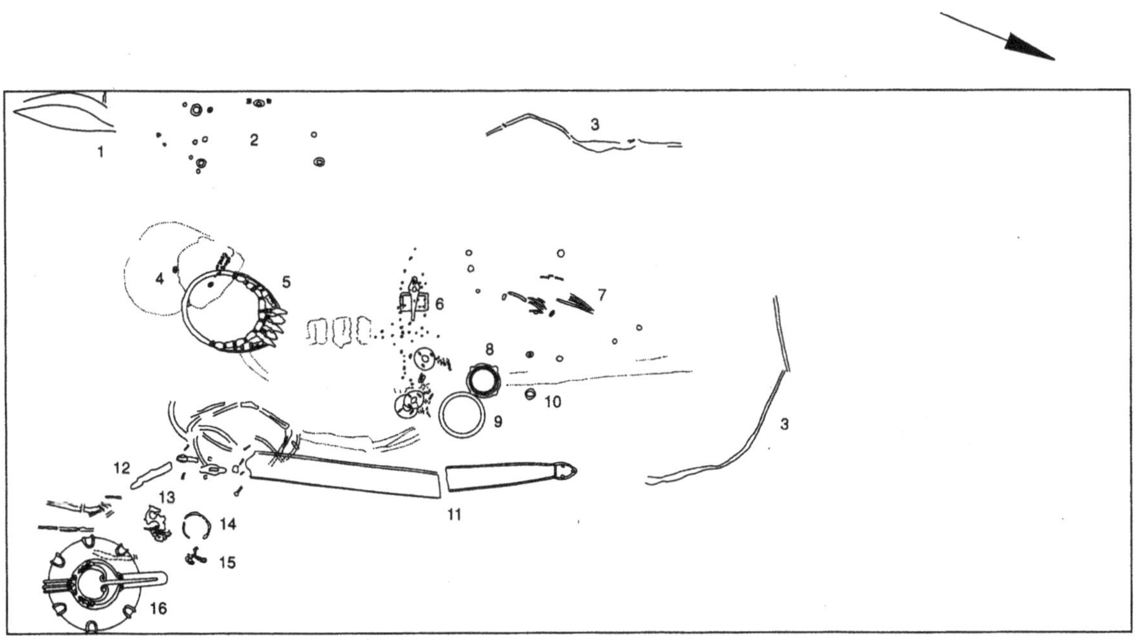

Figure 12.11. *Excavation plan of the Glauberg 1 grave, drawn from an X-ray photo. [1]head of a lance made of iron; [2]small bronze rings and rivets, maybe from a quiver; [3]iron fittings from the rim of a wooden shield; [4]two small gold rings; [5]gold neck-ring; [6]belt-hook and other trimmings for a belt, all made of bronze; [7]iron fragments; [8]bronze rings; [9]gold armring; [10]gold finger-ring; [11]iron sword in a sheath with bronze trimmings; [12]iron fragments; [13]bronze brooch shaped like an animal; [14]bronze ring; [15]two bronze rings; [16]jug with the mead-residue. Scale 1:15. After Bartel et al. 1998: Abb.7. Drawn by M. Pantelius after a model by P. Will.*

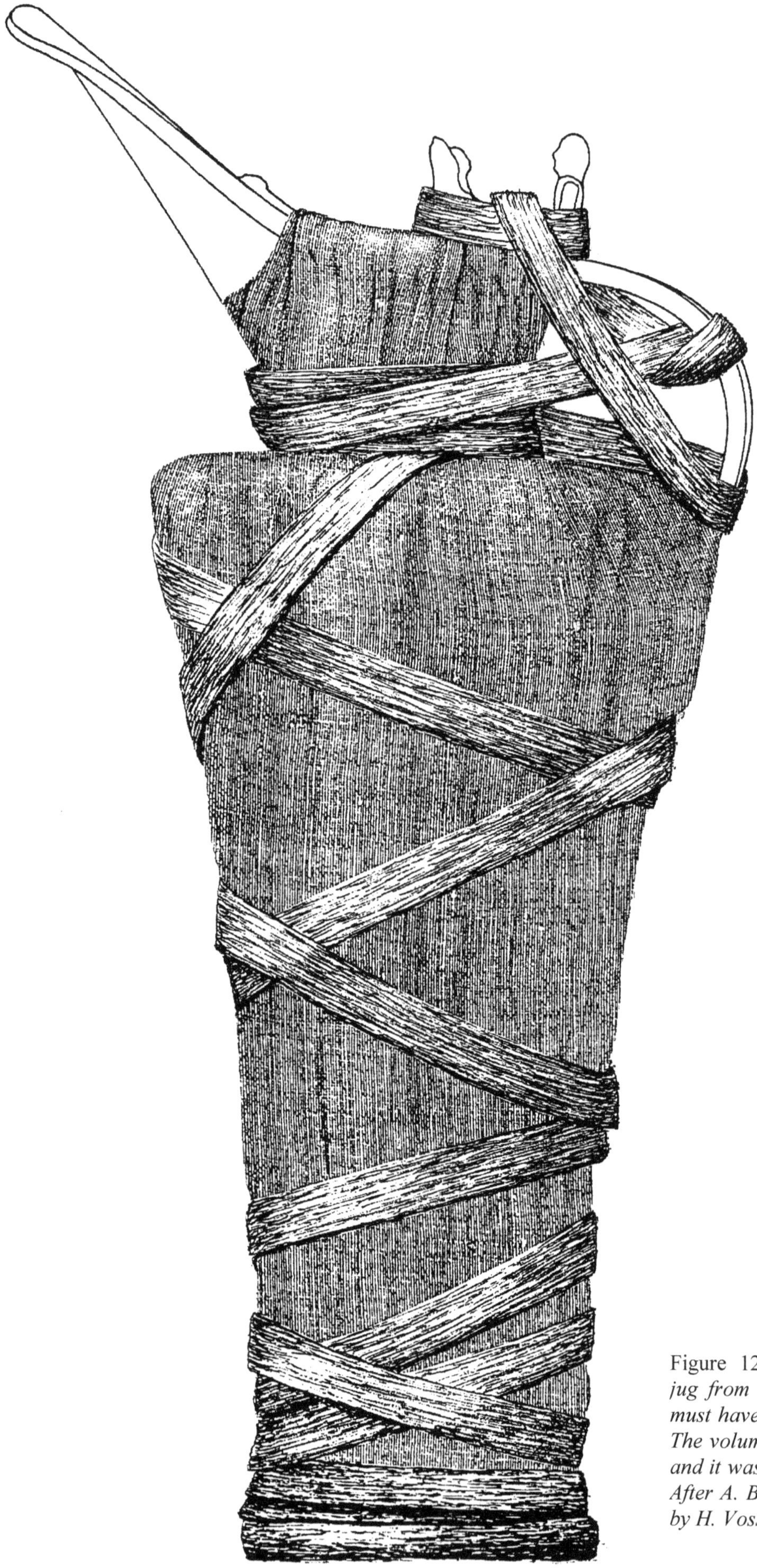

Figure 12.12. *Reconstruction of the jug from the Glauberg 1 grave, as it must have looked in its final wrapping. The volume of the jug is about 4 litres, and it was about 50 cm tall. Scale 1:2. After A. Bartel et al. 1998: 68. Drawn by H. Voss.*

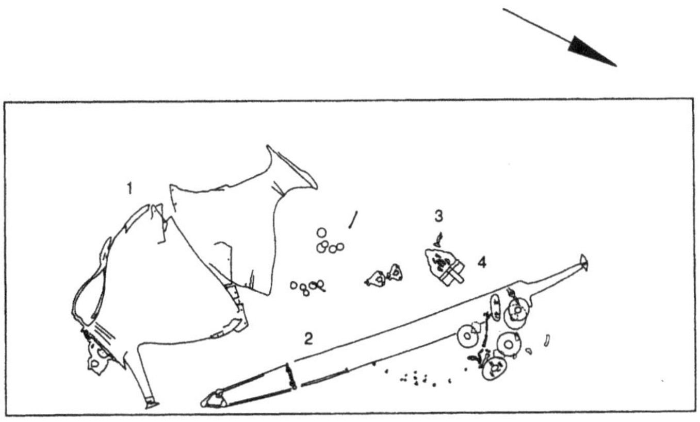

Figure 12.13. *Excavation plan of the Glauberg 2 grave, drawn from an X-ray photo. [1]jug with the honey residue; [2]iron sword in a sheath with bronze trimmings; [3]bronze brooch; [4]belt-hook and other trimmings for a belt, all made of bronze. Scale 1:15. After Bartel et al. 1998: Abb.9. Drawn by M. Pantelius after a model by S. Martins.*

The second grave in the mound at Glauberg contained a somewhat less rich grave than the first (Figure 12.13). It was a cremation burial, and the man's burnt bones were accompanied only by his weapons - there was no golden jewellery like that with the 'prince' in the chamber-grave. Apart from the weapons the only grave gift was a jug, similar to the one found in the rich grave. Analyses of its residue have proved that it contained only 0.15 litres of honey (Rösch 1999: table 1); this low concentration is likely to have derived from a mixed drink, with only a small amount of honey added.

The Hohmichele Grave

Another princely grave from the same period as the Eberdingen-Hochdorf grave is the famous double grave from within the burial mound of Hohmichele, found close to the hilltop stronghold of Heuneburg. The mound was excavated in 1937/38, and a number of graves were found. Although the main grave had been plundered in prehistoric times, a secondary grave chamber remained intact. It contained a man and a woman, buried with a luxurious wagon, jewels and weapons. A large bronze cauldron with a volume of about 20 litres was placed on the wagon.

In the cauldron lay a bronze scoop, and there was a dark residue on the bottom of the cauldron. Lately, a sample of this residue has been examined and it could be that this cauldron also contained a drink made with honey. Unfortunately only a part of the residue was preserved, so it has not been possible to calculate the amount of honey in this drink (Rösch 1999: table 1). Therefore it is uncertain whether the drink was mead or merely a mixed drink containing honey.

The evidence from the graves: overview

Of the nine graves discussed above, six were male graves, and rather rich ones for their time. All appeared to have contained mead (or at least unmixed honey-drinks). The three graves with evidence of a mixed drink were two female graves from the first half of the Bronze Age, and a male grave from the Iron Age (but somewhat less rich than the contemporary male graves with mead).

The differences between these two categories of graves may have been due to several factors. Mead, with its higher honey content, must have been the more costly, perhaps reserved for the chief, his closest followers and special diplomatic guests. The mixed drink was less costly and was probably enjoyed by people of lesser status. Because of its higher sugar content, the mead could also reach a higher alcohol percentage than the mixed drink - and that may be another reason why it might have been reserved for the male élite.

Andrew Sherratt (1991) has written of communal drinking of intoxicants as a universal part of prehistoric rituals that strengthened ties between the male members of social élites. By the end of the Iron Age, later than the burials discussed above, the communal drinking of mead by chiefs and their warriors and guests was described by Classical writers as a ritual of great social importance (Enright 1988: 184-5).

The feasts of the Bronze Age

Indications of strong drinks served at feasts

Mead was never a drink for ordinary consumption; even during Medieval times, when beekeeping was very common because of the great demand for wax to make candles for the Church, mead was a drink to be enjoyed at special occasions and at the courts of kings (Husberg 1994: 321). Even honey-beer was regarded as a special treat in the Medieval period. During the Bronze Age, the two kinds of honey-drink were hardly everyday drinks but instead were luxuries, only to be served on special festive occasions that may very well have been of a cultic nature.

Large, imported bronze buckets have been found in several bog offerings from the Bronze Age. One of the most famous finds of this kind is that from Mariesminde Mose on the island of Fyn (Madsen 1876: 34, figs 25-27; Broholm 1949: 231, 248) where a large bucket was accompanied by 11 golden cups with handles ending in horse heads (Figure 12.14). The bucket's contents - and what was drunk out of the cups - are now unknown but we might guess that it was mead, served for special members of society at big cultic feasts. Perhaps these servings were also of a redistributive kind, in which the common people enjoyed the gifts of the chief: either would fit with our present view of Bronze Age society in northern Europe. Similar gold cups and bronze buckets are known from a number of bog finds both elsewhere in Denmark and in more southern parts of Europe.

Figure 12.14. *Votive find from Mariesminde on the island of Fyn, consisting of a 34 cm tall bronze bucket, into which 11 golden cups had been packed. (Two of the cups are not shown on the photo). The bucket is 34 cm tall, and the cups are 6.3 cm tall without the handle. Photo: Nationalmuseet, Copenhagen.*

The finds from the offering well on the settlement of Berlin-Lichterfelde (Müller 1964: 25; Lehmann 1966: 81) also fit this picture of large-scale cultic celebrations involving the drinking of beverages made with honey. This well contained up to 100 small pots. They were placed in rows and in layers on top of each other, packed in grass, and with grass and stones between each layer (Figure 12.15). Pollen samples from the grass packing make it reasonable to assume that, at the time of deposition, the pots were filled with a drink made from honey and cereals. An interpretation of this find as an offering made in connection with a major celebration, when some kind of honey-drink was enjoyed, would not be unreasonable.

Feasts and dancing

The rock carvings of Scandinavia (Figure 12.16) and finds of small bronze statues may indicate the occurrence of cultic feasts, where the drinking of alcohol and dancing was a part of the celebration.

The Egtved girl, mentioned above, was buried in her dress: a light skirt made of strings worn on the hips and a short shirt. For a girdle, she wore a belt made of wool, ornamented with a large disc of bronze. A modern Danish dancer, Anni Brøgger Kjeldsen, has argued that this dress is well suited for dancing whilst it is impractical for most everyday tasks. Anything carried on the arm, such as a small child, gets tangled in the strings of the skirt.

Her idea accords well with the evidence of a number of small figurines of females from the Bronze Age, showing women in different cultic poses. One, from Faardal (Kjær 1927: 242), is holding up one of her breasts, perhaps to a snake, as a snake-figure was part of the same find. Another figurine, from Grevensvænge (Djupedal and Broholm 1953) is in an acrobatic position, probably as a part of a dance (Figure 12.17). A third one is unfortunately only known from a rather primitive drawing from 1756 (Tanderup and Ebbesen 1979: 20, fig.5). The woman shown here seems also to dance, with hands on her hips and bowed knees (Figure 12.18). The fourth one, from Itzehoe[1] (Figure 12.19), is presenting a vessel. Was the content of the vessel a honey-drink, and is she serving it, or has she just taken a hearty draught to put fire into her dancing? We will never know but it seems to be a reasonable possibility that the dress of the Egtved girl distinguished her as some kind of cultic dancer or priestess, performing her duty at cultic feasts.

On the rock carvings, the dress of the various persons cannot be seen but several of the pictures seem to show dancing. Some of these dances appear to be of an acrobatic kind, like the one shown by the little figurine in Figure 12.17. Other pictures show people jumping - sometimes jumping over ships or ship-figures (Wanke 1997) (Figure 12.20a). Other people seem to dance with lifted arms and intricate movements of the legs (Figures 12.20b, c and d), others dance in a long row like a chain dance (Figures 20e and f), and yet others seem to be part of more complicated dance formations (Figure 12.20g).

[1] *Antiquariske Annaler* 4: 243 (1827).

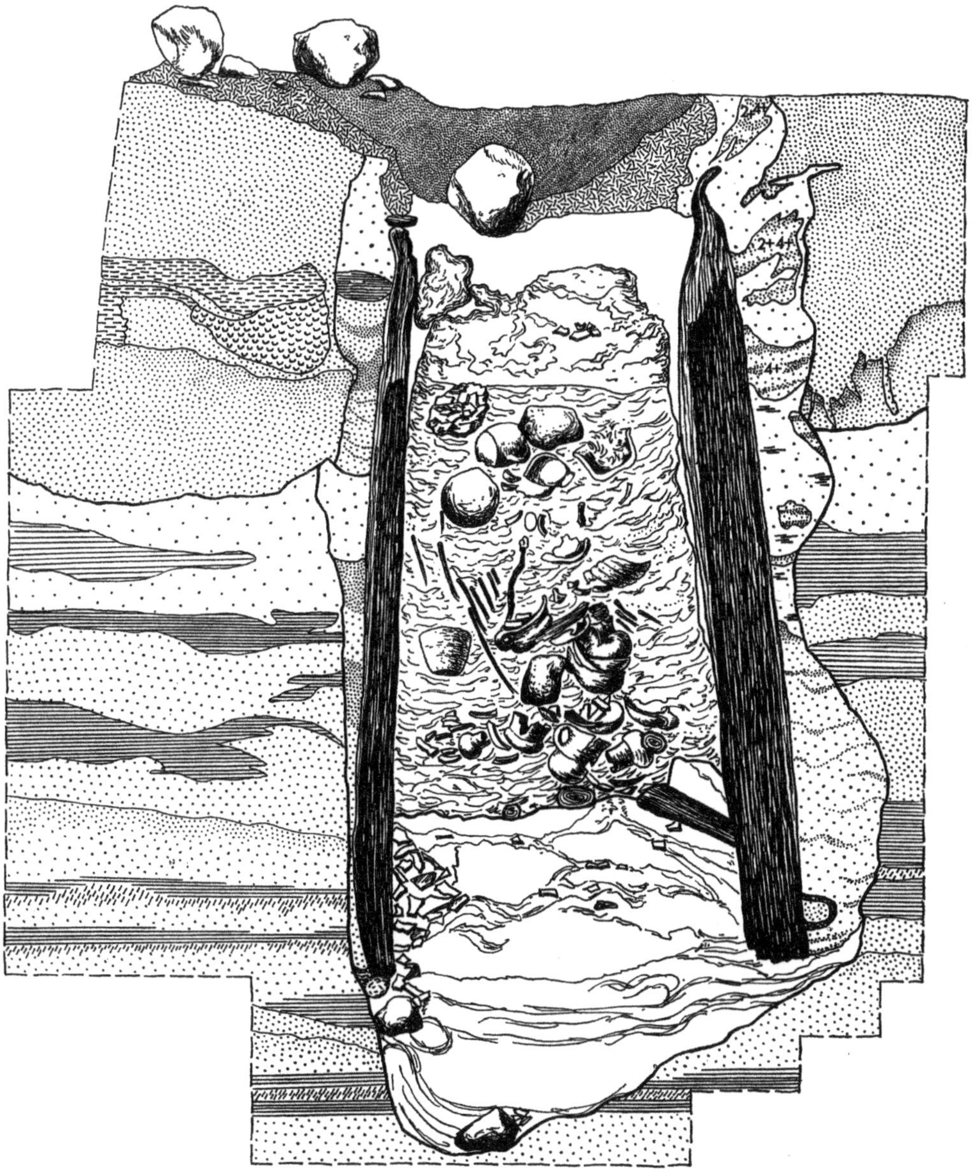

Figure 12.15. Excavation profile of the well found on the settlement of Berlin-Lichterfelde. Its content of small clay pots packed in straw is clearly shown. Scale 1:15. After Müller 1964: fig.2.

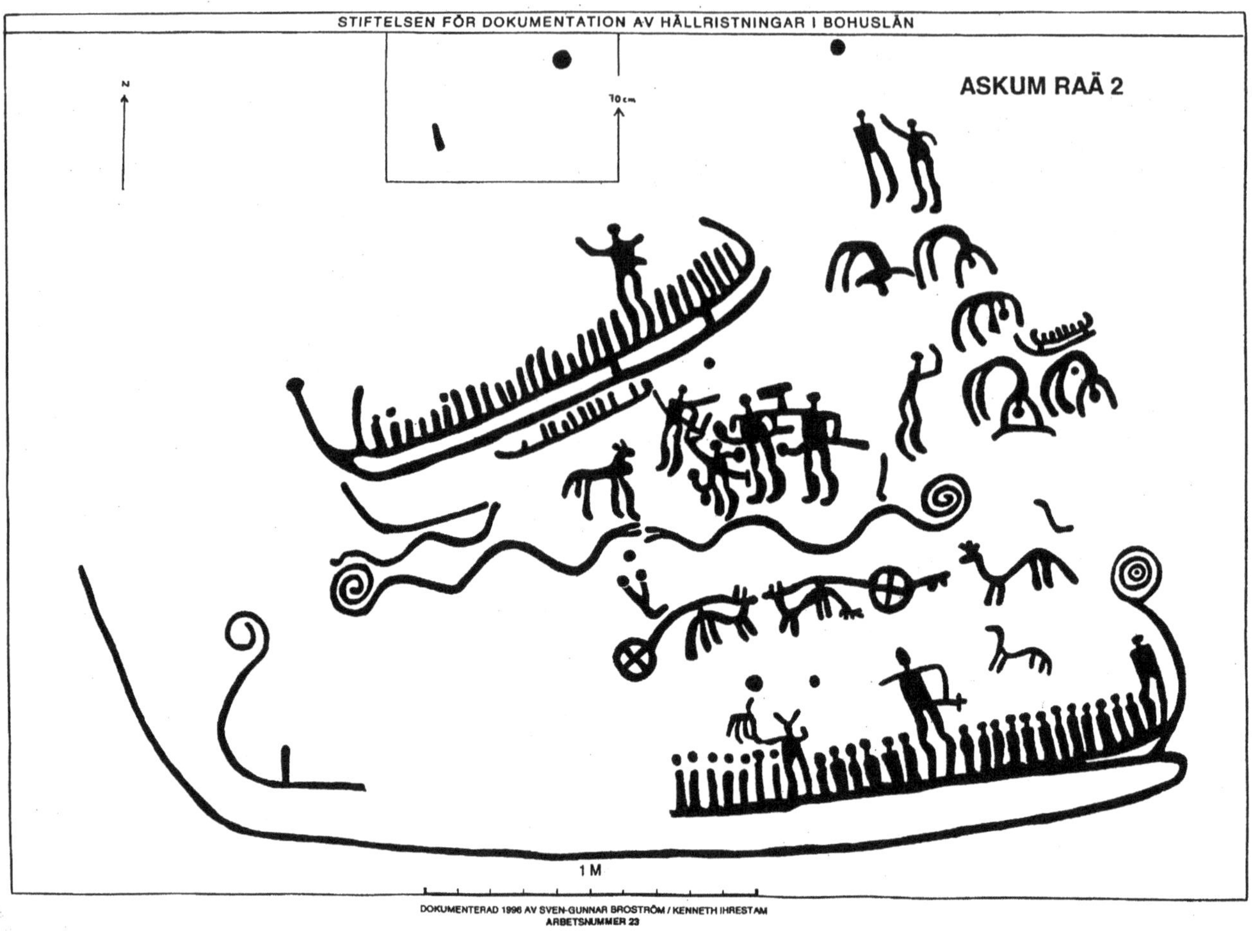

Figure 12.16. *The rock carving Askum Raä 2, showing acrobats among other things. This panel has also been called 'The feast in Askum'. After Bengtsson 1997. Drawn by S.-G. Broström and K. Ihrestamm.*

Figure 12.17. *The woman from Grevensvænge, dressed like the Egtved girl in a skirt made of strings, and in an acrobatic pose. Scale 1:1. Photo: Nationalmuseet, Copenhagen.*

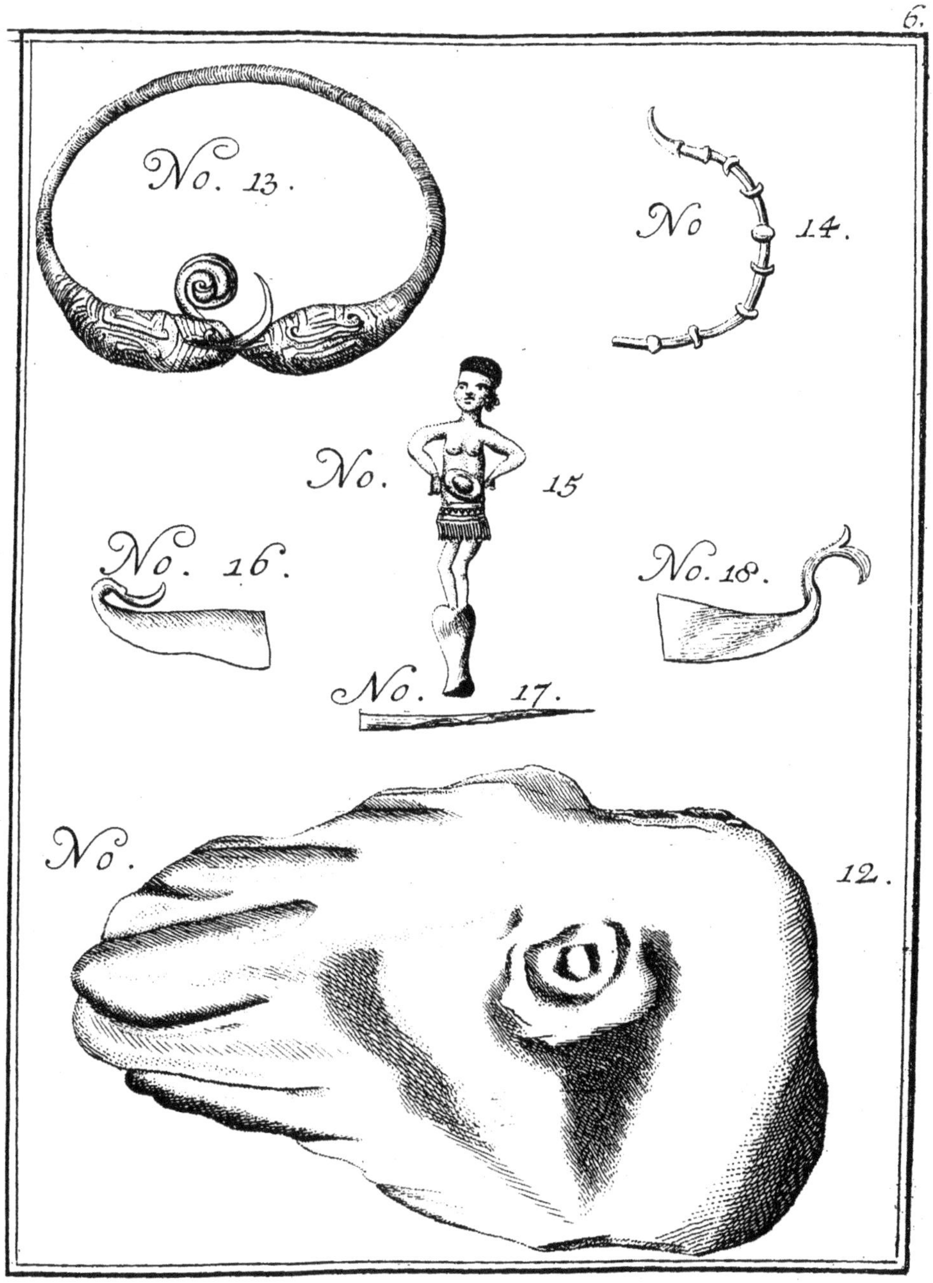

Figure 12.18. *A figurine of unknown size, recorded by Camerer in 1756. It was found in a grave beneath a mound at the village of Beringstädt, parish of Schenefeldt, Rendsburg, Germany.*

Figure 12.19. *Small figurine of a woman dressed like the Egtved girl in a skirt made of strings, and holding a pot. Grip of a small knife found at Itzehoe in Holstein. Scale 2:1. Photo: Nationalmuseet, Copenhagen.*

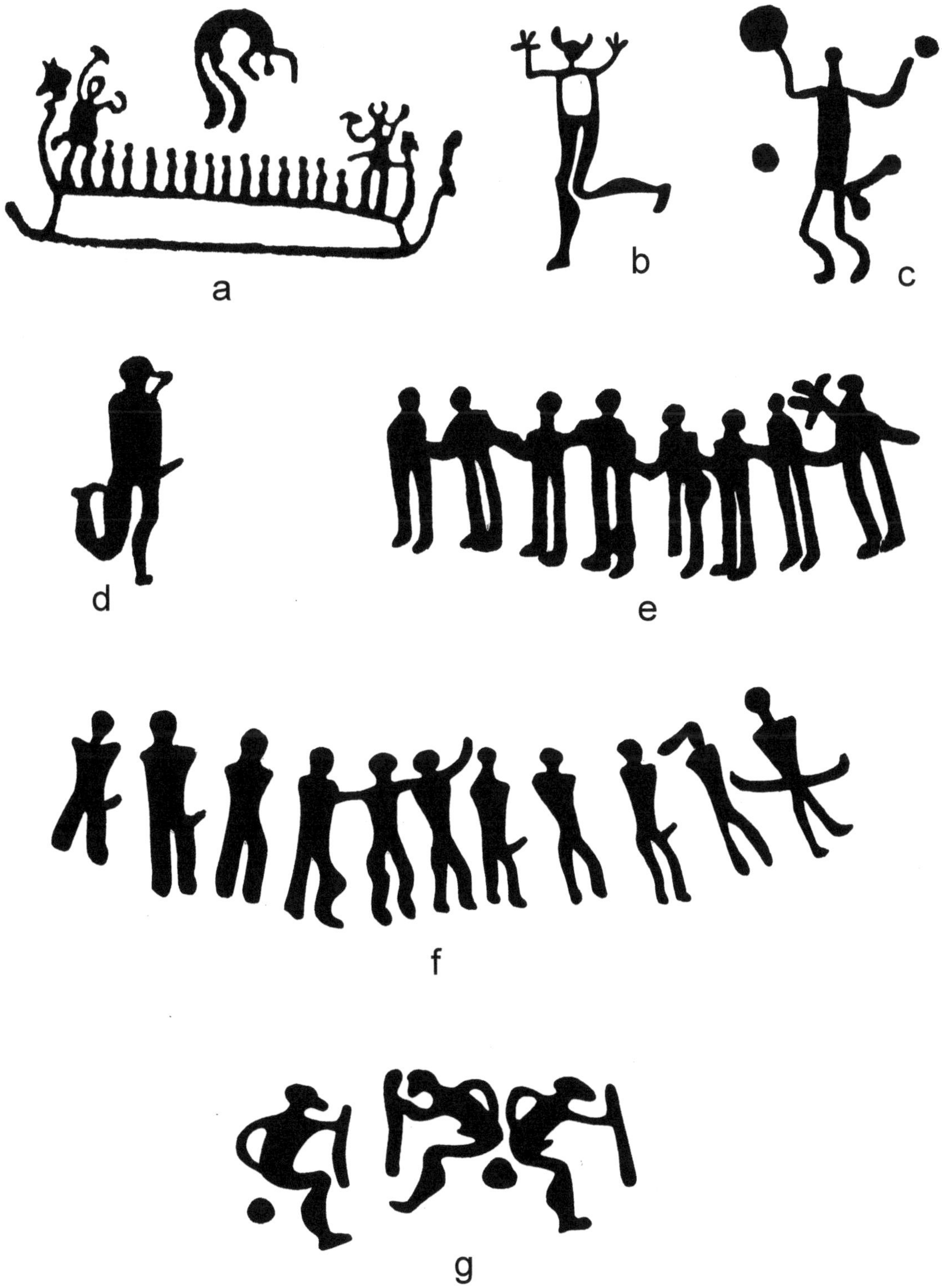

Figure 12.20. *A number of rock carvings with dancing persons of both sexes. The men can be recognised by the penis, the sword-sheath or the horned helmet, the women by the pony-tail hairdo or a cup-mark placed between their legs.*
a. A person jumping over a ship, from Sotorp, Bohuslän, T 355-362, after Wanke 1997: 58; *b - d.* persons dancing with lifted arms and intricate movements of the legs: *b.* from Bro T-192, after photo in *Adoranten* 1986; *c.* from Askum Raä 43:1, after Bengtsson 1998; *d.* from Fossum T-255, after a rubbing in Milstreu 1985; *e - f.* people in a chain-dance: *e.* from Rished, Bohuslän, after Sprockhoff 1957; *f.* from Tegneby, Aspeberget T-25, after photo in Bengtsson 1997: 70; *g.* three men in a figure of a dance, from Fossum 1:2, Hageberget T 255, after photo in Bengtsson 1997. The rock carvings are shown in different scales on this figure. Unless otherwise mentioned, the drawings are by the author.

Conclusion

Finds from the Bronze Age and Early Iron Age show that two kinds of drinks made with honey existed in later European prehistory. Both of them were probably fermented to a degree as alcoholic beverages. We also have evidence of large-scale cultic feasts and evidence of dancing and celebrating; the finds from Berlin-Lichterfelde, the Egtved burial, and the little figurine from Itzehoe seem to connect these two phenomena. I do not think it is going too far to conclude that the obtaining and distribution of honey and honey products played an important social role in the chiefdom societies of northern Europe during the Bronze Age and Early Iron Age.

References

Aner, E. and Kersten, K. 1990. Vejle amt. *Die Funde der älteren Bronzezeit des nordischen Kreises aus Dänemark, Schleswig-Holstein und Niedersachsen* Vol. 9. Neumünster.

Aner, E. and Kersten, K. In press. Thisted amt. *Die Funde der älteren Bronzezeit des nordischen Kreises aus Dänemark, Schleswig-Holstein und Niedersachsen* Vol. 11. Neumünster.

Armbruster, L. 1931. Klassische Bienenzuchtgebiete im Lichte der historischen Betriebslehre und Völkerkunde. *Archiv für Bienenkunde* 12 (1): 1-36.

Barclay, G.J. 1983. Sites of the third millennium BC to the first millennium AD at North Mains, Strathallan, Perthshire. *Proceedings of the Society of Antiquaries of Scotland* 113: 122-281.

Bartel, A., Frey, O.-H., Herrmann, F.-R., Kreuz, A. and Rösch. M. 1998. Ein frühkeltischer Fürstengrabhügel am Glauberg im Wetteraukreis, Hessen: Bericht über die Forschungen 1994-1996. *Germania* 75 (1997): 459-550.

Bengtsson, L. ed. 1997. *Arkeologisk Rapport 3 från Vitlykkemuseet*. Tanumshede.

Bengtsson, L. ed. 1998. *Arkeologisk Rapport 4 från Vitlykkemuseet*. Tanumshede.

Biel, J., Hundt, H.-J., Körber-Grohne, U., Gauer, W. and Hartmann, A. 1985. Das Grab des Keltenfürsten. In *Der Keltenfürst von Hochdorf: Katalog zur Ausstellung Stuttgart, Kunstgebäude vom 14.August bis 13.Oktober 1985*. Stuttgart. 78-163.

Bohncke, S. 1983. The pollen analysis of deposits in a food vessel from the henge monument at North Mains. In G.J. Barclay, Sites of the third millennium BC to the first millennium AD at North Mains, Strathallan, Perthshire. *Proceedings of the Society of Antiquaries of Scotland* 113: 178-80.

Broholm, H.C. 1949. *Danmarks Bronzealder IV: Danmarks kultur i den yngre Bronzealder*. København.

Broholm, H.C. and Hald, M. 1939. Skrydstrupfundet. *Nordiske Fortidsminder* 3 (2:6): 215-347.

Christensen, K. and Jensen, J. 1991. Egtvedpigens alder. *Nationalmuseets Arbejdsmark* 1991: 11-19.

Clarke, D.L. 1970. *Beaker Pottery of Great Britain and Ireland*. Vols I and II. Cambridge: Cambridge University Press.

Crane, E. 1999. *The World History of Beekeeping and Honey Hunting*. London: Duckworth.

Dickson, J.H. 1978. Bronze Age mead. *Antiquity* 52: 108-13.

Dickson, C. and Dickson, J.H. 2000. *Plants and People in Ancient Scotland*. Stroud: Tempus.

Djupedal, R. and Broholm, H.C. 1953. Marcus Schnabel og Bronzealderfundene fra Grevensvænge. *Aarbøger for Nordisk Oldkyndighed og Historie* 1952: 5-59.

Enright, M.J. 1988. Lady with a mead cup: ritual, group cohesion and hierarchy in the Germanic warband. *Frühmittelalterliche Studien* 22: 170-203.

Feinman, G. and Neitzel, J. 1984. Too many types: an overview of sedentary prestate societies in the Americas. *Advances in Archaeological Method and Theory* 7: 39-102.

Gerloff, S. 1975. The Early Bronze Age daggers in Great Britain and a reconsideration of the Wessex culture. *Prähistorische Bronzefund* 6 (2).

Hammer, O., Jørgensen, E.G. and Mikkelsen, V.M. 1948. Studier over danske Honningprøvers Indhold af Blomsterstøv. *Tidskrift for Planteavl* 52 : 293-350.

Henshall, A.S. 1964. A dagger-grave and other cist burials at Ashgrove, Methilhill, Fife. *Proceedings of the Society of Antiquaries of Scotland* 97: 166-79.

Hertz, O. 1996. Biavl i troperne. *Tidsskrift for Biavl* saernummer:16-31.

http://www.pastforward.co.uk/vikings/mead.html

Husberg, E. 1994. Honnung, vax och mjöd. Biodlingen i Sverige under medeltid och 1500-tal. *Avhandlingar från Historiska institutionen i Göteborg* 7.

Hvass, L. 1981. *Egtvedpigen*. København.

Jensen, J. 1982. *The Prehistory of Denmark*. London and New York.

Kjersgaard, E. 1978. Mad og Øl i Danmarks Middelalder. København.

Kjær, H. 1927. To votivfund fra yngre Bronzealder. *Aarbøger for nordisk Oldkyndighed og Historie* 1927: 233-76.

Körber-Grohne, U. 1985. Die biologische Reste aus dem hallstattzeitlichen Fürstengrab von Hochdorf, Gemeinde Eberdingen (Kreis Ludwigsburg). *Hochdorf* 1: 87-164.

Lehmann, H. 1966. Ein dreitausindjähriger 'Klotzstülper' aus Berlin-Lichterfelde. *Berliner Blätter für Vor-und Frühgeschichte* 11: 45-98.

McGovern, P. 1999. A funerary feast fit for King Midas. *Nature*.402 (6764): 863-4.

Madsen, A.P. 1876. *Afbildninger af danske Oldsager og Mindesmærker II. Broncealderen. Samlede Fund.* København

Milstreu, G. 1985. Rapport fra Museets Arbejdsmark. Tema: Menneskefigurer i bevægelse. *Adoranten, Årsskrift för Scandinavian Society for Prehistoric Art* 1985: 19-23.

Müller, A. von, 1964. Die Jungbronzezeitliche Siedlung von Berlin Lichterfelde. *Berliner Beiträge zur Vor- und Frühgeschicht* 9.

Nielsen, S. 1977. Fyldskifter. *Skalk* 2: 32.

Nielsen, S. 1978. Mjød. *Skalk* 2: 32.

Nielsen, S. 1988. Bronzealdergravene fra Bregninge. *Antikvariske Studier* 2: 15-34.

Rösch, M. 1999. Evaluation of honey residues from Iron Age hill-top sites in southwestern Germany: implications for local and regional land use and vegetation dynamics. *Vegetation History and Archaeobotany* 8: 105-12.

Sahlins, M.D. 1968. *Tribesmen*. Englewood Cliffs, NJ: Prentice Hall.

Service, E.R. 1962. *Primitive Social Organisation*. New York: Random House.

Sherratt, A. 1991. Sacred and profane substances: the ritual use of narcotics in later Neolithic Europe. In P. Garwood, D. Jennings, R. Skeates and J. Toms (eds) *Sacred and Profane: proceedings of a conference on archaeology, ritual and religion.* Oxford: Oxford University Committee for Archaeology monograph 32. 50-64.

Sprockhoff, H. 1957. Das bronzerne Zierband von Kronshagen bei Kiel. *Offa* (1955) 14: 5-114.

Tanderup, R. and Ebbesen, K. 1979. *Forhistoriens Historie*. Århus.

Thomsen, Th. 1929. Egekistefundet fra Egtved. *Nordiske Fortidsminder* 2.

Vorwohl, G. 1985. Quantitave honigkundliche Untersuchungen am Sediment des Hochdorfer Mischkessels und an heutigen Honig. *Hochdorf* 1: 126-8.

Wanke, L. 1997. Die Schiffshebung und die Sprünge über das Boot. *Adoranten, Årsskrift för Scandinavian Society for Prehistoric Art* 1997: 52-61.

www.ingramcontent.com/pod-product-compliance
Lightning Source LLC
LaVergne TN
LVHW071218210926
848482LV00015B/836